# LONDON MATHEMATICAL SOCIETY LECTURE NOTE SERIES

Managing Editor: Professor J.W.S. Cassels, Department of Pure Mathematics and Mathematical Statistics, University of Cambridge, 16 Mill Lane, Cambridge CB2 1SB, England

The books in the series listed below are available from booksellers, or, in case of difficulty, from Cambridge University Press.

# LONDON MATHEMATICAL SOCIETY LECTURE NOTE SERIES

Managing Editor: Professor J.W.S. Cassels, Department of Pure Mathematics and Mathematical Statistics, University of Cambridge, 16 Mill Lane, Cambridge CB2 1SB, England

The books in the series listed below are available from booksellers, or, in case of difficulty, from Cambridge University Press.

London Mathematical Society Lecture Note Series. 167

# Stochastic Analysis

Proceedings of the Durham Symposium on Stochastic Analysis, 1990

Edited by
M.T. Barlow
*Trinity College, Cambridge*
and
N.H. Bingham
*Royal Holloway and Bedford New College,
University of London*

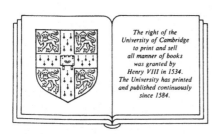

*The right of the
University of Cambridge
to print and sell
all manner of books
was granted by
Henry VIII in 1534.
The University has printed
and published continuously
since 1584.*

CAMBRIDGE UNIVERSITY PRESS

Cambridge

New York  Port Chester  Melbourne  Sydney

CAMBRIDGE UNIVERSITY PRESS
Cambridge, New York, Melbourne, Madrid, Cape Town, Singapore, São Paulo

Cambridge University Press
The Edinburgh Building, Cambridge CB2 8RU, UK

Published in the United States of America by Cambridge University Press, New York

www.cambridge.org
Information on this title: www.cambridge.org/9780521425339

© Cambridge University Press 1991

First published 1991
Re-issued in this digitally printed version 2008

A catalogue record for this publication is available from the British Library

ISBN 978-0-521-42533-9 paperback

# CONTENTS

# PREFACE

This volume contains the proceedings of the Durham Symposium on Stochastic Analysis, held at the University of Durham 11–21 July 1990, under the auspices of the London Mathematical Society.

The core of the Symposium consisted of courses of lectures by six keynote speakers (Aldous, Dawson, Kesten, Meyer, Sznitman and Varadhan), three of which appear here in written form. In addition, there were twenty-six talks by invited speakers; the written versions of eleven of these make up the remainder of the volume.

All the papers in the volume have been refereed.

It is a pleasure to thank here all those individuals and institutions who contributed to the success of the Symposium, and to these Proceedings. We thank the London Mathematical Society for the invitation to organize the meeting, and the Science and Engineering Research Council for financial support under grant GR/F18459. We are grateful for the University of Durham for use of its facilities in Gray College and the Department of Mathematics, to our local organizers, John Bolton and Lyndon Woodward, and to Mrs Susan Nesbitt for her devoted secretarial help. Our main debt is to the speakers and participants at the Symposium, and to the contributors and referees for the Proceedings, and we thank them all. Last but by no means least, we thank David Williams, the organizer of the 1980 Durham Symposium on Stochastic Integrals, for much valuable advice during the planning of this meeting.

Martin Barlow
Nick Bingham

June 1991

# List of participants

RJ Adler (Technion)

DJ Aldous (Berkeley)

MT Barlow (Cambridge)

DJ Balding (Queen Mary Westfield)

FG Ball (Nottingham)

JA Bather (Sussex)

JD Biggins (Sheffield)

MS Bingham (Hull)

NH Bingham (Royal Holloway)

K Burdzy (Seattle)

G Burstein (Imperial )

EA Carlen (MIT)

TK Carne (Cambridge)

T Chan (Cambridge and Heriot-Watt)

JMC Clark (Imperial)

NJ Cutland (Hull)

IM Davies (Swansea)

DA Dawson (Carleton )

RA Doney (Manchester)

EB Dynkin (Cornell)

DA Edwards (Oxford)

RGE Elliott (Alberta)

KD Elworthy (Warwick)

M Emery (Strasbourg)

AM Etheridge (Oxford)

PJ Fitzsimmons (UC San Diego)

L Foldes (London School of Economics )

CM Goldie (Queen Mary and Westfield)

I Goldsheid (Swansea)

PE Greenwood (University of British
Columbia)

DR Grey (Sheffield)

B Hambly (Cambridge)

J Hawkes (Swansea)

RL Hudson (Nottingham)

P Hunt (Cambridge)

SD Jacka (Warwick)

M Jacobsen (Imperial)

J Jacod (Paris VI)

DG Kendall (Cambridge )

WS Kendall (Warwick)

J Kennedy (Cambridge)

H Kesten (Cornell)

PE Kopp (Hull)

H Kunita (Kyushu)

GF Lawler (Duke)

H-L Le (Cambridge)

R Leandre (Strasbourg )

JT Lewis (IAS, Dublin)

TM Liggett (UC Los Angeles)

M Lindsay (Kings, London)

T Lindstrom (Oslo)

TJ Lyons (Edinburgh)

Z Ma (Academia Sinica)

D Mannion (Royal Holloway)

P March (Ohio State)

P McGill (UC Irvine)

JM McNamara (Bristol)

PA Meyer (Strasbourg)

JR Norris (Cambridge)

B Øksendal (Oslo)

Papangelou (Manchester)

E Pardoux (Marseille)

EA Perkins (University of British
Columbia)

M Pinsky (Northwestern)

S Pitts (University College London)

P Protter (Purdue )

O Raimond (Paris VI)

GO Roberts (Nottingham)

M Röckner (Edinburgh and Bonn)

A Rogers (Kings, London )

A-S Sznitman (Courant and Zurich)

SJ Taylor (Virginia)

B Toth (Heriot-Watt)

A Truman (Swansea)

SRS Varadhan (Courant)

JB Walsh (University of British
Columbia)

D Williams (Cambridge)

M Yor (Paris VI)

I Ziedins (Heriot Watt)

# An Evolution Equation for the Intersection Local Times of Superprocesses[1]

ROBERT J. ADLER and MARICA LEWIN

Faculty of Industrial Engineering and Management
Technion—Israel Institute of Technology
Haifa 32000, Israel

## 1  Introduction

The primary aim of this paper is to establish evolution equations for the *intersection local time* (ILT) of the super Brownian motion and certain super stable processes. We shall proceed by carefully defining the requisite concepts and giving all of our main results in the Introduction, while leaving the proofs for later sections. The Introduction itself is divided into four sections, which treat, in turn, the definition of the superprocesses that will interest us, the definition of ILT and some previous results, our main result – a Tanaka-like evolution equation for ILT – and an Itô formula for measure-valued processes along with a description of how to use it to derive the evolution equation. Some technical lemmas make up Section 2 of the paper, while Section 3 is devoted to proofs.

In order to conserve space, we shall motivate neither the study of super-processes *per se* – other than to note that they arise as infinite density limits of infinitely rapidly branching stochastic processes – nor the study of ILT – other than to note that this seems to be important for the introduction of an intrinsic dependence structure for the spatial part of a superprocess. Good motivational and background material on superprocesses can be found in Dawson (1978, 1986), Dawson, Iscoe and Perkins (1989), Ethier and Kurtz (1986), Roelly-Coppoletta (1986), Walsh (1986) and Watanabe (1968), as well as other papers in this volume. Material on ILT can be found in Adler, Feldman and Lewin (1991), Adler and Lewin (1991), Adler and Rosen (1991), Dynkin (1988) and Perkins (1988).

[1]Research supported in part by US-Israel Binational Science Foundation (89-298), Air Force Office of Scientific Research (AFOSR 89-0261) and the Israel Academy of Sciences (702-90).

**(a)    Super Brownian Motion and Super Stable Processes.** We require some notation.

$$M \;=\; M(\Re^d) \;=\; \{\mu : \mu \text{ is a Radon measure on } \Re^d\}.$$

$$M_q \;=\; M_q(\Re^d) \;=\; \{\mu : \mu \in M, \; \textstyle\int_{\Re^d}(1 + \|x\|)^{-q}\mu(dx) < \infty\}.$$

$$C_o \;=\; C_o(\Re^d) \;=\; \{f : \Re^d \to \Re, \; f \text{ continuous}, \; \lim_{\|x\|\to\infty} f(x) = 0\}.$$

$$C_K \;=\; C_K(\Re^d) \;=\; \{f : \Re^d \to \Re, \; f \text{ continuous with compact support}\}.$$

The Sobolev space of functions whose $k$-th derivatives are in $\mathcal{L}^p$ is denoted by $W^{k,p}$. The $d$-dimensional Laplacian is denoted by $\Delta$, and the fractional Laplacian by $\Delta_\alpha = -(-\Delta)^{\alpha/2}$, $\alpha \in (0,2)$. (c.f. Yosida (1965).) With some abuse of notation, we shall let $\Delta_2 \equiv \Delta$. The domain of an operator $A$ is denoted by $D(A)$, $\mathcal{S}_d$ is the Schwartz space of rapidly decreasing functions on $\Re^d$, and $\mathcal{S}_d'$ is the space of tempered distributions on $\Re^d$. $(\Omega, \mathcal{F}, \mathcal{F}_t, P)$ is a filtered probability space.

The *super Brownian motion*, starting at $\mu \in M_q$, is a $M_q$-valued, $\mathcal{F}_t$-adapted, strong Markov process with $X_0 = \mu$ and satisfying

$$\begin{aligned}
\langle \varphi, X_t \rangle &= X_t(\varphi) \\
&= \mu(\varphi) + Z_t(\varphi) + \int_0^t X_s(\Delta\varphi)\,ds,
\end{aligned} \tag{1.1}$$

for every $\varphi \in D(\Delta) \cap \mathcal{S}_d$, where $Z_t$ is a continuous $\mathcal{F}_t$ martingale measure with increasing process

$$[Z(\varphi)]_t = \int_0^t X_s(\varphi^2)\,ds. \tag{1.2}$$

The *super (symmetric) stable process* is defined via the same recipe, but with $\Delta_\alpha$ replacing $\Delta$ and $Z^\alpha$ replacing $Z$ throughout.

For the remainder of this paper, we shall take $X_0 = \mu = $ Lebesgue measure, so we shall implicitly assume that $q > d$ throughout.

**(b)    Intersection Local Time (ILT).** At a heuristic level, the (self) intersection local time of a measure-valued process is a set indexed functional of the form

$$L(B) = \int_B ds\,dt \int_{\Re^d \times \Re^d} \delta(x - y)\,X_s(dx)X_t(dy), \tag{1.3}$$

where $B$ is a finite rectangle in $[0, \infty) \times [0, \infty)$ and $\delta$ is the Dirac delta function. A more precise definition of $L(B)$ is obtained by replacing the delta function by an approximate delta function $f_\epsilon \in S_d$, and then taking an $\mathcal{L}^2$ limit as $\epsilon \to 0$. In particular, let $\{f_\epsilon\}_{\epsilon > 0}$ be a collection of positive $C^\infty$ functions with $\int f_\epsilon(x) dx = 1$ for all $\epsilon$, such that $f_\epsilon \to \delta$ as $\epsilon \to 0$, where convergence is in the sense of distributions.

Replace (1.3) by

$$L_\epsilon(B) = \int_B ds dt \int_{\Re^d \times \Re^d} f_\epsilon(x - y) X_s(dx) X_t(dy). \qquad (1.4)$$

There are two qualitatively different cases that arise in this formulation. The first, and by far the simpler case, arises when the set $B$ does not intersect the diagonal $D = \{(t, t) : t \geq 0\}$. This case has been considered in detail in Dynkin (1988) to whom we refer the reader for details, noting here merely the fact that Dynkin has established the existence of $L(B)$ as an $\mathcal{L}^2$ limit of $L_\epsilon(B)$ for all dimensions $d \leq 7$, and that he treats a wider class of superprocesses than that considered in this paper.

The more difficult case, in which $B \cap D \neq \emptyset$, requires a renormalization argument, since any attempt at a direct approach to (1.3) as a limit of $L_\epsilon(B)$ yields a plethora of infinities. In order to state what is known in this case, we adopt the notation

$$\langle \varphi, X_t \rangle = \langle \varphi(x), X_t(dx) \rangle = \int_{\Re^d} \varphi(x) X_t(dx)$$

$$\langle \psi, X_s \times X_t \rangle = \langle \psi(x, y), X_s(dx) X_t(dy) \rangle = \int_{\Re^d \times \Re^d} \psi(x, y) X_s(dx) X_t(dy).$$

Then Rosen (1990) has proven the following result, which, contrary to the assumption made above, holds only if the initial measure $\mu = X_0$ is finite.

**Theorem 1.1 (Rosen (1990))** . *Let $\hat{\gamma}_\epsilon(T)$ be the approximate, renormalized, intersection local time defined by*

$$\hat{\gamma}_\epsilon(T) = \int_0^T dt \int_0^t ds \left( \langle f_\epsilon(x - y), X_s(dx) X_t(dy) \rangle \right.$$
$$\left. - E\{ \langle f_\epsilon(x - y), X_s(dx) X_t(dy) \rangle | \mathcal{F}_s \} \right), \qquad (1.5)$$

*where the particular sequence $\{f_\epsilon = \epsilon^{-d} f(\cdot / \epsilon)\}$, for some symmetric, positive $f \in C_K$ with $\int f(x) dx = 1$ is chosen for the approximate $\delta$-functions. If $X_t$ is a super Brownian motion and $d = 4$ or $5$, then $\hat{\gamma}_\epsilon(T)$ converges in $\mathcal{L}^2$, to a limit independent of $f$, as $\epsilon \to 0$. $\mathcal{L}^2$ convergence also holds if $X_t$ is a super stable process of index $\alpha$ and $d/3 < \alpha \leq d/2$. The limit process is denoted by $\hat{\gamma}(T)$ and is called Rosen's ILT process for $X$.*

Rosen's results also describe what happens when the conditions on $d$ and $\alpha$ are not satisfied, and cover superprocesses defined over smooth elliptic diffusions.

It is worthwhile to note at this point that, as we shall see later, the renormalization in (1.5) obtained from the conditional expectation is not unique. Rosen's choice of renormalization arises naturally from his style of proof, which is heavily based on moment arguments. Our approach, which is stochastic analysis based, will yield a slightly different renormalization. Note also that if $d \leq 3$ in the Brownian case, or $\alpha > d/2$ in the stable case, the renormalization is not necessary, as both terms in (1.5) have finite second moment for all $\epsilon > 0$. Furthermore, in both of these cases, the super process has a well-defined local time, which can be used to both define the ILT and to develop an evolution equation for it. (cf. the discussion in Adler and Rosen (1991) regarding a similar situation for the Brownian and stable density processes.)

The primary aim of this paper is to go beyond the existence results of Theorem 1.1, and derive an evolution equation describing the spatial (via a parameter to be introduced below) and temporal development of $\gamma(T)$.

**(c)   An Evolution Equation**   Let $p_t$ and $p_t^\alpha$, $\alpha \in (0,2)$ be, respectively, the transition densities of Brownian and symmetric stable processes. Thus, with a slight abuse of notation:

$$p_t^2(x,y) = p_t(x,y) = p_t(x-y)$$
$$= \frac{1}{(4\pi t)^{d/2}} e^{-\|x-y\|^2/4t},$$

$$p_t^\alpha(x,y) = p_t^\alpha(x-y)$$
$$= \frac{1}{(2\pi)^d} \int_{\Re^d} \exp(-ip \cdot (x-y) - t\|p\|^\alpha)\,dp.$$

The corresponding Green's functions, for $\alpha \in (0,2]$ and $\lambda \geq 0$, are defined by

$$G_\alpha^\lambda(x-y) = G_\alpha^\lambda(x,y) = \int_0^\infty e^{-\lambda t} p_t^\alpha(x,y)\,dt \qquad (1.6)$$

where we allow ourselves the luxury of writing $G_\alpha^\lambda$ as a function of either one or two parameters at will. Note that, depending on the values of $\alpha$ and $d$, $G_\alpha^0$ may be identically infinite. For all $\alpha$ and $d$, however, $G_\alpha^\lambda(x,y)$ is finite if $\lambda > 0$ and $x \neq y$. Furthermore, $G_\alpha^\lambda \in \mathcal{L}^1$ for all $\lambda > 0$. We shall list a number of properties of $G_\alpha^\lambda$ in the following section. The property of central importance to us is the fact that the distributional equation

$$(-\Delta_\alpha + \lambda)u = \delta, \qquad (1.7)$$

where $\delta$ is the Dirac delta function, is solved by $u = G_\alpha^\lambda$. That is, for every test function $\varphi \in \mathcal{S}_d$,

$$\int_{\Re^d} ((-\Delta_\alpha + \lambda) G_\alpha^\lambda)(x) \varphi(x) dx = \varphi(0).$$

The fact that $G_\alpha^\lambda \in \mathcal{L}^1$ for $\lambda > 0$ implies that $G_\alpha^\lambda$ can also be treated as an $\mathcal{S}_d'$ distribution. Thus, there exists a family $\{G_\epsilon\}_{\epsilon>0}$ of $C_K$ functions such that $G_\epsilon \to G_\alpha^\lambda$ as $\epsilon \to 0$, in $\mathcal{S}_d'$. Since $\Delta_\alpha$ is a continuous operator on $\mathcal{S}_d'$ (cf. Hörmander (1985), page 70) it follows that

$$G_\epsilon^{\alpha,\lambda} := (-\Delta_\alpha + \lambda) G_\epsilon \to \delta \tag{1.8}$$

as $\epsilon \to 0$, where convergence is again in $\mathcal{S}_d'$. Thus it is not unreasonable to attempt to define a new approximate, renormalized ILT by setting, for every $\varphi \in \mathcal{S}_d$,

$$\begin{aligned} \gamma_\epsilon^\lambda(T, \varphi) &= \int_0^T dt \int_0^t ds \langle \varphi(x) G_\epsilon^{\alpha,\lambda}(x-y), X_s(dx) X_t(dy) \rangle \\ &\quad - \int_0^T \langle \varphi(x) G_\epsilon(x-y), X_t(dx) X_t(dy) \rangle dt. \end{aligned} \tag{1.9}$$

There are a number of differences between the approximate ILT, $\gamma_\epsilon^\lambda(T, \varphi)$, and the approximate ILT $\gamma_\epsilon(T)$, of Rosen. Firstly, the addition of the test function $\varphi$ gives information on the spatial dispersion of self-intersections that is not available otherwise. (The addition of this parameter is also necessitated by the fact that we work with an infinite initial measure.) Secondly, $\gamma_\epsilon^\lambda$ is not so much *an* ILT as a *family* of ILT's indexed by the parameter $\lambda$. This emphasizes the non-uniqueness of the renormalization. The main difference, however, lies in the structure of the renormalization. As the proofs below will show, the renormalizing second term of (1.9) arises naturally from stochastic analysis considerations. The fact that it involves the "doubling" of the measure $X$ – i.e. it involves $X_s \times X_t$ only for $t = s$ – is highly reminiscent of the renormalized, self-intersection local time for $\Re^d$ valued processes for which the renormalization involves the removal of "local double points" of the process. For this reason, we find our renormalization esthetically more appealing than that of (1.5).

**Theorem 1.2** *Let $X_t$ be a super Brownian motion or super stable process, and let $\gamma_\epsilon^\lambda(T, \varphi)$ be its approximate renormalized intersection local time as defined by (1.9). If $d = 4$ or $5$ in the Brownian case, or $d/3 < \alpha \leq d/2$ in the* *stable case, then for all $\lambda > 0$, all $T \in (0, \infty)$ and all $\varphi \in S_d$, $\gamma_\epsilon^\lambda(T, \varphi)$ converges in $\mathcal{L}^2$ to a finite limit $\gamma^\lambda(T, \varphi)$ as $\epsilon \to 0$ which we call a renormalized ILT of $X_t$, and which is independent of the approximating sequence $\{G_\epsilon\}$. Furthermore, $\gamma^\lambda$ has the following representation in terms of the process $X_t$ and the martingale measure $Z_t$:*

$$\gamma^\lambda(T, \varphi) = \int_0^T \int_{\Re^d} \left\{ \int_0^t \langle G_\alpha^\lambda(x - y)\varphi(x), X_s(dx) \rangle \, ds \right\} Z(dt, dy)$$

$$+ \lambda \int_0^T dt \int_0^t ds \, \langle G_\alpha^\lambda(x - y)\varphi(x), X_s(dx) \, X_t(dy) \rangle$$

$$- \int_0^T \langle G_\alpha^\lambda(x - y)\varphi(x), X_t(dx) \, X_T(dy) \rangle \, dt. \qquad (1.10)$$

**(d)  About the Proof.** As is the case with most evolution equations of the above kind, the derivation hinges on finding an appropriate Itô formula and applying it carefully. The following Itô formula, which is actually slightly more general than we require, and is modelled on a similar result of Dawson (1978), will be proven in Section 3.

**Lemma 1.3** (Itô formula). *Let $X_t$ be a Brownian or stable superprocess. Fix $k \geq 1$, $\Psi \in C^2(\Re_+ \times \Re^k)$, and assume $\Psi$ and its first and second order derivatives satisfy a polynomial growth condition at infinity. Let $\varphi_1, \ldots, \varphi_k$ belong to the space $W^{2,p} \cap W^{2,1}$ for some $p \geq 2$. Let $\hat{\varphi} = (\varphi_1, \ldots, \varphi_k)$, and write $\langle \hat{\varphi}, X_t \rangle$ for the vector $(\langle \varphi_1, X_t \rangle, \ldots, \langle \varphi_k, X_t \rangle)$. Then, for all $t > 0$,*

$$\Psi(t, \langle \hat{\varphi}, X_t \rangle) = \Psi(0, \langle \hat{\varphi}, X_0 \rangle) + \int_0^t \Psi_t(s, \langle \hat{\varphi}, X_s \rangle) \, ds$$

$$+ \int_0^t \sum_{i=1}^k \Psi_{x_i}(s, \langle \hat{\varphi}, X_s \rangle) \cdot \langle \Delta_\alpha \varphi_i, X_s \rangle \, ds$$

$$+ \frac{1}{2} \int_0^t \sum_{i=1}^k \sum_{j=1}^k \Psi_{x_i x_j}(s, \langle \hat{\varphi}, X_s \rangle) \cdot \langle \varphi_i \varphi_j, X_s \rangle \, ds$$

$$+ \int_0^t \int_{\Re^d} \sum_{i=1}^k \Psi_{x_i}(s, \langle \hat{\varphi}, X_s \rangle)\varphi_i(x) \, Z(ds, dx) \qquad (1.11)$$

*where*

$$\Psi_t = \partial\Psi(t, x)/\partial t,$$
$$\Psi_{x_i} = \partial\Psi(t, x)/\partial x_i,$$
$$\Psi_{x_i x_j} = \partial^2\Psi(t, x)/\partial x_i \partial x_j.$$

As is usual, the Itô formula also holds if $\Psi$ is a non-anticipative functional of the process $X_t$. We shall be interested in the particular functional

$$\Psi(t, x) = x \int_0^t \langle \psi, X_s \rangle \, ds, \tag{1.12}$$

where $\psi \in W^{2,p} \cap W^{2,1}$ and $x \in \Re^d$. Note that

$$\begin{aligned}
\Psi(0, x) &\equiv 0, \\
\Psi_t(t, x) &= x \langle \psi, X_t \rangle, \\
\Psi_x(t, x) &= \int_0^t \langle \psi, X_s \rangle \, ds, \\
\Psi_{xx}(t, x) &\equiv 0.
\end{aligned}$$

Apply (1.11) with this choice of $\Psi$ to obtain

**Lemma 1.4** *Let $\varphi, \psi \in W^{2,p} \cap W^{2,1}$ for some $p \geq 2$. Then for every finite $T > 0$,*

$$\begin{aligned}
&\int_0^T \langle \psi(x)\varphi(y), \ X_t(dx) X_T(dy) \rangle \, dt \\
&= \int_0^T dt \int_0^t ds \, \langle \psi(x)\Delta_\alpha\varphi(y), \ X_s(dx) X_t(dy) \rangle \\
&\quad + \int_0^T \langle \psi(x)\varphi(y), \ X_t(dx) X_t(dy) \rangle \, dt \\
&\quad + \int_0^T \int_{\Re^d} \int_0^t ds \, \langle \psi(x)\varphi(y), \ X_s(dx) \rangle \ Z(dt, dy). \tag{1.13}
\end{aligned}$$

This lemma is of importance in so far as it leads us to the following Lemma 1.5, which almost implies the central Theorem 1.2. Lemma 1.5, itself, follows in a reasonably straightforward fashion from Lemma 1.4 by first extending the latter to functions of the form $\Phi(x, y) = \sum_{i=1}^N \varphi_i(x)\psi_i(y)$ and then to general $\Phi(x, y) \in C^2(\Re^d \times \Re^d)$ of compact support.

**Lemma 1.5** *Let $\Phi(x, y) \in C^2(\Re^d \times \Re^d)$ have compact support. Then (1.13) continues to hold if $\psi(x)\varphi(y)$ is replaced, throughout, by $\Phi(x, y)$, and $\psi(x) \times \Delta_\alpha\varphi(y)$ by $\Delta_\alpha^{(y)}\Phi(x, y)$, where we use $\Delta_\alpha^{(y)}$ to denote $\Delta_\alpha$ operating on the second variable of $\Phi(x, y)$. That is, after reordering*

$$\begin{aligned}
&\int_0^T dt \int_0^t ds \, \langle \Delta_\alpha^{(y)}\Phi(x, y), \ X_s(dx) X_t(dy) \rangle \\
&\quad + \int_0^T \langle \Phi(x, y), \ X_t(dx) X_t(dy) \rangle \, dt \\
&= \int_0^T \langle \Phi(x, y), \ X_t(dx) X_T(dy) \rangle \, dt \\
&\quad - \int_0^T \int_{\Re^d} \int_0^t ds \, \langle \Phi(x, y), \ X_s(dx) \rangle \ Z(dt, dy). \tag{1.14}
\end{aligned}$$

To derive the evolution equation (1.10) for the ILT, $\gamma(T, \varphi)$, "all" that has to be done is to replace the function $\Phi(x, y)$ of (1.14) by $G_\alpha^\lambda(x - y)\varphi(x)$, and to note the relationship (1.7) between $G_\alpha^\lambda$, $\Delta_\alpha$ and the delta function $\delta$ as well as the definition of the renormalized ILT as the limit of the $\gamma_\epsilon(T, \varphi)$ of (1.9). Unfortunately, however, this simple recipe requires justification at a large number of stages. Hence the long proof of Section 3.

**Acknowledgements:** Our path, through the calculations of the following two sections, was made substantially easier by being able to obtain guidance from the Master of the Moment, Jay Rosen. Steve Krone pointed out some minor errors in an earlier version of the paper.

## 2   On Evaluating Moments

An important component in virtually all the proofs of the following section will be the evaluation of moments of the kind $E \prod_{i=1}^{k} \langle \varphi_i, X_{t_i} \rangle$, when $X_t$ is a Brownian or stable super process. An algorithm for calculating these, based exclusively on the transition densities $p_t(x, y)$ and $p_t^\alpha(x, y)$ of Section 1(c), has been given by Dynkin (1988)[2].

$$
E\left(\prod_{i=1}^{k} \langle \varphi_i, X_{t_i} \rangle \right)
$$
$$
= \sum_{D_k} \int \prod_{v \in V_-} dy_v \prod_{\ell \in L} p_{s_{f(\ell)} - s_{i(\ell)}}^\alpha (y_{f(\ell)} - y_{i(\ell)})
$$
$$
\times \prod_{v \in V_o} ds_v dy_v \prod_{i=1}^{k} \varphi_i(z_i) dz_i \qquad (2.1)
$$

where $p_t^\alpha(x) \equiv 0$ if $t < 0$, and $D_k$ is the set of directed binary graphs with $k$ exits marked $1, 2, \ldots, k$. The convention that $p_t^\alpha(x) \equiv 0$ if $t < 0$ is important, and will be used throughout the remainder of this paper without further comment. Given such a graph, $L$ is the set of directed links, and if the link $\ell \in L$ goes from vertex $v$ to vertex $w$, we write $v = i(\ell)$, $w = f(\ell)$. To each vertex $v$ we associate two variables.

$$
(s_v, y_v) \in \Re_+ \times \Re^d ,
$$

---

[2]Dynkin's formulae were acually proven for finite initial measures, and any positive, measurable $\varphi$. That the formulae also hold for a Lebesgue initial measure, and $\varphi \in \mathcal{L}^p$, $p > 1$, is a consequence of Dynkin's proof and Theorem 2.3 of Adler and Lewin (1991).

which we refer to as the time and space coordinates of $v$. $V_-$ denotes the set of entrances for our graph, and if $v \in V_-$ we set $s_v = 0$ and $y_v = x_v$. If $v$ is the exit labelled by $j$, $i \leq j \leq n$, we set

$$(s_v, y_v) \equiv (t_j, z_j).$$

Finally, $V_o$ denotes the set of internal vertices, i.e. those vertices that are neither entrances nor exits.

We shall, in what follows, be interested only in third and fourth order moments. In the latter case, the set $D_4$ of (2.1) consists of the following six basic graphs, and their various combinatorial rearrangements. The contribution of graph – i.e. the integral appearing in (2.1) – is written after the graph. Since we shall be concerned only with finiteness of moments, we shall not bother with the combinatorial factors associated with each graph.

### Graph 1

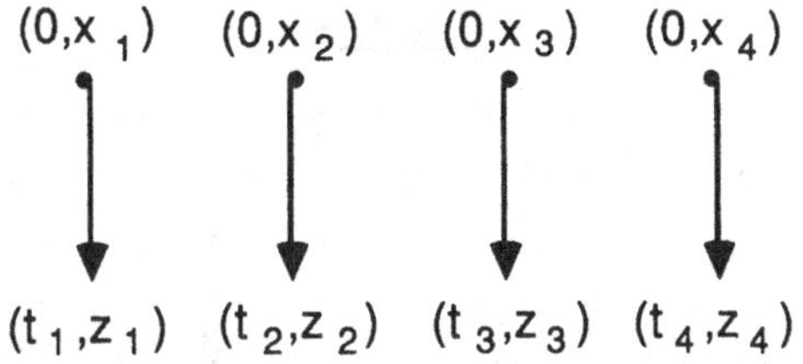

$$(0, x_1) \quad (0, x_2) \quad (0, x_3) \quad (0, x_4)$$

$$(t_1, z_1) \quad (t_2, z_2) \quad (t_3, z_3) \quad (t_4, z_4)$$

$$\prod_{i=1}^{4} \left\{ \int p_{t_i}^{\alpha}(x_i, z_i) \varphi_i(z_i) \, dx_i dz_i \right\}.$$

## Graph 2

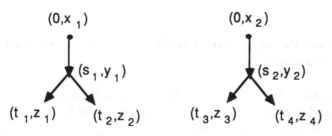

$$\int dx_1 dx_2 \int dy_1 dy_2 \int ds_1 ds_2 \, p^\alpha_{s_1}(x_1, y_1) p^\alpha_{s_2}(x_2, y_2)$$

$$\times \int p^\alpha_{t_1-s_1}(y_1, z_1) p^\alpha_{t_2-s_1}(y_1, z_2) p^\alpha_{t_3-s_2}(y_2, z_3) p^\alpha_{t_4-s_2}(y_2, z_4) \prod_{i=1}^{4} \varphi(z_i) dz_i$$

## Graph 3

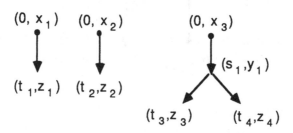

$$\int dx_1 dx_2 dx_3 \int ds_1 \int dy_1 \, p_{s_1}(x_3, y_1)$$

$$\times \int p^\alpha_{t_1}(x_1, z_1) p^\alpha_{t_2}(x_2, z_2) p^\alpha_{t_3-s_1}(y_1, z_3) p^\alpha_{t_4-s_1}(y_1, z_4) \prod_{i=1}^{4} \varphi(z_i) dz_i$$

## Graph 4

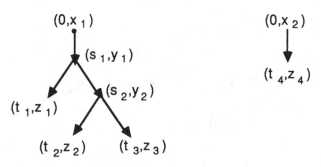

$$\int dx_1 dx_2 \int ds_1 \int dy_1 \, p_{s_1}^\alpha(x_1, y_1) \int ds_2 \int dy_2 \, p_{s_2-s_1}(y_1, y_2)$$

$$\times \int p_{t_1-s_1}^\alpha(y_1, z_1) p_{t_2-s_2}^\alpha(y_2, z_2) p_{t_3-s_2}^\alpha(y_2, z_3) p_{t_4}^\alpha(x_1, z_4) \prod_{i=1}^{4} \varphi(z_i) dz_i$$

## Graph 5

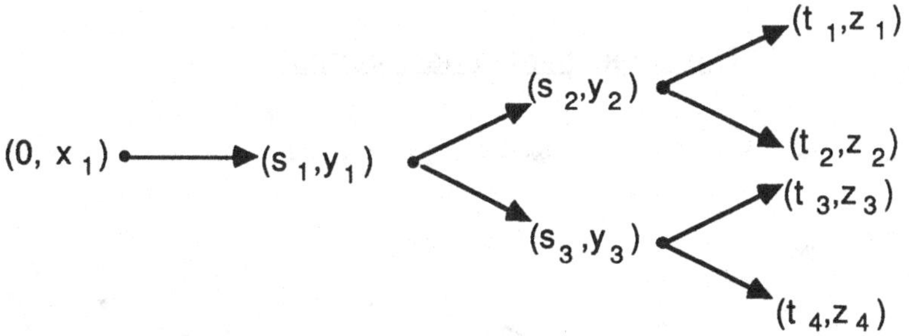

$$\int dx_1 \int ds_1 \int dy_1 \, p_{s_1}^\alpha(x_1, y_1) \int ds_2 ds_3 \int dy_2 dy_3 \, p_{s_2-s_1}^\alpha(y_1, y_2) p_{s_3-s_1}^\alpha(y_1, y_3)$$

$$\times \int p_{t_1-s_2}^\alpha(y_2, z_1) p_{t_2-s_2}^\alpha(y_2, z_2) p_{t_3-s_3}^\alpha(y_3, z_3) p_{t_4-s_3}^\alpha(y_3, z_4) \prod_{i=1}^{4} \varphi(z_i) dz_i$$

### Graph 6

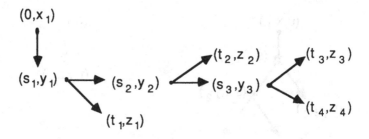

$$\int dx_1 \int ds_1 \int dy_1\, p_{s_1}^{\alpha}(x_1, y_1) \int ds_2 \int dy_2\, p_{s_2-s_1}^{\alpha}(y_1, y_2) \int ds_3 \int dy_3\, p_{s_3-s_2}^{\alpha}(y_2, y_3)$$

$$\times \int p_{t_1-s_1}^{\alpha}(y_1, z_1) p_{t_2-s_2}^{\alpha}(y_2, z_2) p_{t_3-s_3}^{\alpha}(y_3, z_3) p_{t_4-s_3}^{\alpha}(y_3, z_4) \prod_{i=1}^{4} \varphi(z_i) dz_i$$

Similar considerations apply in establishing the graphs and their contributions required for calculating third order moments. Since the pattern should, by now, be clear, we shall merely list the graphs, leaving the calculation of their contributions to the reader, and the discussion of the following section.

### The three third order graphs

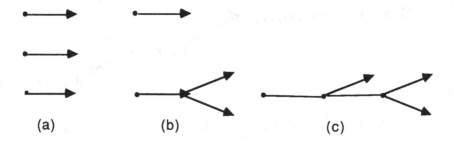

(a)                    (b)                              (c)

## 3   Proofs

**(a)    Proof of Lemma (1.3) – The Itô Formula.** The Itô formula

(1.11) follows in a reasonably straightforward fashion from a standard Itô formula for continuous semi-martingales, and so we shall only sketch the proof, highlighting the technical dificulties. We shall also restrict ourselves to the one-dimensional case ($k = 1$ in (1.11)) which is all we really need in this paper.

Fix a test function $\varphi \in W^{2,p} \cap W^{2,1}$, $p \geq 2$, and note that $\langle \varphi, X_t \rangle$, where $X_t$ is either a super Brownian motion or super stable process, is a continuous semi-martingale (cf., for example Roelly-Coppoletta (1986) for this result for a more restricted class of $\varphi$, and Adler and Lewin (1991) for $\varphi \in W^{2,p} \cap W^{2,1}$). The associated finite variation process is given by $\langle \Delta_\alpha \varphi, X_t \rangle$, and the martingale part by $\langle \varphi, Z_t \rangle$, with associated increasing process $[\langle \varphi, Z \rangle]_t = \int_0^t \langle \varphi, X_s \rangle^2 ds$, (cf.(1.2)).

It thus follows from a standard Itô formula (e.g. Karatzas and Shreve (1988)), that for $\Psi \in C^2(\Re_+ \times \Re)$ and $\varphi \in W^{2,p} \cap W^{2,1}$, the formula (1.11) of Lemma 1.3 holds, with the singular exception that the last term there (remember that $k = 1$) is replaced by

$$\int_0^t \Psi_x(s, \langle \varphi, X_s \rangle)\, d\langle \varphi, Z_s \rangle. \tag{3.1}$$

To complete the proof, we have to show that the above expression is equivalent to

$$\int_0^t \int_{\Re^d} \Psi_x(s, \langle \varphi, X_s \rangle)\varphi(x)\, Z(ds,\, dx).$$

Recall that, by assumption, there exists an integral $a \geq 0$ such that $\Psi_x(s, x) \leq C(s)\|x\|^a$ for all $x \in \Re^d$. It thus follows from Dynkin's moment formulae that $E(\Psi_x(s, \langle \varphi, X_s \rangle))^2 < \infty$. Hence the integral (3.1) is also in $\mathcal{L}^2(P)$.

We can therefore approximate $\Psi_x(s, \langle \varphi, X_s \rangle)$ in $\mathcal{L}^2(P)$ by a simple function of the form

$$\sum_{i=1}^n \sum_{A_i} c_{A_i} I_{A_i}(\omega) I_{(s_i, t_i)}(s)$$

where

$$(s_i, t_i] \cap (s_j, t_j] = \emptyset$$

if $i \neq j$, $\bigcup_i (s_i, t_i] = \Re_+$, and $A_i \in \mathcal{F}_{s_i}$, $\bigcup_{A_i} A_i = \Omega$. Thus the integral (3.1) can be approximated in $\mathcal{L}^2(P)$ by

$$\sum_i \sum_{A_i} c_{A_i} I_{(0,t]}(t_i) [\langle \varphi, Z_{t_i} \rangle - \langle \varphi, Z_{s_i} \rangle]. \tag{3.2}$$

But, according to Walsh's (1986) formulation of stochastic integration with respect to martingale measures, if we also take an approximation via simple functions to $\psi(x) = \Psi_x(s, \langle \varphi, X_s \rangle)\varphi(x)$, we immediately have that (3.2) is also an $\mathcal{L}^2(P)$ approximation to

$$\int_0^t \int_{\Re^d} \Psi_x(s, \langle \varphi, X_s \rangle)\, \varphi(x)\, Z(ds,\, dx).$$

Going to the $\mathcal{L}^2(P)$ limit on both sides of this equivalence establishes the required result.

**(b)     Proof of Lemma 1.4.**  As in the previous proof, one can apply a regular Itô formula for the continuous semi-martingale $\langle \varphi, X_t \rangle$ to the non-anticipative functional $\Psi(t, x) = x \int_0^t \langle \varphi, X_s \rangle ds$, and evaluate it at $\Psi(t, \langle \psi, X_t \rangle)$. Noting the various identities for $\Psi$ and its derivatives following (1.12), we thus obtain (1.13) but for the fact that the last term is replaced by

$$\int_0^T \int_0^t ds\, \langle \psi(x), X_s \rangle\, d\langle \varphi, Z_t \rangle.$$

As before, we have to rewrite this as a stochastic integral against the martingale measure $Z(dt,\, dx)$ to obtain the required form of (1.13). The argument, however, is precisely as before.

**(c)     Proof of Lemma 1.5.**  We commence with functions $\Phi_N(x, y)$ of the form

$$\Phi_N(x, y) = \sum_{k=1}^N \varphi_k(x)\psi_k(y) \tag{3.3}$$

where both $\varphi, \psi \in W^{2,p} \cap W^{2,1}$. It follows immediately from Lemma 1.4, with the notation of our lemma, that

$$\int_0^T \langle \Phi_N(x, y),\, X_t(dx)\, X_T(dy) \rangle\, dt$$

$$= \int_0^T dt \int_0^t ds\, \langle \Delta_\alpha^{(y)} \Phi_N(x, y),\, X_s(dx)\, X_t(dy) \rangle$$

$$+ \int_0^T \langle \Phi_N(x, y),\, X_t(dx)\, X_t(dy) \rangle\, dt$$

$$+ \int_0^T \int_{\Re^d} \int_0^t ds\, \langle \Phi_N(x, y),\, X_s(dx) \rangle\, Z(dt,\, dy). \tag{3.4}$$

This, of course, is the required (1.14) but for the restriction of the form (3.3). We need (3.4) for $\Phi(x, y) \in C^2(\Re^d \times \Re^d)$ of compact support.

Note, however, that every such function can be represented as the limit of functions $\Phi_N$ of the form (3.3), with convergence in the supremum norm

$\|\varphi\|_\infty = \sup_x |\varphi(x)|$. Thus, let $\Phi$ be as required, and let $\{\Phi_N\}_{N\le 1}$ be an appropriate approximating sequence. We shall proceed by showing that each of the four terms in (3.4) is Cauchy in $\mathcal{L}^2(P)$, and that the limit variable is the corresponding expression with $\Phi_N$ replaced by $\Phi$. We shall give details only for the first and last terms. The first three are all similar, but the last, being a stochastic integral, is somewhat different.

(A)   The First Term:   Because of the form of $\Phi_N$ as a sum, it is clear that $\mathcal{L}^2(P)$ Cauchy convergence of $\int_0^T \langle \Phi_N(x,y),\ X_t(dx)\, X_T(dy)\rangle\, dt$ will result from

$$E|\langle \Phi_N(x,y),\ X_t(dx)\, X_T(dy)\rangle|^2 \le C \cdot \|\Phi_N\|_\infty^q \qquad (3.5)$$

for some $q > 0$, where $C$ is a constant that may depend on $T$ and $\Phi$. In particular, if $D \subset \Re^d \times \Re^d$ is the union of the supports of the $\Phi_N$, we expect $C$ to depend on $D$. ($D$, of course, can be assumed compact.) The left-hand side of (3.5) is equivalent to

$$E\left(\sum_{k=1}^N \langle \varphi_k,\ X_t\rangle \cdot \langle \psi_k,\ X_T\rangle\right)^2$$
$$= \sum_{j=1}^N \sum_{k=1}^N E\left\{\langle \varphi_k,\ X_t\rangle\langle \varphi_j,\ X_t\rangle\langle \psi_k,\ X_T\rangle\langle \psi_j,\ X_T\rangle\right\}. \qquad (3.6)$$

Each expectation here can be treated via the graph formulae of Section 2. Since these are fourth order moments, there are six different graphs to consider. While the requisite calculations are not difficult, they are tedious, so we shall look, in detail, at only two cases. Consider the case arising from Graph 1, the simplest of the six. By the formula following the graph, the corresponding contribution to (3.6) is

$$\sum_{j=1}^N \sum_{k=1}^N \int_{\Re^{8d}} \varphi_k(z_1)\varphi_j(z_2)\psi_k(z_3)\psi_j(z_4)$$
$$\times\ p_t^\alpha(x_1, z_1)\, p_t^\alpha(x_2, z_2)\, p_T^\alpha(x_3, z_3)\, p_T^\alpha(x_4, z_4) \prod_{i=1}^4 dx_i dz_i.$$

Note that the $p^\alpha$ are symmetric, and do the $x_i$ integrals. The resulting expression is

$$\sum_{j=1}^N \sum_{k=1}^N \int_{\Re^{4d}} \varphi_k(z_1)\psi_k(z_2)\varphi_j(z_3)\psi_j(z_4)\, dz_1 dz_2 dz_3 dz_4$$
$$= \int_{\Re^{4d}} \Phi_N(z_1, z_2)\Phi_N(z_3, z_4)\, dz_1 dz_2 dz_3 dz_4. \qquad (3.7)$$

Since the $\Phi_N$ all have support within a common compact set $D$, this last integral is bounded above by $|D|^4\|\Phi_N\|_\infty^2$, where $|D|$ denotes the Lebesgue measure of $D$.

As a second example, consider the cases corresponding to the evaluation of the summands of (3.6) via Graph 2. There are really two different subcases here, depending on whether the two disjoint subgraphs there end with $\{(t,\varphi_k),(t,\varphi_j)\}$ and $\{(T,\psi_k),(T,\psi_j)\}$ or with $\{(t,\varphi_k),(T,\psi_j)\}$ and $\{(t,\varphi_j),(T,\psi_k)\}$. Consider the first of these. Their contribution to (3.6) is of the form

$$\sum_{j=1}^{N}\sum_{k=1}^{N}\int dx_1 dx_2 \int dy_1 dy_2 \int ds_1 ds_2\, p_{s_1}^\alpha(x_1,y_1)\, p_{s_2}^\alpha(x_2,y_2)$$

$$\times\ \int p_{t-s_1}^\alpha(y_1,z_1)\, p_{t-s_1}^\alpha(y_1,z_2)\, p_{T-s_2}^\alpha(y_2,z_3)\, p_{T-s_2}^\alpha(y_2,z_4)$$

$$\times\ \varphi_j(z_1)\varphi_k(z_2)\psi_j(z_3)\psi_k(z_4)\prod_{i=1}^{4} dz_i. \tag{3.8}$$

As before, note the symmetry $p_t^\alpha(u,v) = p_t^\alpha(v,u)$, and thus integrate out $x_1$ and $x_2$. Perform the $y_1, y_2$ integrals using the Chapman-Kolmogorov formula to obtain that (3.8) is equivalent to

$$\sum_{j=1}^{N}\sum_{k=1}^{N}\int ds_1 ds_2\int p_{2(t-s_1)}^\alpha(z_1,z_2)p_{2(T-s_2)}^\alpha(z_3,z_4)$$

$$\times\ \varphi_j(z_1)\varphi_k(z_2)\psi_j(z_3)\psi_k(z_4)dz_1 dz_2 dz_3 dz_4$$

$$=\int ds_1 ds_2\int p_{2(t-s_1)}^\alpha(z_1,z_2)p_{2(T-s_2)}^\alpha(z_3,z_4)$$

$$\times\ \Phi_N(z_1,z_3)\Phi_N(z_2,z_4)dz_1 dz_2 dz_3 dz_4. \tag{3.9}$$

Applying the same argument as before, and noting that $s_1, s_2 < T$, we obtain that (3.9) is bounded above by $T^2|D|^2\|\Phi_N\|_\infty^2$.

The remaining graphs are treated similarly, as are the other three non-stochastic integrals in (3.4). It remains to show, however, that the Cauchy, $\mathcal{L}^2(P)$, convergence is to the correct limit.

In the current case, however, this is easy. Since the convergence of the $\Phi_N$ to $\Phi$ is in the supremum norm, convergence is also uniform. For each $s, t \geq 0$, $X_t \times X_s$ is a.s. a finite measure on $D$, and so $\langle \Phi_N, X_s \times X_t\rangle \overset{a.s.}{\to} \langle \Phi, X_s \times X_t\rangle$. We have just shown that $\langle \Phi_N, X_s \times X_t\rangle$ is $\mathcal{L}^2(P)$ convergent. Since the $\mathcal{L}^2(P)$ and a.s. limit must agree, we are done.

(B) <u>The Stochastic Integral Term</u>: We must now apply a similar argument to the last term of (3.4). We start with Cauchy convergence, for which we take the $\Phi_N$ as above. We shall show that

$$E\left(\int_0^T \int_{\Re^d} \int_0^t ds \, \langle \Phi_N(x,y), X_s(dx)\rangle \, Z(dt, dy)\right)^2 \le C.\|\Phi_N\|_\infty^2 \qquad (3.10)$$

where $C = C(T, D)$. Since $Z$ is a martingale measure with associated increasing process $[\langle \varphi, Z. \rangle]_T = \int_0^T \langle \varphi^2, X_t \rangle dt$, (cf. Walsh (1986)), the left-hand side of (3.10) is equivalent to

$$E \int_0^T \int_{\Re^d} \left(\int_0^t \langle \Phi_N(x,y), X_s(dx)\rangle ds\right)^2 X_t(dy) \, dt. \qquad (3.11)$$

Note the form of $\Phi_N$, and expand the square to obtain that this is equivalent to

$$\int_0^T dt \int_0^t ds \int_0^t ds' \sum_{i=1}^N \sum_{j=1}^N E\left\{\langle \varphi_j, X_s\rangle \langle \varphi_k, X_{s'}\rangle \langle \psi_j \psi_k, X_t\rangle\right\} . \qquad (3.12)$$

Evaluation of each summand here hinges on the three graphs at the end of Section 2. Consider, for example, the third and most complex of these. The corresponding contributions to (3.12) are of the form

$$\int_0^T dt_3 \int_0^{t_3} dt_1 \int_0^{t_3} dt_2 \sum_{i=1}^N \sum_{j=1}^N \int_{\Re^d} dx \int_0^T ds_1 \int_{\Re^d} dy_1 \, p_{s_1}^\alpha(x, y_1)$$

$$\times \int_0^T ds_2 \int_{\Re^d} dy_2 \, p_{s_2-s_1}(y_1, y_2) \int_{\Re^{3d}} p_{t_1-s_1}^\alpha(y_1, z_1) p_{t_2-s_2}^\alpha(y_2, z_2) p_{t_3-s_2}^\alpha(y_2, z_3)$$

$$\times \varphi_j(z_1)\varphi_k(z_2)\psi_j(z_3)\psi_k(z_3) \, dz_1 dz_2 dz_3. \qquad (3.13)$$

(There are, of course, a number of contributions of this kind with the rôles of the various pairs of times and test functions interchanged.) As in the previous case, use the symmetry of $p_t^\alpha$ to integrate out $x$, and Chapman-Kolmogorov to integrate over $y_1$. After rearrangement, (3.13) becomes

$$\int_{\Re^{3d}} \Phi_N(z_1, z_3)\Phi_N(z_2, z_3)$$

$$\times \int_0^T dt_3 \int_0^{t_3} dt_1 \int_0^{t_3} dt_2 \int_0^T ds_1 \int_0^T ds_2 \int_{\Re^d} dy_2 \, p_{t_1+s_2-2s_1}^\alpha(y_2, z_1)$$

$$\times p_{t_2-s_2}^\alpha(y_2, z_2) p_{t_3-s_2}^\alpha(y_2, z_3) dz_1 dz_2 dz_3. \qquad (3.14)$$

Bound $\Phi_N(z_1, z_3)\Phi_N(z_2, z_3)$ by $\|\Phi_N\|_\infty^2$, and then perform the $z_3$ integral over all of $\Re^d$ rather than just $D$. Then integrate out $y_2$ via Chapman-Kolmogorov, to bound (3.14) by

$$\|\Phi_N\|_\infty^2 \int_0^T dt_3 \int_0^{t_3} dt_1 \int_0^{t_3} dt_2 \int_0^T ds_1 \int_{D \times D} p_{(t_1+t_2-2s_1)}^\alpha(z_1, z_2) \, dz_1 dz_2. \qquad (3.15)$$

Integrate $z_2$ over $\Re^d$, and then $z_1$ over $D$ to finally find a bound of the form $|D|T^4\|\Phi\|_\infty^2$, as required. All other terms can be handled similarly.

As before, to complete the proof, we must show that the $\mathcal{L}^2(P)$ limit, assured by Cauchy convergence is, in fact, the stochastic integral

$$\int_0^T \int_{\Re^d} \int_0^t ds\, \langle \Phi(x,y),\, X_s(dx)\rangle\, Z(dt,dz). \qquad (3.16)$$

This, however, is relatively straightforward by a subsequencing argument. Implicit in the above calculations is the fact that, for every $y \in D_Y = \{y : (x,y) \in D\}$ and $t \in [0,T]$, $\langle \Phi_N(x,y), X_t(dx)\rangle$ is Cauchy in $\mathcal{L}^2(P)$. Thus there is an a.s. convergent subsequence $\langle \Phi_{N_k}(x,y), X_t(dx)\rangle$, which, since $X_t(D_x) < \infty$, where $D_X = \{x : (x,y) \in D\}$, and the convergence of $\Phi_N$ to $\Phi$ is uniform, implies $\langle \Phi_{N_k}(x,y), X_t(dx)\rangle \overset{a.s.}{\to} \langle \Phi(x,y), X_t(dx)\rangle$ as $k \to \infty$. Furthermore, both $\langle \Phi_N(x,y), X_t(dx)\rangle$ and $\langle \Phi(x,y), X_t(dx)\rangle$ are uniformly continuous in $y \in D_Y$ and $t \in [0,T]$. Hence, along the subsequence $\{N_k\}_{k \geq 1}$ we must have

$$\int_0^T \int_{\Re^d} \int_0^t ds\, \langle \Phi_{N_k}(x,y),\, X_s(dx)\rangle\, Z(dt,dy)$$

$$\overset{a.s.}{\to} \int_0^T \int_{\Re^d} \int_0^t ds\, \langle \Phi(x,y),\, X_s(dx)\rangle\, Z(dt,dy).$$

Thus the $\mathcal{L}^2(P)$ convergence along the full sequence must be to the appropriate limit.

**(d)   Proof of Theorem 1.2**   We can now, finally, turn to the proof of Theorem 1.2. We shall concentrate on establishing the evolution equation (1.10). Since the proof is of an $\mathcal{L}^2$ nature, the convergence of $\gamma_\epsilon^\lambda(T,\varphi)$, which forms the first part of the theorem, will be proven *en passant*.

To start, let $\{G_\epsilon\}_{\epsilon > 0}$ be a sequence of $C^\infty$ functions of bounded support such that $G_\epsilon \overset{\mathcal{L}^1}{\to} G_\alpha^\lambda$ as $\epsilon \to 0$. Take the evolution equation (1.14), and replace the function $\Phi(x,y)$ appearing there by $G_\epsilon(x-y)\varphi(x)$. Subtract

$$\lambda \int_0^T dt \int_0^t ds\, \langle G_\epsilon(x-y)\varphi(x),\, X_s(dx)\, X_t(dy)\rangle$$

from both sides of the resultant equation. Note the definition (1.9) of $\gamma_\epsilon^\lambda(T,\varphi)$, as well as (1.8), to obtain

$$\begin{aligned}
\gamma_\epsilon^\lambda(T,\varphi) &= \int_0^T \int_{\Re^d} \int_0^t ds\, \langle G_\epsilon(x-y)\varphi(x),\, X_s(dx)\rangle\, Z(dt,dy)\\
&+ \lambda \int_0^T dt \int_0^t ds\, \langle G_\epsilon(x-y)\varphi(x),\, X_s(dx)\, X_t(dy)\rangle\\
&- \int_0^T \langle G_\epsilon(x-y)\varphi(x),\, X_t(dx)\, X_T(dy)\rangle\, dt. \qquad (3.17)
\end{aligned}$$

If we now show that each of the three terms on the right-hand side of this equation converges in $\mathcal{L}^2$, as $\epsilon \to 0$, to a limit that is independent of the sequence $\{G_\epsilon\}$, then the same must be true of the left-hand side, and so the central Theorem 1.2 will be established.

The proof hinges on the moment formulae given in Section 2, and very often closely parallels similar calculations in Rosen (1990). Because of the fact that we have three terms to consider, each one involving a number of graphs, and because the examples given in Rosen's paper show how these calculations must be carried out, we shall now carry out only one of the moment calculations in detail.

The term that we choose is the stochastic integral with respect to $Z$. There are two reasons for this. Firstly, Rosen's calculations do not involve an expression of this kind, and secondly, it will become clear during the calculations where the main conditions of the theorem (i.e. $d = 4, 5$ in the Brownian case and $d/3 < \alpha \le d/2$ in the stable case) come from.

Thus, we now turn to proving that for any sequence $\{G_\epsilon\}$, of the form described above, converging in $\mathcal{L}^1$ to $G_\alpha^\lambda$,

$$\lim_{\epsilon, \epsilon' \to 0} E\left\{ \int_0^T \int_{\Re^d} \int_0^t ds \, \langle (G_\epsilon - G_{\epsilon'})(x - y)\varphi(x), X_s(dx) \rangle \, Z(dt, dy) \right\}^2 = 0 .$$
$$(3.18)$$

To save on notation, fix $\epsilon, \epsilon' > 0$ and set $D(x) = G_\epsilon(x) - G_{\epsilon'}(x)$. From Section 2, three different graphs arise in computing (3.18), described there as (a)–(c). We claim that (a) is easy, and so leave it to the reader. A typical term arising from (b) is

$$\int_{\Re^d} dx_1 \int_0^T ds_1 \int_{\Re^d} dy_1 \, p_{s_1}^\alpha(x_1, y_1) \int_0^T dt_1 \int_{\Re^d} dz_1 \, p_{t_1 - s_1}^\alpha(y_1, z_1)\varphi^2(z_1)$$
$$\times \int_0^T dt_2 \int_{\Re^d} dz_2 \, p_{t_2 - s_1}^\alpha(y_1, z_2)D(z_1 - z_2)$$
$$\times \int_{\Re^d} dx_2 \int_0^T dt_3 \int_{\Re^d} dz_3 \, p_{t_3}^\alpha(x_2, z_3)D(z_1 - z_3) .$$
$$(3.19)$$

(If this expression is not "obvious", then write $D(x, y)$ as $\sum_{j=1}^N d_j(x)d_j'(y)$, and consider a term in this expansion corresponding to the graph

$$(t_1, d_j(z_1) d_k(z_1) \varphi^2(z_1))$$

$$(0,x_1) \longrightarrow (s_1,y_1) $$

$$(t_2, d'_k(z_2))$$

$$(0,x_2) \longrightarrow (t_3, d'_j(z_3))$$

Then sum over $j, k = 1, \ldots, N$ to obtain (3.19) for this $D$, following the style of argument of the previous proof.)

In order to simplify (3.19), use the symmetry of $p^\alpha$ to integrate out $x_1$ and $x_2$. Then use the Chapman-Kolmogorov equation to integrate out $y_1$. This leaves

$$\int_0^T ds_1 \int_0^T dt_3 \int_{s_1}^T dt_1 \int_{s_1}^T dt_2 \int_{\Re^{3d}} dz_1 dz_2 dz_3$$
$$\times \ p^\alpha_{t_1+t_2-2s_1}(z_1, z_2) D(z_1 - z_2) D(z_1 - z_3) \varphi^2(z_1), \qquad (3.20)$$

where the lower bound of $s_1$ on $t_1$ and $t_2$, which was implicit in (3.19), now needs to be noted explicitly. Furthermore, by ignoring a factor of 2, we can assume that $t_1 \leq t_2$.

Noting now that $D(z) = G_\epsilon(z) - G_{\epsilon'}(z)$, that the $G_\epsilon$ are $\mathcal{L}^1$ approximations to $G_\alpha^\lambda$, and that since $\alpha < 2d$ we have $G_\alpha^\lambda \in \mathcal{L}^1$ for all $\lambda > 0$, we can integrate out $z_3$ in (3.20) to obtain

$$\int_0^T ds_1 \int_0^T dt_3 \int_{s_1}^T dt_1 \int_{t_1}^T dt_2 \int_{\Re^{2d}} dz_1 dz_2 \ p^\alpha_{t_1+t_2-2s_1}(z_1 - z_2) D(z_1 - z_2) \varphi^2(z_1)$$

$$= \int_0^T ds_1 \int_0^T dt_3 \int_{s_1}^T dt_1 \int_{t_2}^T dt_2 \int_{\Re^{2d}} dz_1 dz_2 \ (t_1 + t_2 - 2s_1)^{-d/\alpha}$$
$$\times p_1^\alpha((t_1 + t_2 - 2s_1)^{-1/\alpha}(z_1 - z_2)) D(z_1 - z_2) \varphi^2(z_1), \qquad (3.21)$$

where the second line follows from the scaling relationship

$$p_t^\alpha(x) = t^{-d/\alpha} p_1^\alpha(t^{-1/\alpha} x).$$

Now use the fact that stable and Gaussian densities are uniformly bounded to bound (3.21) by a constant multiple of

$$\int_0^T ds_1 \int_0^T dt_3 \int_{s_1}^T dt_1 \int_{t_1}^T dt_2 \ (t_1 + t_2 - 2s_1)^{-d/\alpha} \times \int_{\Re^{2d}} \varphi^2(z_1) D(z_1 - z_2) dz_1 dz_2.$$
(3.22)

Using the finite support of $\varphi$ and the fact that $G_\alpha^\lambda \in \mathcal{L}^1$, we have that the integral over $z_1$ and $z_2$ is finite, and, by dominated convergence, can be made arbitrarily small by taking $\epsilon, \epsilon' \downarrow 0$. Thus, to show that the contribution of the graph under consideration to the left-hand side of (3.18) is asymptotically zero, we need only establish the finiteness of the time integrals in (3.22). In fact, there are only three time variables – $s_1, t_1, t_2$ – of interest. The integration is elementary, and the integral is easily seen to converge if $d < 3\alpha$.

To conclude, we consider a typical term corresponding to the third type of graph, viz.

$$\int_{\Re^d} dx \int_0^T ds_1 \int_{\Re^d} dy_1 \ p_{s_1}^\alpha(x, y_1) \int_{s_1}^T dt_1 \int_{\Re^d} dz_1 \ p_{t_1 - s_1}^\alpha(y_1, z_1)$$

$$\times \int_{s_1}^T ds_2 \int_{\Re^d} dy_2 \ p_{s_2 - s_1}^\alpha(y_1, y_2) \int_{s_2}^T dt_2 \int_{\Re^d} dz_2 \int_{s_2}^T dt_3 \int_{\Re^d} dz_3$$

$$\times \ p_{t_2 - s_2}^\alpha(y_2, z_2) p_{t_3 - s_2}^\alpha(y_2, z_3) D(z_1 - z_2) D(z_1 - z_3) \varphi^2(z_1) \ . \quad (3.23)$$

As before, integrate out $x$, and then use the Chapman-Kolmogorov equation to integrate out $y_1$. The remaining integrand is

$$p_{t_1 + s_2 - 2s_1}^\alpha(z_1, y_2) p_{t_2 - s_2}^\alpha(y_2, z_2) p_{t_3 - s_2}^\alpha(y_2, z_3) D(z_1 - z_2) D(z_1 - z_3) \varphi^2(z_1) \ . \quad (3.24)$$

Use the scaling relationship as before, but only on the first transition probability, and then integrate out $y_2$, again applying Chapman-Kolmogorov. Up to a constant multiple, the integrand becomes

$$(t_2 + t_3 - 2s_2)^{-d/\alpha} (t_1 + s_2 - 2s_1)^{-d/\alpha} D(z_1 - z_2) D(z_1 - z_3) \varphi^2(z_1) \ . \quad (3.25)$$

Arguing as before completes the calculation as long as $d < 3\alpha$, in order to do the time integrals, and if $\alpha < 2d$, which is used to ensure that $G_\alpha^\lambda \in \mathcal{L}^2$. This completes the proof.

# References

[1] Adler, R.J., Feldman, R.E. and Lewin, M. (1991), Intersection local times for infinite systems of planar Brownian motions and the Brownian density process, *Annals of Probability*, **19**, in press.

[2] Adler, R.J. and Lewin, M. (1991), Local time and Tanaka formulae for super Brownian motion and super stable processes, *Stochastic Processes and Their Applications*, to appear.

[3] Adler, R.J. and Rosen, J.S. (1991), Intersection local times of all orders for Brownian and stable density processes—construction, renormalization and limit laws, *Ann. Probability*, to appear.

[4] Dawson, D. (1978), Geostochastic calculus, *Canadian J. Statistics*, **6**, 143–168.

[5] Dawson, D. (1986), Measure-valued processes: construction, qualitative behaviour and stochastic geometry, *Proc. Workshop on Spatial Stochastic Models, Lecture Notes in Math.*, **1212**, 69–93, Springer-Verlag.

[6] Dawson, D., Iscoe, I. and Perkins, E.A. (1989), Super Brownian motion: path properties and hitting probabilities, *Prob. Theory and Related Topics*, **83**, 135–205.

[7] Dynkin, E.B. (1988), Representation for functionals of superprocesses by multiple stochastic integrals, with applications to self-intersection local times, *Astérisque*, **157–158**, 147–171.

[8] Ethier, S. and Kurtz, T. (1986), *Markov Processes*, Wiley, New York.

[9] Hörmander, L. (1985), *The Analysis of Linear Partial Differential Operators III*, Springer-Verlag, Berlin.

[10] Karatzas, I. and Shreve, S.E. (1988), *Brownian Motion and Stochastic Calculus*, Springer-Verlag, New York.

[11] Perkins, E.A. (1988), Polar sets and multiple points for super Brownian motion, *Trans. Amer. Math. Soc.*, **305**, 743–795.

[12] Roelly-Coppoletta, S. (1986), A criterion of convergence of measure-valued processes: application to measure-branching processes, *Stochastics*, **17**, 43–65.

[13] Rosen, J.S. (1990), Renormalization and limit theorems for self-intersections of super processes, preprint.

[14] Walsh, J. (1986), An introduction to stochastic partial differential equations, *Lecture Notes in Math.*, **1180**, 265–437, Springer-Verlag.

[15] Watanabe, S. (1968), A limit theorem for branching processes and continuous state branching processes, *J. Math. Kyoto Univ.*, **8**, 141–167.

[16] Yoshida, K.Y. (1980), *Functional Analysis*, Springer-Verlag, Berlin.

# The Continuum Random Tree II: An Overview

David Aldous[*]

University of California, Berkeley

## 1 INTRODUCTION

Many different models of random trees have arisen in a variety of applied setting, and there is a large but scattered literature on exact and asymptotic results for particular models. For several years I have been interested in what kinds of "general theory" (as opposed to *ad hoc* analysis of particular models) might be useful in studying asymptotics of random trees. In this paper, aimed at theoretical probabilists, I discuss aspects of this incipient general theory which are most closely related to topics of current interest in theoretical stochastic processes. No prior knowledge of this subject is assumed: the paper is intended as an introduction and survey.

To give the really big picture in a paragraph, consider a tree on $n$ vertices. View the vertices as points in abstract (rather than $d$-dimensional) space, but let the edges have length ($= 1$, as a default) so that there is metric structure: the distance between two vertices is the length of the path between them. Consider the average distance between pairs of vertices. As $n \to \infty$ this average distance could stay bounded or could grow as order $n$, but almost all natural random trees fall into one of two categories. In the first (and larger) category, the average distance grows as order $\log n$. This category includes supercritical branching processes, and most "Markovian growth" models such as those occurring in the analysis of algorithms. This paper is concerned with the second category, in which the average distance grows as order $n^{1/2}$. This occurs with Galton-Watson branching processes conditioned on total population size $= n$ (in brief, CBP(n)). At first sight that seems an unnatural model, but it turns out to coincide (see section 2.1) with various combinatorial models, and is similar to more general models of *critical* branching processes conditioned to be large (in any reasonable way). The fundamental

[*]Research supported by N.S.F. Grants MCS87-01426 and MCS 90-01710

fact is that, by scaling edges to have length $n^{-1/2}$, these random trees converge in distribution as $n \to \infty$ to a limit we call the CCRT (for *compact continuum random tree*). This was treated explicitly in Aldous [2] in a special case and in Aldous [3] in the natural general case, though (as we shall see) many related results are implicit in recent literature. Thus asymptotic distributions for these models of discrete random trees can be obtained immediately from distributions associated with the limit tree. The limit tree is closely connected with *Brownian excursion*. In fact two different 1-parameter processes associated with the tree – the *search depth* process and the *height profile* process – are intimately connected with Brownian excursion (sections 2.4 and 3.2). Section 2 is a chatty account of 4 different ways of looking at the CCRT. In section 3 I take natural distributional questions about CBP(n) asymptotics (with known or unknown answers), which can be expressed in terms of the CCRT and see what can be said about the limit distributions, using the Brownian excursion representation in particular. Nothing I say is essentially new: I use the word "novel" (intended to be weaker than "new") to refer to results about CBP(n) asymptotics obtainable from known Brownian excursion results (e.g. Corollaries 3 and 6, and Proposition 12) and vice versa (e.g. (41) as a fact about Brownian excursion). One could conversely pick haphazardly some facts about Brownian excursion and apply them to random trees, but that somehow seems less interesting.

Scaling the edges of CBP(n) to have length $n^{-\alpha}$ ($0 < \alpha < 1/2$) gives (section 2.5) another limit tree I call the SSCRT (*self-similar continuum random tree*). Further, the same limit tree is obtained whether we root at the progenitor or whether we re-root at a uniform random individual in the population. This limit tree – which relates to the 3-dimensional Bessel process BES(3) in the same way that the CCRT relates to Brownian excursion – is less natural from the combinatorial viewpoint. But being more tractable (from the self-similarity inherited from BES(3)) it is useful in the theoretical stochastic process investigations below.

Sections 5 and 6 are speculative. There has been recent theoretical interest in existence, uniqueness and properties of "Brownian motion" whose state space is some deterministic fractal set in $d$ dimensions, the set typically constructed by some recursive procedure giving strong regularity properties. Our limit trees are "dimension 2" (inherited from Brownian sample paths), and it is intuitively clear that "Brownian motion" can be defined with these trees as its state space. Unlike other exotic state spaces, we can actually do some simple distributional calculations with these Brownian motions, and the purpose of section 5 is to present these back-of-an-envelope calculations. To develop rigorously a theory of Brownian motion on general continuum trees would be an interesting project, and some thoughts are presented in section 5.2.

Section 6 is a quixotic venture into superprocesses. It is trivial to construct Markov processes *indexed by* a continuum tree. Making the index set the particular CCRT or SSCRT gives variants of the usual superprocess. This is the idea developed by theoreticians under the name "historical process", but the theoretical literature makes this appear a deep and sophisticated object. I assert one should start from scratch and regard a superprocess as a tree-indexed process rather than as a measure-valued process. My purpose is to indicate (section 6.1) how this leads to insights which seem simpler or different from those obtained in the traditional approach.

*Acknowledgements.* I thank Jim Pitman for much help with and information about Brownian excursion and BES(3), Steve Evans for help with superprocesses, and Martin Barlow for discussions on diffusions on fractals.

## 2   THE BIG PICTURE
The first four subsections elaborate on the following four fundamental facts.

- Conditioned Galton-Watson branching processes correspond to a natural and well-studied class of combinatorial models of random trees.

- One particular model can be constructed from simple random walk conditioned on first return to 0 at time $2n$, and so its asymptotics can be expressed in terms of Brownian excursion.

- Another particular model can be constructed from a direct (i.e. not involving conditioning) algorithm, and by taking limits one gets a direct algorithm for global construction of a limit tree.

- By considering asymptotics of subtrees spanned by a fixed number of randomly chosen vertices, one sees that the limit random tree must be the same (up to a scale factor) for all models in the class.

Foundational work giving rigorous definitions and proofs concerning existence of "continuum trees" (without any specific probability model present) and abstract convergence results is in Aldous [3], and it is not worth repeating such "general abstract nonsense" here.

### 2.1   CBP(n) and Combinatorial Models
Let $\xi \geq 0$ be integer-valued and satisfy

$$E\xi = 1$$

$$0 < \text{var } \xi = \sigma^2 < \infty. \tag{1}$$

Such a $\xi$ is $d$-lattice, for some $d \geq 1$. We want to allow $d > 1$ for natural combinatorial examples (e.g. binary trees). Associate with $\xi$ the distribution $\hat{\xi}$ defined by

$$P(\hat{\xi} = i) = (i+1)P(\xi = i+1), \ i \geq 0 \tag{2}$$

and note that

$$E\hat{\xi} = E\xi(\xi - 1) = \text{var } \xi = \sigma^2.$$

Consider the simple Galton-Watson branching process with offspring distribution $\xi$, starting with 1 individual in generation 0. Write $\mathcal{T}$ for the "family tree" of this branching process. Let $\mathcal{T}_n$ have the distribution of $\mathcal{T}$ conditioned on the total population size $|\mathcal{T}| = n$. This CBP(n) (for "conditioned branching process") distribution is our object of study.

*Tangential remarks.* 1. If $\xi$ and $\eta$ come from the same exponential family, i.e. for some $(c, \theta)$

$$P(\xi = i) = c\theta^i P(\eta = i); \ i \geq 0$$

then the conditioned branching processes constructed from $\xi$ and from $\eta$ are identical. Thus we lose no generality by considering only *critical* branching processes. The chance that the total population size is exactly $n$ decreases exponentially fast for sub- and super-critical branching processes, but only polynomially fast in the critical case: in this sense the critical case is most natural as a model for $n$-trees.

2. In the language of freshman statistics, if $\xi$ is "number of daughters of a randomly-picked mother", then $\hat{\xi}$ at (2) is "number of sisters of a randomly-picked girl". The two distributions are identical iff they are the Poisson(1) distribution.

3. I use "Galton-Watson process" to mean the family tree of the process. Old-fashioned textbooks use it to mean the process of population sizes in successive generations, which I call the "height profile" of the Galton-Watson process.

*Simply generated trees.* Results about CBP(n) appear in the combinatorial literature under this name (introduced by Meir and Moon [35], apparently unaware of the branching process connection). Though the identification has subsequently become well known, there seems no convenient "translation guide" in existence, so I give one here.

A "rooted" tree simply has one vertex distinguished and called the root: imagine a family tree of descendants of a single progenitor, the root. We consider only rooted trees. Such a tree is called *ordered* if we distinguish birth order: if an individual (vertex) has 3 offspring then these are distinguished as "first", "second" and "third". Consider the family tree $\mathcal{T}$ of the unconditioned Galton-Watson branching process with offspring $\xi$. Write $p_i = P(\xi = i)$. Then the distribution of $\mathcal{T}$ on rooted ordered trees $t$ is

$$
\begin{aligned}
P(\mathcal{T} = t) &= \prod_{v \in t} p_{d(v,t)} \\
&= \prod_{i \geq 0} p_i^{D_i(t)} \\
&= \omega(t) \text{ say}
\end{aligned}
\tag{3}
$$

where $d(v,t)$ is the out-degree (number of children) of vertex $v$ in $t$, and $D_i(t)$ is the number of vertices in $t$ with out-degree $i$. Thus the distribution of the CBP(n) tree $\mathcal{T}_n$ is specified by

$$P(\mathcal{T}_n = t) \text{ is proportional to } \omega(t) \text{ on } \{t : |t| = n\} \tag{4}$$

where $|t|$ denotes the number of vertices in $t$.

One can get to (4) without explicitly mentioning Galton-Watson processes. Let $(c_i; i \geq 0)$ be non-negative constants with $c_0 = 1$, and let

$$\phi(y) = \sum_i c_i y^i$$

be the associated generating function. Let $\hat{\omega}(t)$ be some collection of non-negative "weights" for trees. Define

$$y_n = \sum_{t:|t|=n} \hat{\omega}(t)$$

and let $Y(x) = \sum_n y_n x^n$ be the associated generating function. Then it is easy to see the following are equivalent.

$$Y(x) \equiv x\phi(Y(x)) \tag{5}$$

$$\hat{\omega}(t) = \prod_{i \geq 0} c_i^{D_i(t)} \tag{6}$$

A combinatorial definition of "simply generated tree" is "a family of weights satisfying (5), or equivalently (6)". So a random simply generated tree $\mathcal{T}_n$ is defined as

$$P(\mathcal{T}_n = t) \text{ is proportional to } \hat{\omega}(t) \text{ on } \{t : |t| = n\}.$$

To see why this is really the same as the CBP(n) model, note that for any $\tau$ with $\phi(\tau) < \infty$ we can define a probability distribution

$$P(\xi = i) = p_i = c_i \tau^i / \phi(\tau), \ i \geq 0. \tag{7}$$

Choose the $\tau$ which makes $E\xi = 1$. For $\omega(t)$ defined at (3), we see

$$\omega(t) = \hat{\omega}(t) \ \tau^{|t|-1} / \phi^{|t|}(\tau).$$

Thus on $\{t : |t| = n\}$, $\omega$ is proportional to $\hat{\omega}$, and so the two models for $\mathcal{T}_n$ are identical.

Elementary calculations from (7) show that the condition "$E\xi = 1$" specifying $\tau$ is the condition

$$\tau \phi'(\tau) = \phi(\tau)$$

and that the variance $\sigma^2 \equiv \mathrm{var}(\xi)$ is

$$\sigma^2 = \tau^2 \phi''(\tau) / \phi(\tau). \tag{8}$$

The right-side expression appears in combinatorial papers without mention of its simple interpretation as "offspring variance".

*Examples.* The idea of all the combinatorial examples is that all $n$-vertex trees of a certain type should be equally likely. One aspect of "type" is that we can place restrictions on out-degrees. Another aspect is that sometimes we want to distinguish birth-order (*ordered trees*) and sometimes we don't. In the set-up above, ordered trees become the case

$$c_i = 1 \text{ if } i \text{ is an allowed out-degree}, \ = 0 \text{ if not}$$

and unordered trees become the case

$$c_i = 1/i! \text{ if } i \text{ is an allowed out-degree}, \ = 0 \text{ if not }.$$

Various offspring distributions $p_i = P(\xi = i)$ are recorded below as a handy reference: the values of $\sigma$ are needed to connect our results with those in the combinatorial literature on special models. To reiterate the point: the uniform distribution on the following "types of $n$-vertex tree" coincides with the CBP(n) description with the stated offspring distribution.

ordered (= planar) trees.

Unrestricted degree: shifted geometric distribution $p_i = 2^{-i}, i \geq 0$; $\sigma^2 = 1$.
Strict binary (0 or 2 offspring): $p_0 = p_2 = 1/2$; $\sigma^2 = 1$.

Strict $t$-ary (0 or $t$ offspring): $p_0 = 1 - 1/t, p_t = 1/t;\ \sigma^2 = t - 1$.
Unary-binary (0, 1 or 2 offspring): $p_0 = p_1 = p_2 = 1/3;\ \sigma^2 = 2/3$.

unordered labelled trees.
Unrestricted degree: Poisson distribution $p_i = e^1/i!, i \geq 0;\ \sigma^2 = 1$.
Unary-binary: $p_0 = \frac{1}{2+\sqrt{2}}, p_1 = \frac{\sqrt{2}}{2+\sqrt{2}}, p_3 = \frac{1}{2+\sqrt{2}};\ \sigma^2 = \frac{2}{2+\sqrt{2}}$.
Strict $t$-ary: same as ordered case.

*Remark.* I have slid over one issue: in the combinatorial story the trees are regarded as rooted and *labelled,* i.e. the $n$ vertices are distinguishable. The distinction between labelled and unlabelled is irrelevant for ordered rooted trees (because the ordering serves to distinguish vertices anyway) but relevant for unordered trees. The model "all unordered unlabelled trees equally likely" does *not* fit into this set-up, and no simple probabilistic description is known.

## 2.2   Ordered Trees and Brownian Excursion
With a finite rooted ordered tree $t$ on $n$ vertices we can associate the following two sequences (the terminology is not standard).

The *height profile* $(h(j); j \geq 0)$, where $h(j)$ is the number of vertices at distance $j$ from the root.

The *search depth* $(x(i); 1 \leq i \leq 2n - 1)$ defined as follows. At each vertex $v$, suppose the edges at $v$ leading away from the root are *ordered* as "first", "second", etc. Then depth-first search of the tree is the following deterministic walk $(v(i) : 1 \leq i \leq 2n - 1)$ around the vertices. Let $v(1) = $ root. Given $v(i)$ choose (if possible) the first (in the ordering) edge at $v(i)$ leading away from the root which has not already been traversed, and let $(v(i), v(i+1))$ be that edge. If not possible, let $(v(i), v(i+1))$ be the edge from $v(i)$ leading towards the root.

This walk terminates with $v(2n - 1) = $ root, having traversed each edge exactly once in each direction. Finally, define the search depth $x(i) = $ distance from root to $v(i)$.

There is a connection between the two sequences: for $j \geq 1$

$$h(j) = \text{ number of upcrossings of } (j - 1, j) \text{ by the sequence } x(i). \quad (9)$$

For a random tree distributed as CBP(n) these become random sequences

$(H(j))$ and $(X(i))$, say. Define the *rescaled cumulative height profile process*

$$H_t^n = n^{-1} \sum_{j \leq n^{1/2}t} H(j), \ t \geq 0 \tag{10}$$

and the *rescaled search depth process*

$$X_t^n = n^{-1/2}X([2nt]), \ 0 \leq t < 1. \tag{11}$$

Conventions about rescaling constants are awkward – e.g. one might want to rescale by $(2n)^{-1/2}$ in (11) – but my conventions are chosen to make rescaled edge-lengths $= n^{-1/2}$ consistently.

Returning to the unscaled process $X(i)$, set $X(0) = X(2n) = -1$. For any model, the process $X(i)$ has steps $\pm 1$ and first returns to the starting level after step $2n$. The simplest model for such a random process would be "simple symmetric random walk, conditioned on first return to starting level at time $2n$". The key fact is that this describes the depth search process in one particular model of random trees: the combinatorial model of "uniform ordered trees", which is the CBP(n) model with shifted geometric $(1/2)$ offspring distribution.

Various forms of this fact have been known to combinatorialists for a long time But its significance for probabilistic asymptotics was overlooked until recently (I learned it from Durrett et al [17], who attribute it to Harris). It is intuitively obvious (and true [16] – see also [13] and [9] p. 104 for references and history) that conditioned random walk rescales to Brownian excursion, and so (for this special model of random trees) the rescaled search depth process converges to Brownian excursion. It is equally intuitively obvious from (9) that the rescaled height profile process converges to the total occupation density of Brownian excursion.

On a finite tree, the search depth process determines the ordered tree in a simple way: each $+1$ step draws a new edge, and each $-1$ step retraces an existing edge toward the root. So it is intuitively clear that, for the special "uniform ordered $n$-tree" model, there is a limit tree whose realizations can be constructed from realizations of sample paths $f(t)$ of Brownian excursion. In non-standard terms, an infinitesimal positive increment of $f$ draws an infinitesimal new edge, and an infinitesimal negative increment of $f$ retraces an existing edge toward the root. In standard terms, given $0 < t_1 < t_2 < \ldots < t_k < 1$, let $s_i = \min_{t_i < t < t_{i+1}} f(t)$. Draw an edge of length $f(t_1)$, and label one end "root" and the other end "$t_1$". Inductively, from $t_i$ move back distance $f(t_i) - s_i$ toward the root, then make a new edge of length $f(t_{i+1}) - s_i$ and label its endpoint "$t_{i+1}$". The shapes of these trees are consistent as $(t_i)$ varies, and define a "continuum tree" with vertices labelled by $0 < t < 1$.

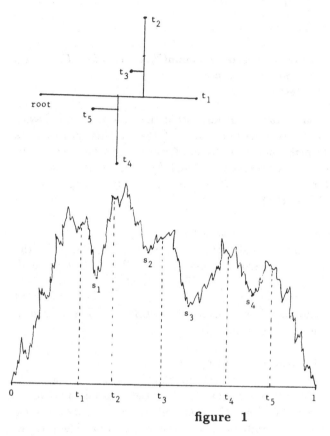

**figure 1**

A rigorous treatment of constructing continuum trees from continuous functions is given in [3], Theorem 13: it turns out that distributional properties of Brownian excursion are irrelevant, and that any continuous function $f_0$ with certain qualitative properties (e.g. local minima are dense) can be used. Later we shall use $(\mathcal{S}_0, \mu_0)$ to denote the continuum tree constructed from such a $f_0$. Regard $\mathcal{S}_0$ as the vertex-set, labelled by $0 < t < 1$, and regard $\mu_0$ as the "uniform probability distribution" on $\mathcal{S}_0$ induced from Lebesgue measure. Write $(\mathcal{S}, \mu)$ for the particular continuum random tree ("the CCRT") constructed from ($2\times$) Brownian excursion.

## 2.3    The Limit Trees: Global Constructions

Another special case of CBP(n) is where the offspring distribution is Poisson(1). Combinatorially, this is the uniform random unordered labelled tree. Many algorithms for simulating this random tree are known: the following was discovered in Aldous [5].

**Algorithm 1** *Fix $n \geq 2$.*
*Take a root vertex 1.*
*For $2 \leq i \leq n$ connect vertex $i$ to vertex $V_i = \min(U_i, i-1)$, where $U_2, \ldots, U_n$ are independent and uniform on $1, \ldots, n$.*
*Randomly permute the labels.*

The advantage of this particular algorithm is that the $n \to \infty$ limit behavior is intuitively easy to see. It is proved in Aldous [2] that the first process (the CCRT) described informally below is the limit when edges are rescaled to length $n^{-1/2}$, and the second process (the SSCRT) is the limit when edges are rescaled to length $n^{-\alpha}$, $0 < \alpha < 1/2$, or more generally to length $1/a(n)$ where $a(n) \to \infty, a(n) = o(n^{1/2})$.

**The compact continuum random tree $(\mathcal{S}, \mu)$.**
Take a half-line $[0, \infty)$, and cut-and-paste as follows. Let $C_1, C_2, \ldots$ be the times of a non-homogeneous Poisson process of rate $r(t) = t$. Cut the half-line into intervals $[C_i, C_{i+1})$ . Start with the line segment $[0, C_1)$, and make 0 the root. Grow a tree inductively by adding $[C_i, C_{i+1})$ as a branch connected to a random point $J_i$, chosen uniformly over the existing tree. The process is the closure of the union of all branches. $\square$

**The self-similar continuum random tree $(\mathcal{R}, \nu)$.**
Start at time 0 with an infinite continuous line $[0, \infty)$, and make 0 the root. At time $0 < t < \infty$ there is a tree composed of the original line and of finite line segments connected with each other; only a finite number of such segments connecting with each finite interval of the original line. The process grows according to the rules
(i) in each time increment $(t, t + dt)$, in each segment $(x, x + dx)$ of the tree constructed at time $t$, there is chance $dt\,dx$ of a "birth";
(ii) if a birth occurs at time $t$ and place $x$, then a new branch with random exponential(rate $t$) length is instantly attached at $x$.
The process is the closure of the tree at time infinity. $\square$

In these limit processes, regard $\mathcal{S}$ and $\mathcal{R}$ as random sets, indicating the spatial position of the limit "continuum tree". Then $\mu$ and $\nu$ are random measures supported by $\mathcal{S}$ and $\mathcal{R}$, representing how the vertices are spread over the tree.

In other words, with the tree $\mathcal{T}_n$ constructed by Algorithm 1 we associate the empirical distribution $\mu_n$ of the vertices: $\mu_n$ puts mass $1/n$ on each vertex. As space-rescaled $\mathcal{T}_n$ converges to $\mathcal{S}$, so does $\mu_n$ (with the induced space-rescaling) converge to $\mu$. Similarly, when edge-lengths of $\mathcal{T}_n$ are rescaled to length $1/a(n)$ to get the limit $\mathcal{R}$, let $\nu_n$ be the measure putting mass $1/a^2(n)$ on each vertex: then $\nu_n$, with the induced space-rescaling, converges to $\nu$.

As the notation suggests, we shall see below that the CCRT $(\mathcal{S}, \mu)$ constructed above is the same as that constructed from ($2\times$) Brownian excursion in the previous section.

## 2.4   The Convergence Result for CBP(n)

The results in the previous two sections depended on exact combinatorial relations for finite $n$, in the two special cases. A natural first step in seeking to generalize is to consider the general CBP(n) model. Neither of the previous methods works: there is in general no constructive algorithm like Algorithm 1 known, and while any tree can be coded as a walk with steps $\pm 1$ as in section 2.2, the process obtained from general CBP(n) does not have any standard dependence structure which makes convergence to Brownian excursion look easy to prove. But using different techniques (outlined below), an abstract result "rescaled general CBP(n) converges to the CCRT" is proved in Aldous [3] Theorem 23. Without setting up the precise statement of the abstract result, let us state the concrete consequence (which actually turns out to be equivalent to the *a priori* stronger abstract result – c.f. [3]) for the rescaled search depth process $X^n$ at (11).

**Theorem 2** *For CBP(n), as $n \to \infty$*

$$(X_t^n; 0 \le t \le 1) \xrightarrow{d} (2\sigma^{-1}W_t; 0 \le t \le 1)$$

*where $W = (W_t; 0 \le t \le 1)$ is standard Brownian excursion.*

Here "convergence in distribution" is the usual weak convergence of processes. Note we use "Brownian excursion" to mean Brownian excursion *of duration* 1.

An immediate corollary is the result for the rescaled cumulative height profile process $H^n$ at (10).

**Corollary 3** *For CBP(n), as $n \to \infty$*

$$(H_s^n; s \ge 0) \xrightarrow{d} (H_{\sigma s/2}; s \ge 0) \tag{12}$$

*where*

$$H_s = \int_0^1 1_{(W_t \le s)} dt.$$

To use an old-fashioned term, the abstract result behind Theorem 2 is an *invariance principle*: the distribution of the limit tree $S$ doesn't depend on the offspring distribution $\xi$, except through the s.d. $\sigma$ as a scale factor. (This may be thought surprising – one's first guess might be that $\sigma$ would affect the *shape* of the tree). As with the classical invariance principle (convergence of i.i.d. partial sums to Brownian motion) one might expect the result to be true for much more general models, and we discuss this briefly in section 4.

A final ingredient of the big picture is an "intrinsically tree-ish" description of the CCRT $S$. To give this, we need to introduce a different species of tree $t$. Let $t$ have $k$ labelled leaves, a root with degree 1, and binary branchpoints (and hence $2k - 1$ edges). Let the edge-lengths be positive reals, and regard the tree as unordered. Such a tree $t$ can be specified by its topological shape $t^*$, say, and by the $2k - 1$ edge-lengths $(l_i)$. Define $\mathcal{R}(k)$ to be a random tree of this type with density

$$f(t^*, l_1, \ldots, l_{2k-1}) = s \exp(-s^2/2), \quad s = \sum_{i=1}^{2k-1} l_i. \tag{13}$$

In other words, the edge-lengths are independent of the shape of the tree, which is uniform on all shapes; moreover the edge-lengths are exchangeable (and hence we didn't need to specify exactly which edge was edge $i$). These random trees satisfy the natural consistency condition in $k$. It turns out that the distribution of $(S, \mu)$ is specified by the fact that the subtree $\mathcal{R}(k)$ spanned by $k$ "uniform" (i.e. chosen according to $\mu$) random vertices has density (13). More generally, just as ordinary stochastic processes can be specified via consistent families of f.d.d.'s, so ([3] Theorem 3) a random continuum tree can be specified by "random f.d.d.'s", the subtrees spanned by randomly-chosen vertices. The point is that in general there is no "canonical" way of labelling vertices of continuum trees, so random f.d.d.'s are a natural substitute for ordinary f.d.d.'s.

The proof in [3] that rescaled CBP($n$) converges is based upon convergence of random f.d.d.'s. Fix $k$, choose at random $k$ vertices from CBP($n$), consider the subtree spanned by these $k$ vertices and the root, rescale and let $n \to \infty$. Using classical asymptotics for sizes of critical BPs it can be shown that the limit tree is (a scale factor $\sigma^{-1}$ times) $\mathcal{R}(k)$. In [3] we develop such "exchangeability and weak convergence" techniques as a hopefully useful way of establishing convergence of more general models to more general limit continuum random trees. In principle one could seek to prove Theorem 2 directly, by first proving convergence of finite dimensional distributions $(X_{t_1}^n, \ldots, X_{t_k}^n)$. In a special model of binary trees this was recently carried out by Gutjahr and Pflug [26], based on exact combinatorial formulas, but the general CBP(n) model seems less tractable. Direct approaches to Corollary 3 are easier, but not powerful enough to establish the Theorem.

We now tie this up with the special constructions of $S$ in sections 2.2 and 2.3. The construction in section 2.2 of a tree from a function $f$ and points $(t_i)$, applied to a sample path of $2W$ and to $k$ uniform random points, plainly must give a tree isometric to $\mathcal{R}(k)$. The connection with the global construction in section 2.3 is more surprising. Let $\hat{\mathcal{R}}(k)$ be the subtree obtained from the first $k$ branches $[C_{i-1}, C_i], i \leq k$ in the global construction. Then a direct computation ([3] section 4.3) shows that $\hat{\mathcal{R}}(k)$ has distribution (13):

$$\hat{\mathcal{R}}(k) \stackrel{d}{=} \mathcal{R}(k) \tag{14}$$

In other words, $\hat{\mathcal{R}}(k)$ is isometric to the subtree $\mathcal{R}(k)$ of $S$ spanned by $k$ *randomly* chosen vertices of $S$.

It is intuitively clear that the natural "local" result associated with Corollary 3 should be true. Write $H(j)$ for the number of vertices at height $j$ from the root, and define the rescaled height profile process

$$h_s^n = n^{-1/2} H([n^{1/2} s]).$$

**Conjecture 4** *For CBP(n), as $n \to \infty$,*

$$(h_s^n; s \geq 0) \stackrel{d}{\to} (\frac{\sigma}{2} l_{\sigma s/2}; s \geq 0)$$

*where*

$$l_s = \frac{d}{ds} \int_0^1 1_{(W_t \leq s)} dt$$

*is the total occupation density of Brownian excursion $W$.*

The "weak convergence" methods of [3] are too weak to be used here. In some special cases there are exact combinatorial expressions for means and moments of $H(j)$, and in these special cases one could no doubt establish Conjecture 4, but the general case seems to require delicate analytical asymptotics. Distributional properties are discussed in section 3.2.

*Remark: local time convention.* Above I use total occupation density for Brownian excursion, and later I use total occupation density for BES(3). These are of course "local times as space-indexed processes", up to normalization conventions. Occasionally I use "local time at a point" as a time-indexed process, still using the occupation density normalization.

*Technical note.* Using the construction of $S$ from Brownian excursion, we get a random measure $\mu$ on $S$ induced from Lebesgue measure on $[0, 1]$: this is the same measure $\mu$ which occurs as the limit empirical distribution of vertices (section 2.3). The same applies to $(\mathcal{R}, \nu)$, the SSCRT: in the construction below from 2-sided BES(3), $\nu$ is the measure induced from Lebesgue measure on the line.

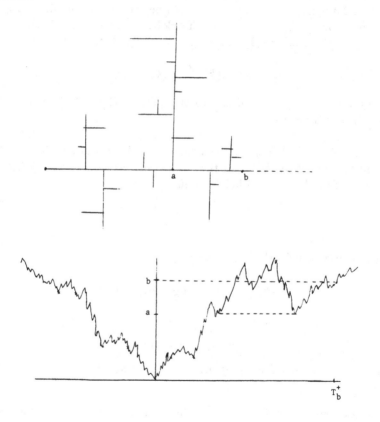

**figure 2**

## 2.5 The Self-Similar Continuum Random Tree

Sketched above is the SSCRT $(\mathcal{R}, \nu)$ given by the global construction in section 2.3. (The "baseline" is drawn horizontally.)

Recall that standard BES(3) is the process distributed as the radial part of 3-dimensional standard Brownian motion started at 0. We shall be concerned with 2-sided standard BES(3) $B = (B_s; -\infty < s < \infty)$. Here *2-sided* means that $(B_t; t \geq 0)$ and $(B_{-t}; t \geq 0)$ are independent copies of standard BES(3). It turns out that we can construct a realization of $\mathcal{R}$ from a realization of $2B$, analogous to the construction of $\mathcal{S}$ from $2W$ in section 2.2. In brief, we construct a tree labelled by $\{t : t \geq 0\}$ from $(2B_t; t \geq 0)$ and separately construct another tree labelled by $\{-t : t \geq 0\}$ from $(2B_{-t}; t \geq 0)$; then we join the trees by identifying (for each $b \geq 0$) the points labelled $T_b^-$ and $T_b^+$, where

$$T_b^+ = \max\{t : 2B_t = b\}, \ T_b^- = \min\{t < 0 : 2B_t = b\}.$$

This becomes the point **b** on the baseline at distance $b$ from the root. In figure 2, we regard positive-time BES(3) as tracing out the part of the tree above the baseline, and negative-time BES(3) tracing out the part below the baseline.

Here is a verbal description of how $\mathcal{R}$ arises as a limit of rescaled CBP(n). Rescale edges to have length $1/a(n)$, where throughout this section

$$a(n) \to \infty, \ a(n) = o(n^{1/2}).$$

Let $\nu_n$ be the measure putting mass $1/a^2(n)$ on each vertex. Then the rescaled random set $\mathcal{T}_n$ of vertices of CBP(n) converges in distribution to $\mathcal{R}$, and the random measure $\nu_n$ converges to $\nu$. From this limit procedure (or from the BES(3) construction) we see that the SSCRT has a self-similarity property: multiplying distances by a constant $c$ doesn't affect the distribution of the random set $\mathcal{R}$, though it does take the measure $\nu$ to $c^{-2}\nu$.

These results could be formalized and proved in the same way as was done in [3] for the CCRT. Here are the concrete results analogous to Theorem 2 and Corollary 3.

Reconsider the search depth $(x(i); 1 \leq i \leq 2n - 1)$ associated with a tree $t$ in section 2.2. The search starts and ends at the root: $x(1) = x(2n - 1) = 0$. For present purposes we want to center at the root, so we define $(x^*(i); -n < i < n)$ by

$$x^*(i) = x(i), 1 \leq i < n; \ x^*(-i) = x(2n - i), 1 \leq i < n$$

with $x^*(0) = -1$. For a random tree distributed as CBP(n) this becomes a random sequence $(X^*(i))$; we also have the height profile process $(H(j))$ as in section 2.2. Rescale as

$$\hat{H}_s^n = a^{-2}(n) \sum_{j \le a(n)s} H(j), \ s \ge 0 \tag{15}$$

and

$$\hat{X}_s^n = a^{-1}(n) X^*([2a^2(n)s]), -\infty < s < \infty. \tag{16}$$

**Theorem 5** *For CBP(n), as $n \to \infty$*

$$(\hat{X}_s^n; -\infty < s < \infty) \ \xrightarrow{d} \ (2\sigma^{-1} B_s; -\infty < s < \infty)$$

*where B is 2-sided standard BES(3).*

**Corollary 6** *For CBP(n), as $n \to \infty$*

$$(\hat{H}_s^n; s \ge 0) \ \xrightarrow{d} \ (Q_{\sigma s/2}; s \ge 0) \tag{17}$$

*where*

$$Q_s = \int_{-\infty}^{\infty} 1_{(B_t \le s)} dt.$$

Here is the analog of Conjecture 4 for the (local) height profile process.

**Conjecture 7** *For CBP(n), as $n \to \infty$,*

$$(a^{-1}(n) H([a(n)s]); \ s \ge 0) \ \xrightarrow{d} \ (\frac{\sigma}{2} q_{\sigma s/2}; \ s \ge 0) \ \overset{d}{=} \ (\frac{\sigma^2}{4} q_s; \ s \ge 0)$$

*where*

$$q_s = \frac{dQ_s}{ds}$$

*is total occupation density for 2-sided BES(3).*

Kolchin [33] Theorem 2.5.4 and Kennedy [30] Theorem 1 have given the 1-dimensional convergence results implicit in Conjecture 7, but I have not seen the full weak convergence result published explicitly.

It is well known that the total occupation density of one-sided BES(3) is the diffusion with drift and variance rates

$$\mu(x) = 2, \quad \sigma^2(x) = 4x,$$

or equivalently $|B_2(s)|^2$, where $B_d$ is standard $d$-dimensional Brownian motion. It follows that $(q_s)$ is $|B_4(s)|^2$, or equivalently the diffusion with

$$\mu(x) = 4, \quad \sigma^2(x) = 4x.$$

The marginal distributions are Gamma$(2, \cdot)$:

$$f_{q(s)}(x) = \frac{x}{4s^2} \exp(-\frac{x}{2s}). \tag{18}$$

See Pitman and Yor [40, 41] for extensive accounts of related properties of Bessel processes. As mentioned above, this Gamma limit distribution for generation size in conditioned Galton-Watson processes was known, but analytic proofs give little insight into why this particular limit distribution holds. The BES(3) representation gives one: the positive-time and negative-time occupation densities are obviously i.i.d. exponentials.

## 2.6   Discrete limits of CBP(n)

As a final piece of background, one can take limits in CBP(n) without rescaling edge-lengths. In this setting, the limit process $\mathcal{T}_\infty$ (described below) depends on the entire distribution of $\xi$. This result is in Grimmett [24] and in [2] Theorem 2, in the special case of Poisson(1) offspring; and the general case is implicit in Kesten [31].

**The discrete infinite tree $\mathcal{T}_\infty$.**
For each $k = 0, 1, 2, \ldots$ create independently branching processes, whose first generation size has distribution $\hat{\xi}$ but whose subsequent offspring distribution is $\xi$. Regard these as trees with root $i_k$ and other vertices unlabelled. Then connect $i_0, i_1, i_2, \ldots$ as a path, deem $i_0$ the root and delete labels. $\square$

This is a convenient place to introduce the idea of *random re-rooting*. A random tree $\mathcal{T}_n$ distributed as CBP(n) is normally considered as rooted at the progenitor of the branching process. We may, however, choose another vertex $v$ of $\mathcal{T}_n$ and declare that to be the root. (To avoid discussing ordering of the re-rooted tree, regard trees as unordered). If $v$ is chosen uniformly at random from the $n$ vertices, call this procedure "random re-rooting". In the combinatorial model "uniform random labelled unordered tree", i.e. the Poisson(1) special case of CBP(n), it is immediate from the combinatorial description that random re-rooting does not change the distribution of the random tree. For general CBP(n), the distribution does change. However, the discrete limit distribution $T^*_\infty$ is almost the same as $T_\infty$ above, except for one change:

the branching process rooted at $i_0$ has first generation

offspring distribution $\xi$ instead of $\hat{\xi}$.                    (19)

This result, implicit in earlier work, is given explicitly in Aldous [1]. As an aside, the idea of taking discrete limits in randomly re-rooted trees works for almost all the larger class of "height $O(\log n)$ trees" mentioned in the introduction, whereas for those trees looking at limits around the original root is not interesting – this topic is the subject of [1].

## 2.7  Symmetries of Trees, and the Arrow of Time

We now have four ways to look at the CCRT $S$ (Brownian excursion, the global construction, limits of CBP(n), and the random f.d.d.'s (13)). An audience from theoretical stochastic processes is likely to concentrate on the first way, and think the whole subject is just a corner of Brownian excursion theory. But I hope to show that misses the point: all four ways are useful in doing calculations.

As an illustration, the fact that in a special case CBP(n) is exactly invariant under random re-rooting implies immediately that

the distribution of the CCRT is invariant under random re-rooting.   (20)

By considering the search depth process, we could write this as a statement about Brownian excursion $W$. Fix $u$ and define

$$\hat{W}_u(s) \;=\; W_u + W_{u+s} - 2 \inf_{u \le t \le u+s} W_t, \; 0 \le s \le 1 - u$$

$$=\; W_u + W_{u+s-1} - 2 \inf_{u+s-1 \le t \le u} W_t, \; 1 - u \le s \le 1.$$

Then (20) becomes:

$$W \stackrel{d}{=} \hat{W}_U, \; \text{where } U \text{ is uniform on } [0,1].   \tag{21}$$

This is much less helpful than (20). I find it conceptually helpful to think of trees as purely **spatial** objects, without any notion of "time" involved. The point is that here are many different ways to associate "time" with a tree: the "intrinsic" time mentioned below if different from the notions of time induced by the Brownian excursion construction in section 2.2 and different again from the notions of time in the global constructions in section 2.3. Further, in section 5 we will consider trees as range spaces for random processes, in which setting having a notion of "time" attached to the tree itself is really confusing.

Having said this, recall that in a discrete-time branching process such as CBP(n) we would normally think of the vertices at distance $d$ from the root as "the individuals alive at time $d$", since we are drawing the family tree with edges of unit length. Analogously, in a continuum tree we may consider "time" to be "distance from the root" – I call this *intrinsic time*. Loosely, we may think of the CCRT as a family tree for individuals with infinitesimal lifetimes, the vertices at distance $t$ from the root representing the individuals alive at time $t$. Thus the processes $(l_t)$ and $(q_t)$ in the previous sections represent population sizes at time $t$ in the CCRT and SSCRT. But the interesting symmetries of our trees, such as (20), involve changes in direction of intrinsic time, and this is why it helps to think of the trees as purely spatial objects.

As illustration, consider the *interpretation of the SSCRT as an ancestor process*. In section 2.5 the SSCRT was presented as a limit of rescaled CBP(n) as seen from the progenitor. Here the direction of time is indicated in the top diagram in figure 3. But as at (19) and (20) we can look at CBP(n) from the standpoint of a uniform random individual. Then rescaling as in section 2.5 gives the same limit SSCRT. Here the interpretation of the baseline is as the ancestral line back from the random individual $V$ towards the progenitor, and a bush branching off the baseline at b indicates relatives of $V$ whose last common ancestor with $V$ was at (rescaled) time $b$ in the past. See the middle diagram in figure 3. (Incidently, the bottom diagram arises in a context discussed in section 6.)

*Relations between the limits.* There are several relations amongst these processes.

1. $\mathcal{R}$ is the "large-scale" limit of $\mathcal{T}_\infty$ (the discrete infinite tree), and the "small-scale" local (i..e. around the root) limit of $\mathcal{S}$ ([2] Theorem 11). The latter fact is a translation of the fact that BES(3) is the rescaled limit of Brownian excursion near 0.

2. $\mathcal{R}$ can be obtained by attaching to the baseline a $\sigma$- finite process of (mostly

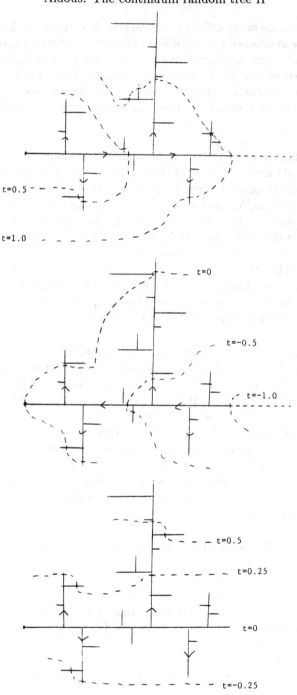

Figure 3

small) rescaled copies of $S$ ([2] section 6). In figure 2, the "bush" attached at $a$ arises from the excursion of $2B$ above $a$, drawn over the dashed line. This translates to a "last-exit" decomposition of BES(3) into excursions above levels $b$ ending at the last exit time from $b$. See section 3.5 for applications.

3. The fact that Brownian excursion is "BES(3) bridge" is suggestive, but I see no solid interpretation in terms of our trees.

## 2.8  Discussion
The preceding sections contain my subjective view of "the big picture". But there is much more one could say about related matters.

1. In classical applied probability, there is a branching process description of the total number of customers served in a busy period of a $M/G/1$ queue. For a critical $M/M/1$ queue, this gives a correspondence between the continuous time simple symmetric random walk (number of arrivals − number of departures) and the shifted geometric (1/2) Galton-Watson branching process, and this is exactly the correspondence of section 2.2 translated into continuous time.

2. There is recent theoretical literature on trees associated with Brownian-type processes. Neveu and Pitman, whose work is summarized in [37], discuss trees associated with upcrossings of size $h$ in Brownian excursion conditioned to reach height $h$ (instead of conditioning on duration). The trees they obtain are the family trees of continuous-time critical branching processes where individual lifetimes have exponential($2/h$) distribution and are followed either by death (probability 1/2) or by a split into 2 new individuals (probability 1/2). Their construction resembles that in section 2.2 in that branchpoints correspond to local minima. But fundamental to my set-up is the idea of trees as having distances between vertices, and one really needs to draw the trees as in figure 1 to make this work.

3. Conversely, Waymire et al. ([25],[9] p. 284) start with the continuous-time binary branching process above and show that, conditioning on total population size $= n$ and letting $n \to \infty$, the time to extinction rescales to the maximum of Brownian excursion.

4. For the reader interested in pursuing distributional properties of Brownian excursion, relevant papers include Chung [12]; Knight [32]; Salminen [44]; Imhof [27]; Biane [10]; Biane and Yor [11].

5. As a fanciful analogy, there are two ways to paint a picture on a piece of paper. You can divide the paper into small pixels and paint each in turn; or you can start with broad brush strokes in the middle and then fill in medium and smaller size details. The latter is analogous to the global construction of the CCRT in section 2.3; the former is analogous to its construction from Brownian excursion, where the sample path of the excursion "traces the outline of the tree".

6. Obviously we could replace the CBP(n) model with the model of critical Galton-Watson branching processes conditioned to have *height* (i.e. number of generations before extinction) greater than $h$. Then a rescaled $h \to \infty$ limit is the variant of the CCRT obtained as in section 2.2 from Brownian excursion conditioned to reach height 1 at least. This seems less natural from the viewpoint of discrete random tree models. I do not know if this limit has a global construction like those of section 2.3, or a simple description of random subtrees like (13).

# 3   DISTRIBUTIONAL PROPERTIES

Obviously branching processes are very amenable to study via generating function methods. Various questions about CBP(n) have been studied by combinatorialists (and some probabilists) using exact formulae in special cases and generating function asymptotics for the general case. Kolchin [33] provides a useful summary of the extensive Russian work in this area. We shall see how well the "weak convergence, continuum trees and Brownian excursion" approach does on these questions.

## 3.1   Height

Write $G_n$ for the *height* of CBP(n), i.e. the number of generations before extinction. Since $G_n$ is the maximum of the search depth process, an obvious corollary of Theorem 2 is

**Corollary 8** *For CBP(n), as $n \to \infty$*

$$n^{-1/2} G_n \xrightarrow{d} 2\sigma^{-1} W^*$$

*where $W^* = \sup_{0 \le t \le 1} W_t$ is the maximum of Brownian excursion.*

Expressions for the mean and the distribution of $W^*$ are well known in the stochastic processes literature, e.g. Kennedy [29] or [9] p. 85:

$$EW^* = \sqrt{\pi/2} \tag{22}$$

$$P(W^* \leq x) = 1 - 2 \sum_{k=1}^{\infty} (4x^2 k^2 - 1) \exp(-2x^2 k^2). \tag{23}$$

It is undoubtedly true that all moments converge in Corollary 8, but I did not keep track of moments in [3] so this does not rigorously follow from our approach. The result for means

$$n^{-1/2} E G_n \to \sqrt{2\pi}\, \sigma^{-1} \tag{24}$$

was established via generating function asymptotics by Flajolet and Odlyzko [22], generalizing various special cases known earlier. The general limit distribution result of Corollary 8 is Theorem 2.4.3 of Kolchin [33]. Special cases have been known for a long time: Renyi and Szekeres [42] studied the "uniform random unordered labelled tree" and obtained an expression for the limit distribution which (using Corollary 8) becomes the expression

$$P(W^* \leq x) = 2^{1/2} \pi^{5/2} x^{-3} \sum_{k=1}^{\infty} k^2 \exp(-k^2 \pi^2 / 2x^2). \tag{25}$$

So the right sides of (25) and (23) must be equal. The special case of "uniform random ordered trees" (where of course the result is immediate from the ideas of section 2.2) has also been studied – see Takacs [49] for a recent treatment and references.

*Remark.* Here and elsewhere, combinatorial arguments typically give *local* limit theorems, which are stronger than the convergence in distribution obtained by our methods.

## 3.2 Height Profile

Corollary 3 and Conjecture 4 provide a connection between occupation density $(l_s; s \geq 0)$ of Brownian excursion $W_t$ and asymptotic height profiles of CBP(n). In this section we look at the explicit formulas available in the literature.

Working directly with Brownian excursion, Knight [32] Theorem 2.3 gives the following expression for the marginal density of $l_s$.

$$f_{l_s}(y) = 2^{3/2} \pi^{5/2} s^{-3} \int_0^1 f_{W^*}\left(\frac{\pi^2 (1-t)}{2s^2}\right) f(t, y) dt$$

where $f_{W^*}$ is the density of $W^*$ (i.e. the derivative of (23)), and where

$$f(t, y) = -\frac{(2\pi t)^{-1/2}}{2s} \sum_{i=0}^{\infty} \frac{1}{i!} \frac{d^{i-1}}{dy^{i-1}} \left( y^i \frac{d^2}{dy^2} \exp(-(2t^{-1} s^2 (y+i)^2))) \right)$$

Convergence of 1-dimensional distributions in the setting of Conjecture 4 follow from classical asymptotics for generation sizes and extinction times in critical branching processes: see Theorem 2.5.6 of Kolchin [33] or Kennedy [30] Theorem 3. This approach leads to the following indirect expression for the density.

$$f_{l_s}(y) = \frac{y}{4} \int_0^1 (1-t)^{-3/2} \exp(-\frac{y^2 s^2}{8(1-t)}) g_{2s}(y/2, t) \, dt$$

where $g_s(y,t)$ is the density whose joint characteristic function $\psi_s(\theta_1, \theta_2)$ is given by

$$1/\psi_s(\theta_1, \theta_2) = \frac{\sinh(s\sqrt{-2i\theta_2})}{s\sqrt{-2i\theta_2}} - i\theta_1(\frac{\sinh(s\sqrt{-i\theta_2/2})}{s\sqrt{-i\theta_2/2}})^2.$$

Finally, by combinatorial analysis of the uniform random ordered tree, Takacs [47] obtains the formula

$$f_{l_s}(y) = 2 \sum_{j=1}^{\infty} \sum_{k=1}^{j} \binom{j}{k} e^{-(y+2sj)^2/2} \frac{(-y)^k}{(k-1)!} H_{k+2}(y + 2sj)$$

where $H_k$ are the Hermite polynomials.

There are simpler formulas for moments, e.g. for means

$$E l_s = 4s \exp(-2s^2)$$

but these are best though of as facts about the *distribution* of heights of *random* vertices of $\mathcal{S}$, as in section 3.3. Instead of emphasizing exact formulas (about which I have nothing new to say), let me emphasize some symmetry properties. In terms of CBP(n) with height $G_n$, there is no offspring distribution for which the height profile process exactly satisfies

$$(H(j); 0 \le j \le G_n) \stackrel{d}{=} (H(G_n - j); 0 \le j \le G_n). \tag{26}$$

So from the branching process viewpoint there is no reason to suspect that the occupation time process $(l_s)$ has the <u>height-reversal symmetry</u> property

$$(l_s; 0 \le s \le W^*) \stackrel{d}{=} (l_{W^*-s}; 0 \le s \le W^*). \tag{27}$$

But this is indeed true. Then Corollary 3 gives a sense in which (26) is always asymptotically true as $n \to \infty$. The symmetry (27) and a related identity

$$\sup_s l_s \stackrel{d}{=} 2W^*. \tag{28}$$

have some relevance to interesting questions about CBP(n). Here is one example: others are in the next section.

Odlyzko and Wilf [38] were interested in the maximal height profile

$$H_n^* = \max_j H(j)$$

for CBP(n). This is difficult to analyze by combinatorial methods, and required a lot of work to get a $O(n^{1/2} \log n)$ upper bound for $EH_n^*$. In view of (28), Conjecture 4 would imply

$$n^{-1/2} H_n^* \xrightarrow{d} \sigma W^*$$

and suggest the result for means

$$n^{-1/2} E H_n^* \to \sigma \sqrt{\pi/2}.$$

Finally, one could consider the sum $\sum_j j H(j)$ of heights of all $n$ vertices of CBP(n). Corollary 3 implies

**Corollary 9**

$$n^{-3/2} \sum_j j H(j) \xrightarrow{d} 2\sigma^{-1} I$$

$$where\ I = \int_0^1 W_s ds.$$

Darling [14] gives an expression for the Laplace transform of $I$. Takacs [48] gives a combinatorial proof of a special case of Corollary 9 and gives a complicated expression for the distribution of $I$ in terms of infinite sums and special functions.

*Heuristics for (27) and (28).* These results are a small part of a big picture discussed in detail by Biane and Yor [11]. From my viewpoint they are anomalous because they do not seem to follow from any symmetry property of the continuum tree $S$. Here are heuristics in terms of branching process asymptotics. Let $U$ be the diffusion on state space $[0, \infty)$ with drift rate $\mu(x) = 0$ and variance rate $\sigma^2(x) = x$. This is the continuous limit of the generation size process in a (unconditioned) critical Galton-Watson branching process. More exactly, the limit where the initial population is $u_0 n^{1/2}$, the offspring variance is 1, the population size is divided by $n^{1/2}$ and the inter-generation time is $n^{-1/2}$. Thus the limit process in Conjecture 4 (with $\sigma = 1$) ought to be the conditioned diffusion

$$l_t^* \equiv (U_t; 0 \le t < \infty | \int_0^T U_s\, ds = 1, U_0 = 0) \qquad (29)$$

where $T = \inf\{t > 0 : U_t = 0\}$. Thus we are conditioning on $U$ having an excursion from 0 of *area* 1.

With this description of $l^*$ the "height-reversal symmetry" property (27) becomes intuitively obvious: a 1-dimensional diffusion is reversible, and conditioning on a reversible event preserves reversibility.

From Corollary 3 we have

$$(l_s^*; s \ge 0) \stackrel{d}{=} (\tfrac{1}{2} l_{s/2}; s \ge 0) \qquad (30)$$

where $l$ is the occupation density of Brownian excursion $W$. But there is another way of looking at $l^*$. Being a drift-free diffusion, $U_t$ is a time-change of standard Brownian motion $\beta(t)$. With this time-change representation, a miracle occurs: the conditioning in (29) becomes conditioning $\beta$ to have an excursion of *duration* 1, i.e. to be Brownian excursion $W$. Precisely, we get

$$(l_t^*; t \ge 0) \stackrel{d}{=} (W_{L^{-1}(t)}; t \ge 0) , \text{ where } L(u) = \int_0^u 1/W_s\, ds. \qquad (31)$$

Putting together (31) and (30) gives a result of Jeulin (really a conditional form of the classic Ray-Knight description of local time for Brownian motion – see Bianne [10] Theorem 3) saying that the occupation density for Brownian excursion is a random time change of another Brownian excursion. And this relation gives (28).

### 3.3  Heights of Specified Vertices

Asking about asymptotics of heights $h_n(v)$ of particular vertices $v$ in CBP(n) doesn't quite make sense: one has to specify how the vertex is chosen. Obviously Theorem 2 gives one case. Fix $s$ and let $v_n$ be the $[2ns]$'th vertex visited in the depth search process: then

$$n^{-1/2} h_n(v_n) \stackrel{d}{\to} 2\sigma^{-1} W_s.$$

The limit marginal density of Brownian excursion is given by the formula (Ito-McKean [28] section 2.9 (3a))

$$f_{W_s}(x) = 2^{1/2} \pi^{-1/2} s^{-3/2} (1-s)^{-3/2} x^2 \exp(-x^2/(2s(1-s))) \qquad (32)$$

While the limit result (for general CBP(n)) is novel, this way of picking vertices is not particularly interesting from the viewpoint of discrete random trees. Instead, let us consider $h(V)$, where $V$ is a random vertex of $S$ chosen according to the "uniform" measure $\mu$, and $h$ denotes height. So $h(V) \overset{d}{=} 2W_U$, where $U$ is uniform on $[0,1]$. As explained below, this has density

$$f_{h(V)}(x) = x\exp(-x^2/2) \tag{33}$$

and so

$$Eh(V) = \sqrt{\pi/2}. \tag{34}$$

So Theorem 2 implies the result for uniform vertices $V$ of general CBP(n):

$$n^{-1/2}h_n(V) \overset{d}{\to} \sigma^{-1}h(V) \tag{35}$$

and suggests the result for all moments, in particular for means

$$n^{-1/2}Eh_n(V) \to \sigma^{-1}\sqrt{2/\pi}. \tag{36}$$

These limit results (35,36) for general CBP(n) were proved by generating function methods by Meir and Moon[35] Theorems 4.5 and 4.6 (in special cases, exact formulas are available). In fact they proved the local limit theorem corresponding to convergence of expectations of 1-dimensional distributions in Conjecture 4:

$$Eh_s^n \to \frac{\sigma}{2}El_{\sigma s/2} = \sigma^2 s\exp(-\sigma^2 s^2/2).$$

Note that implicit in (34) and (22) is the fact that the mean height of $S$ is exactly twice the mean height of a random vertex of $S$:

$$Eh(V^*) = 2Eh(V) = \sqrt{2\pi} \tag{37}$$

where $V^*$ denotes a vertex at maximal height. An explanation of "exactly twice" comes from the stronger fact

$$E(h(V)|h(V^*)) = \frac{1}{2}h(V^*). \tag{38}$$

This follows from the height-reversal symmetry property (27) of the limit height profile process $l^*$, because $s \to l_s^*$ is the conditional density of $h(V)$ given $S$.

There are several ways to understand (33), of which integrating (32) over $0 < s < 1$ is the least useful. The most elegant is to use the fact (14) that the subtree $\mathcal{R}(k)$ of $\mathcal{S}$ spanned by $k$ uniform random vertices is distributed as the tree produced by the first $k$ branches in the global construction of section 2.3. So $h(V_1)$ is distributed as the first cut-point in the global construction, which obviously has density (33). Properties of the joint distribution of $(h(V_1), \ldots, h(V_k))$ can in principle by obtained from the explicit distribution (13) of the subtree. For example when $k = 2$, a tree with leaves at heights $y_1, y_2$ has edges of lengths $x, y_1 - x, y_2 - x$ (for some $x \leq y_1 \wedge y_2$), and so we can use (13) to see that $(h(V_1), h(V_2))$ has joint density

$$f(y_1, y_2) = \int_0^{y_1 \wedge y_2} (y_1 + y_2 - x) \exp(-(y_1 + y_2 - x)^2/2) \, dx.$$

Note that $f(s, s) = El_s^{*2}$, for the limit height profile $l_s^*$ (i.e. for Brownian excursion occupation density, up to factors of two (30)). This provides some alternative explanations for formulas in Chung [12] section 6.

## 3.4    Diameter of the Compact Continuum Tree

The *diameter* $\Delta_n$ of CBP(n) is the maximal distance between a pair of vertices. The abstract result behind Theorem 2 (or Theorem 2 itself) implies

$$n^{-1/2}\Delta_n \overset{d}{\to} \sigma^{-1}\Delta \tag{39}$$

where $\Delta$ is the diameter of the CCRT $\mathcal{S}$. Using the representation of $\mathcal{S}$ in terms of Brownian excursion $W$,

$$\Delta = 2 \sup_{0 \leq t_1 < t_2 \leq 1} (W_{t_1} + W_{t_2} - 2 \inf_{t_1 \leq t \leq t_2} W_t). \tag{40}$$

Szekeres [46] gave a generating function proof of the existence of a limit in (39) for the special case "uniform random unordered labelled trees". From his result we obtain the following expression for the density function $f_\Delta(x)$.

$$3/\sqrt{2\pi} \, f_\Delta(x) = \tag{41}$$

$$\sum_{m=1}^{\infty} \{\frac{64}{x^4}(4b_{m,x}^4 - 36b_{m,x}^3 + 75b_{m,x}^2 - 30b_{m,x}) + \frac{8}{x^2}(4b_{m,x}^3 - 10b_{m,x}^2)\} \exp(-b_{m,x})$$

$$\text{where } b_{m,x} = (8\pi m/x)^2.$$

From this Szekeres computes

$$E\Delta = \frac{4}{3}\sqrt{2\pi} = \frac{4}{3}EG \tag{42}$$

where $G$ is the height of the CCRT $\mathcal{S}$.

As facts about Brownian excursion, (41) and (42) are novel. It is an open problem to establish them directly from (40). Incidently, the argument for (41) is similar to the argument giving (25) for the limit height; so it is likely that (41) has an equivalent expression resembling (23) in format.

I want to present an informal "argument by symmetry" which explains the simple relation (42) between mean diameter and mean height. Given the continuum random tree $\mathcal{S}$, choose a point $V$ uniformly in the tree (according to the measure $\mu$) and let $\mathbf{G}^*$ be the height of the tree rooted at $V$. Then $\mathbf{G}^* \stackrel{d}{=} \mathbf{G}$ by re-rooting symmetry (20). I shall argue informally

$$E(\mathbf{G}^*|\Delta) = 3\Delta/4 \qquad (43)$$

which obviously implies (42).

As a preliminary, although $\mu(\mathcal{S}) = 1$ by definition, we can think of "$\mathcal{S}$ conditioned on $\mu(\mathcal{S}) = c$" as the limit of CBP(cn) with $\sigma = 1$ under the $n^{-1/2}$ rescaling.

It is clear that $\mathcal{S}$ has a unique "center" $v$, that is a point such that $\mathcal{S}$ can be regarded as the union of two trees $\mathcal{S}_1, \mathcal{S}_2$ rooted at $v$ and each having height $\Delta/2$. These trees have random sizes $(\mu(\mathcal{S}_1), \mu(\mathcal{S}_2)) = (A_1, A_2)$, say, where $A_1 + A_2 = 1$. The key fact is

conditional on $(\Delta, A_1)$, $\mathcal{S}_1$ and $\mathcal{S}_2$ are independent

and distributed as $\mathcal{S}$ conditioned on having

height $= \Delta/2$ and size $= A_1$ (resp. $A_2$)

One sees this informally by considering the "uniform random unordered labelled tree" model with even diameter, where a "center" exists. In section 3.2 we discussed the height profile process $l_s^*$ for $\mathcal{S}$ in terms of excursions of the diffusion $U$. Writing $l^{1*}$ for the height profile process of $\mathcal{S}_1$ we get in the notation of (29)

conditional on $(\Delta, A_1)$, $(l_s^{1*}; 0 \leq s \leq \Delta/2) \stackrel{d}{=}$

$$(U_s; 0 \leq s \leq \Delta/2)|T = \Delta/2, \int_0^{\Delta/2} l_s^{1*} ds = A_1).$$

But the conditioning preserves the time-reversibility of the excursions of $U$. Thus, conditional on $(\Delta, A_1)$,

$$A_1^{-1} E \int_0^{\Delta/2} s l_s^{1*} ds = \Delta/4$$

(c.f. the argument below (38)). This expression gives the conditional mean distance from the center to a point $V_1$ chosen uniformly in $S_1$. Clearly the height of $S$ rooted at $V_1$ is this distance plus $\Delta/2$. Applying the same result for $S_2$ gives (43).

## 3.5  Processes Associated with the SSCRT

We now turn to the SSCRT $\mathcal{R}$ constructed globally in section 2.3, or from 2-sided BES(3) $B_s$ in section 2.5. In section 2.5 we discussed the height profile process ($q_s$): here I shall discuss some distributions of other processes defined in terms of the tree $\mathcal{R}$. For $b > 0$ it is useful to write $\mathbf{b}$ for the point on the baseline at distance $b$ from the root, and $\mathcal{R}_b$ for the part of $\mathcal{R}$ connected to the initial segment $[\text{root}, \mathbf{b}]$ of the baseline.

*The projection process.* This is the process $(Z(b); b \geq 0)$, where $Z(b) = \nu(\mathcal{R}_b)$, the total "weight" of $\mathcal{R}_b$. There are two interpretations of the process as limits in CBP(n), using either the original root or the random re-rooting procedure of section 2.7, and the latter is more interesting. Let $V_n$ be a uniform random vertex of CBP(n). Let $V_n^*(b)$ be the ancestor of $V$ in the $ba(n)$'th generation before $V$, and let $Z_n(b)$ be the total number of descendants of $V_n^*(b)$. Then from Theorem 5 (for re-rooted CBP(n))

$$\sigma^{-2} a^{-2}(n) Z_n(b) \xrightarrow{d} Z(b).$$

In [2] section 7 the global construction was used to prove

**Lemma 10** $(Z(b), b \geq 0)$ *is the positive stable* $(1/2)$ *process, that is*

$$E \exp(-\theta Z(b)) = \exp(-b\sqrt{2\theta})$$

$$Z(b) \stackrel{d}{=} b^2 Z(1).$$

It is convenient to record an easy calculation here:

$$\int_0^b (b-s) Z(ds) \stackrel{d}{=} (4/9) b^3 Z(1). \tag{44}$$

One can alternatively obtain Lemma 10 from the BES(3) representation. The last exit time process $(T_b^+; b \geq 0)$ for $(2B_t)$ is a positive stable $(1/2)$ process (see e.g. [40]) and then $Z(b) \stackrel{d}{=} T_b^+ + T_b^-$ because $\nu$ is the measure induced by $2B$ from Lebesgue measure.

*Remark.* One could construct BES(3) by starting with the last exit time process $(T_b^+)$ and then filling in excursions above levels $b$. In the global construction, each bush attached to the baseline represents such an excursion.

*The depth process.* Write $F_b$ for the height of $\mathcal{R}_b$, considered rooted at the original root, and write $D_b$ for the height of $\mathcal{R}_b$, considered re-rooted at **b**. Think of $D_b$ as the "depth" of **b**. We can give $D_b$ an interpretation as a limit in CBP(n), using as above a uniform random vertex $V$ of CBP(n). Let $D_n(b)$ be the number of generations until extinction, for the process of descendants of the ancestor of $V$ in the $ba(n)$'th generation before $V$. Then

$$a^{-1}(n)D_n(b) \overset{d}{\to} D_b.$$

A symmetry property baseline reversibility which is obvious from the global construction is the following: the distribution of $\mathcal{R}_b$ is invariant under reflection of the baseline segment [root, **b**] about its midpoint. In particular

$$D_b \overset{d}{=} F_b \quad \text{for each } b. \tag{45}$$

But $(F_b)$ and $(D_b)$ are different as *processes*, e.g. because $D_b - b$ is non-decreasing in $b$ whereas $F_b$ does not have that property.

**Lemma 11** *For each $b$,*

$$P(F_b \leq a) = P(D_b \leq a) = (1 - b/a)^2, \ a > b$$

This can be obtained from the description of $F_b$ in terms of BES(3):

$$F_b = \sup_{T_b^- \leq s \leq T_b^+} 2B_s.$$

For by the hitting probability formula for BES(3)

$$P(\sup_{0 \leq s \leq T_b^+} 2B_s < a) = P(\inf_{s \geq 0} B_s > b/2 \mid B_0 = a/2) = 1 - b/a.$$

*Last common ancestors.* Any two vertices of $\mathcal{R}$ have a last common ancestor, the point at which the paths from the root to the vertices diverge. Questions like the following are natural in terms of the interpretation of the SSCRT as a limit of critical Galton-Watson branching processes conditioned to survive forever. Define $C_b$ to be the distance from the root to the last common ancestor of *all* points at distance $b$ from the root (this last common ancestor must be on the baseline). And define $G_b$ to be the distance from the root to the last common ancestor of two randomly-chosen points $V_1, V_2$ at distance $b$ from the root (chosen according to conditioned $\nu$, the measure with total mass $q_b$) – this last common ancestor need not be on the baseline. These processes inherit the 1-self-similarity property

$$C_b \stackrel{d}{=} bC_1, \quad G_b \stackrel{d}{=} bG_1$$

as does $F_b$ above.

**Proposition 12** *(a)* $P(C_1 > c) = (1 - c)^2$, $0 < c < 1$.
*(b)* $P(G_1 > g) = 2g^{-2}(1 - g)(g + (1 - g)\log(1 - g))$, $0 < g < 1$.

To see (a), consider the point process $\mathcal{P} = \{(s^*, h^*)\}$ recording the heights $h^*$ and positions $s^*$ of bushes branching off the baseline in the global construction. Then $\mathcal{P}$ is a Poisson point process of intensity $\rho(s, h) = 2h^{-2}$. This fact comes out of the argument in [2] section 6, or alternatively from the BES(3) construction using excursions from last exit times. Plainly

$$
\begin{aligned}
P(C_1 > c) &= P(\text{ no points } (s, h) \text{ of } \mathcal{P} \text{ with } s < c \text{ and } h > 1 - s) \\
&= \exp(-\int_0^c \int_{1-s}^\infty \rho(s, h)dhds)
\end{aligned}
$$

giving (a). For (b), we can use the 2-sided BES(3) description to rewrite $G_1$ as follows. Consider local time measure on $\{s : B_s = 1\}$; pick at random two times from this measure (normalized to a probability measure), and write $T_{(1)}, T_{(2)}$ for the order statistics of these two times; define

$$
\begin{aligned}
G_1 &= \min_{t < T_{(1)} \text{ or } t > T_{(2)}} B_t \quad \text{if } T_{(1)} < 0 < T_{(2)} \\
&= \min_{T_{(1)} < t < T_{(2)}} B_t \quad \text{if not.}
\end{aligned}
$$

Routine but tedious calculations with BES(3) lead to (b).

## 4   DIFFERENT MODELS FOR RANDOM TREES.

Notwithstanding the open questions mentioned in sections 2 and 3, I regard the story of the invariance principle for CBP(n) as now well-understood. From the viewpoint of asymptotics for discrete trees, there are two natural research directions.

*Similar limits for more general models.* I am willing to make a bold conjecture:

In any reasonable model for random $n$-trees where the diameter is $\Theta(n^{1/2})$, the rescaled trees converge to limit processes which coincide with, or can be simply derived from, the limit trees discussed here.

Here is an example of a different model, IMST(n). Start with $n$ isolated vertices. Repeatedly, choose a pair of vertices uniformly and join them by an edge, provided they are not already in the same component. Ultimately a tree is obtained.

The discrete infinite tree limit (c.f. section 2.6) for this model was obtained in Aldous [4], This limit is different from the CBP(n) limit, but rescaling the discrete limit tree into a continuum tree gives exactly the SSCRT. This is strong evidence – but nowhere near a proof - that $n^{-1/2}$ rescaling of IMST(n) gives the CCRT. The "exchangeability" formalizations in [3] are designed to help with examples like this. Loosely, all we need is an argument that

$$P(D_n > xn^{1/2}) \to \exp(-x^2/2) \tag{46}$$

where $D_n$ is the distance in IMST(n) between two prespecified vertices, and then the techniques of [3] could bootstrap the argument into a proof of convergence to the CCRT. Unfortunately (46) seems difficult.

Here are some rather different models where we expect the CCRT limit.

1. Uniform random unordered unlabelled trees.

2. Uniform random spanning trees of expander graphs (e.g. hypercubes) – see Aldous [5].

3. Steele's [45] "exponential family" of random $n$-trees, with a parameter determining the mean proportion of leaves.

Another interesting application is to *random mappings*, a well-studied topic surveyed in Kolchin [33]. Here we choose uniformly at random one of the $n^n$ functions $f : \{1, \ldots, n\} \to \{1, \ldots, n\}$ and consider the graph with edges $(i, f(i))$. The component containing 1 consists of a cycle with attached trees: by representing the trees as in section 2.2, one can represent the entire mapping as a walk of length $2n$. In ongoing work with Jim Pitman it is shown that these walks rescale to reflecting Brownian bridge.

*Different limit continuum random trees.* A much harder topic is the study of asymptotics for random trees whose definition involves the geometry of $d$-dimensional space. Two combinatorial examples are the uniform random spanning trees of $Z^d$ studied in Pemantle [39], and the Euclidean minimum spanning trees on random points in $R^d$ studied in Aldous and Steele [7]. And numerous examples such as directed animals and DLA appear in the physics literature. In many of these examples it is natural to conjecture the existence of continuum limit trees after rescaling, but I am not aware of any rigorous proofs.

## 5  BROWNIAN MOTION ON CONTINUUM TREES.

### 5.1  Generalities

I want to discuss distributional properties of "standard Brownian motion" $(\mathbf{X}_t; t \geq 0)$ taking values in the CCRT $(\mathcal{S}, \mu)$ or in the SSCRT $(\mathcal{R}, \nu)$ (by default, random processes start at the root). The discussion is necessarily heuristic, because no rigorous proof of existence of $\mathbf{X}_t$ has been written down. Such processes may be interesting as counterparts to the recent rigorous theory of Brownian motion on regular fractals developed by Barlow and Perkins [8], Lindstrom [34] and others. Loosely, our particular CRTs have "dimension 2", inherited from Brownian sample paths: in view of the rigorous construction of continuum trees from general continuous functions in [3], one can certainly construct continuum trees of any fractional dimension. Despite the large physics literature, rigorous study of diffusions on non-regular fractal sets seems difficult. But obviously a tree structure is a great simplification, and a rigorous theory of Brownian motion on rather general continuum trees seems a natural next step. I outline below the shape that such a theory might take, but do not intend to pursue the topic myself.

It is important to regard $\mathcal{S}$ and $\mathcal{R}$ as *spatial* trees, and downplay their constructions from Brownian excursion and BES(3). For simple symmetric random walk on a discrete tree there is an elementary formula for the mean first passage time from one specified vertex to another. Using these formulas it is easy to see that in CBP(n) (with $\sigma^2 = 1$ for simplicity) the mean passage time between random vertices $\sim \sqrt{\pi/2}\, n^{3/2}$. Results of this type go back to Moon [36]. Thus one way to think of Brownian motion on $\mathcal{S}$ is as a rescaled limit of simple random walk on CBP(n): make the edges have length $n^{-1/2}$ so as to get the limit $\mathcal{S}$, and make the time between steps be $n^{-3/2}$. Similarly, we could start with simple symmetric random walk on the discrete infinite tree of section 2.6: the same rescalings should lead to Brownian motion on $\mathcal{R}$. The latter discrete setting was studied in Kesten [31], who refers to his unpublished work proving the existence of a limit $n^{-3/2}T_{n^{1/2}} \xrightarrow{d} T$ for the first passage time $T_{n^{1/2}}$ from the root to the point at distance $n^{1/2}$ along the infinite path. I interpret $T$ as the time taken by $\mathbf{X}$ on $\mathcal{R}$ to first hit the point on the baseline at distance 1 from the root. The ingredients of a *proof* of the existence of **X** via such a weak convergence construction would plainly be similar to this unpublished work of Kesten.

In the next section I suggest an alternative "sample path" construction which I believe will handle more general continuum trees. But my main purpose is to exhibit concrete calculations of distributions (mostly expectations) associated with **X**, and this is the subject of sections 5.3 and 5.4.

## 5.2   An Occupation Density Construction

In this section let us work with the precise definition of a (deterministic) *continuum tree* $(S_0, \mu_0)$ given in [3] section 2.3. Essentially, $S_0$ is a set which is topologically a tree, i.e. has a unique non-self-intersecting path between any pair of points, and $\mu_0$ is a probability measure on $S_0$, which should be thought of as a "uniform measure". $S_0$ contains a "root" denoted by 0. (As a technical remark, for our purposes here we also assume $S_0 = \text{support}(\mu_0)$, though in [3] a weaker condition was used.)

We now define *Brownian motion* on a compact continuum tree $(S_0, \mu_0)$ (the locally compact case needed for $\mathcal{R}$ is similar) to be a $S_0$-valued process $(\mathbf{X}_t; t \geq 0)$ with the following properties.

(i) Continuous sample paths.

(ii) Strong Markov.

(iii) Reversible with respect to its invariant measure $\mu_0$.

(iv) For each path $[[a, b]] \subset S_0$ and each $x \in [[a, b]]$,

$$P_x(T_a < T_b) = \frac{d(x, b)}{d(a, b)} \tag{47}$$

where $d$ denotes distance.

(v) For points $a, b \in S_0$ let $m_{a,b}(\cdot)$ be the mean occupation measure for the process started at $a$ and run until it first hits $b$. Then

$$m_{a,b}(dx) = 2d(b; c(b : a, x))\mu_0(dx), \ x \in S_0 \tag{48}$$

where $c(b : a, x)$ is the point at which the paths $[[b, a]]$ and $[[b, x]]$ diverge.

The *skeleton* $S_\infty$ of $S_0$ is the set of points $x$ which are in the interior of some path $[[a, b]]$. A consequence of the precise definition of continuum tree is that the skeleton has $\mu_0$-measure zero. Thus the process spends Lebesgue-0 time in the skeleton.

In contrast to the setting of general fractal subsets of $R^d$ [34], there is no difficulty in proving *uniqueness* here, because of the explicit formulas (47,48). A sledgehammer proof could be based upon the result about general Markov processes being determined up to random time-change by their exit place distributions.

To outline an existence argument, let $V_1, \ldots, V_k$ be picked independently from $\mu_0$ and let $\mathcal{R}(k)$ be the subtree of $\mathcal{S}_0$ spanned by the root and the points $V_1, \ldots, V_k$. Then $\mathcal{R}(k)$ is a (random) tree consisting simply of a finite number of edges with positive edge-lengths. Recall (14) that in the particular case of the CCRT $\mathcal{S}$, this $\mathcal{R}(k)$ is distributed as the tree produced by the first $k$ branches in the global construction of section 2.3. Let $\mu_k$ be the natural induced Lebesgue measure on $\mathcal{R}(k)$. Assume the following regularity property (*local homogeneity*): there exist deterministic $a_k \to \infty$ such that

$$\frac{1}{a_k}\mu_k \to \mu_0 \text{ a.s. as k } \to \infty \tag{49}$$

The CCRT has this property with $a_k = \sqrt{2k}$ (this is essentially Theorem 3 (ii) of [2], but also follows easily from the Brownian excursion representation).

On each $\mathcal{R}(k)$ we can define ordinary Brownian motion $X_k(t)$ started at the root 0. One way of viewing $\mathbf{X}$ is as the weak limit of these ordinary Brownian motions as we "fill out" the tree and speed up time:

$$X_k(a_k t) \overset{d}{\to} \mathbf{X}(t) \text{ as } k \to \infty.$$

In fact we can do better and use a sample path construction. The key idea is: if we measure time by a suitable "local time" rather than absolute time, then we can make these Brownian motions consistent as $k$ increases. Fix $\tau > 0$. Run ordinary Brownian motion $X_k(t)$ on $\mathcal{R}(k)$ until the local time at the root reaches $\tau$. Regard the accumulated local time as an occupation density $l_k$:

$$\int_0^\tau 1_{(X_k(t)\in A)} \, dt = \int_A l_k(\tau,\omega,x)\mu_k(dx), \ A \subset \mathcal{R}(k). \tag{50}$$

We can construct $l_{k+1}$ in terms of $l_k$. For $\mathcal{R}(k+1)$ consists of $\mathcal{R}(k)$ plus a new edge $(B_{k+1}, V_{k+1})$ attached at a point $B_{k+1} \in \mathcal{R}(k)$. The occupation density $l_{k+1}(\tau,\omega,x)$ coincides with $l_k(\tau,\omega,x)$ for $x \in \mathcal{R}(k)$, while on the new edge its conditional distribution given $l_k(\tau,\omega,B_{k+1}(\omega)) = l$ is the occupation density for 1-dimensional Brownian motion on a line segment of length $d(B_{k+1}, V_{k+1})$

started at 0 and run until the local time at 0 reaches $l$. Continuing for all $k$, we get a function $l_\infty(\tau, \omega, x), x \in \mathcal{S}_\infty$ defined on the skeleton $\mathcal{S}_\infty$ of $\mathcal{S}_0$ which satisfies (50) for all $k$. If we can show

$$l_\infty(\tau, \omega, \cdot) \text{ extends to a bounded continuous function on } \mathcal{S}_0, \qquad (51)$$

then by (49) $l_\infty$ is the $k \to \infty$ limit time-rescaled occupation density for Brownian motion on $\mathcal{R}(k)$, and this can be regarded as occupation density for some process $(\mathbf{X}(t) : 0 \le t \le t(\tau))$ on $\mathcal{S}_0$ run until its density at the root reaches $\tau$, i.e. until absolute time

$$t(\tau) = \int_{\mathcal{S}_0} l_\infty(\tau, \omega, x) \mu_0(dx).$$

In other words, we are trying to construct $\mathbf{X}$ by specifying its occupation density up to times $\tau$. The argument is rigorous up to (51); and it remains to put together different $\tau$'s and show that $\mathbf{X}$ has the properties (i)-(v). I conjecture that very little more than (49) is required – perhaps only some weak "metric entropy" condition on $\mathcal{S}_0$.

### 5.3  Easy Distributional Properties

We now return to the special cases of the CCRT $(\mathcal{S}, \mu)$ and the SSCRT $(\mathcal{R}, \nu)$, and assume that Brownian motion $\mathbf{X}$ exists as a weak limit of rescaled random walks as in section 5.1 and also via the construction in section 5.2.

The SSCRT case is somewhat more tractable, because it inherits from $\mathcal{R}$ a self-similarity property.

$$(\mathbf{X}_{ct}; t \ge 0) \stackrel{d}{=} (c^{1/3}\mathbf{X}_t; t \ge 0). \qquad (52)$$

One explanation of the "1/3" comes from the rescaling in section 5.1 of discrete random walk. Another comes from a scaling (53) of first hitting times, which we now derive. As in section 3.5, for $b > 0$ write b for the point on the baseline at distance $b$ from the root, and write $Z(b) = \nu(\mathcal{R}_b) = $ the total "weight" of the part $\mathcal{R}_b$ of $\mathcal{R}$ connected to the initial segment [root, b] of the baseline. Appealing to (48)

$$E(T_{\mathbf{b}}|\mathcal{R}) = \int_0^b 2(b-y)Z(dy)$$

where we write $(\cdot|\mathcal{R})$ to mean conditioning on the realization of the random tree $(\mathcal{R}, \nu)$. Now Lemma 10 says that $Z(\cdot)$ is the positive stable 1/2 process, and appealing to (44)

$$E(T_{\mathbf{b}}|\mathcal{R}) \stackrel{d}{=} \frac{8}{9}b^3 Z(1). \qquad (53)$$

Of course $EZ(1) = \infty$ and so the unconditional first hitting time $T_b$ has infinite expectation: this is a disadvantage of working with the SSCRT.

Turning to the CCRT $(\mathcal{S}, \mu)$, consider first hitting times $T_x$ on arbitrary $x \in \mathcal{S}$. Immediate from (48) is

$$E_x(T_y|\mathcal{S}) = \int_{\mathcal{S}} 2d(y, c(y : x, z))\mu(dz). \tag{54}$$

Another consequence of (48) is a simple formula for mean round trip times between leaves (here the root counts as a leaf)

$$E_x(T_y|\mathcal{S}) + E_y(T_x|\mathcal{S}) = 2d(x, y) \text{ for any leaves } x, y \in \mathcal{S}. \tag{55}$$

These have nothing to do with the particular structure of $\mathcal{S}$, but reflect elementary general identities for simple random walks on discrete trees. Now let $V$ and $V_1$ be independent random points of $\mathcal{S}$ chosen according to $\mu$ (which puts mass 1 on the leaves – here continuum trees are simpler than discrete trees!). Using invariance (20) under random re-rooting, $Ed(V_1, V) = Ed(\text{root}, V)$ and $E_{V_1}T_V = ET_V (= E_{\text{root}}T_V)$. It follows from (55) that

$$\begin{aligned} ET_V &= Ed(\text{root}, V) \\ &= \sqrt{\pi/2} \text{ by (34)} \end{aligned} \tag{56}$$

which is the Brownian motion analog of the result of Moon [36] mentioned earlier. A slight elaboration is provided by

$$E(T_V| h(V)) = h(V) \tag{57}$$

where $h(V) = d(\text{root}, V)$. This is based upon another symmetry property of $\mathcal{S}$. Recall from (14) that we may regard $[[\text{root}, V]]$ as arising from the initial line-segment $[0, C_1]$ in the global construction of $\mathcal{S}$. Treating this initial line segment as a baseline, we can define a "projection process" $\hat{Z}(b)$ analogous to the SSCRT case (section 3.5). That is, $\hat{Z}(b)$ is the $\mu$-mass ultimately attached to the first $b$ units of the initial segment. The global construction makes clear the reversibility property

$$(\hat{Z}(b); 0 \le b \le h(V)) \overset{d}{=} (\hat{Z}(h(V) - b); 0 \le b \le h(V)). \tag{58}$$

Rewriting (54) as

$$E(T_V|\mathcal{S}) = \int_0^{h(V)} 2(h(V) - b)d\hat{Z}(b)$$

(58) implies (57).

It is immediate from (55) that, given $\mathcal{S}$, the $v$ which maximizes the mean round trip time from the root to v and back is the vertex $V^*$ at maximal distance from the root. For future reference,

$$E_{\text{root}}T_{V^*} + E_{V^*}T_{\text{root}} = 2\sqrt{2\pi}. \qquad (59)$$

by (22). On the other hand, one can show that the $v$ which maximizes the one-sided hitting time $E_{\text{root}}(T_v|\mathcal{S})$ is $\underline{\text{not}}$ $V^*$.

One could calculate variances of hitting times in similar ways, starting from formulas in the discrete setting. Moon's result in [36] Corollary 7.3.1 (whose proof relies on the special structure of the uniform random unordered tree) implies

$$\text{var}(T_V) = \frac{32}{15}.$$

In principle this variance could be decomposed as the sum of three components – contributions from the choice of $\mathcal{S}$, the choice of $V$ given $\mathcal{S}$, and the choice of Brownian path given $\mathcal{S}$ and $V$ – but the computations look messy.

### 5.4  Hard Distributional Properties

The distribution of baseline hitting times for Brownian motion on the SSCRT turns out to be related to the distribution of inverse local time in the CCRT. Here is our best shot at describing these distributions, though it doesn't qualify as "explicit".

Consider 1-dimensional Brownian motion, started at 0 and run until its occupation density at 0 reaches $a$. It is well known ([43] VI.52) that its occupation density $Z_s$ at position $s > 0$ behaves as the *Ray-Knight diffusion*

$$Z_0 = a, \ dZ_s = \sqrt{2Z_s}d\beta_s, \ s > 0, \qquad (60)$$

where $(\beta_s)$ is another 1-dimensional Brownian motion.

Now consider the global construction of $\mathcal{S}$ via the cut-and-join points $(C_i, J_i)$. Fix $a > 0$ and define $(Z_s^a, s \geq 0)$ by

$$\begin{aligned}
Z_0^a &= a \\
dZ_s^a &= 2\sqrt{Z_s^a}d\beta_s, \text{ on } C_{i-1} \leq s < C_i \\
Z_{C_i}^a &= Z_{J_i}^a
\end{aligned}$$

Clearly $(Z_s^a, 0 \leq s \leq C_k)$ describes the occupation density over $\mathcal{R}(k)$ for Brownian motion on the tree $\mathcal{R}(k)$ constructed from the first $k$ cuts, the motion run until the local time at the root 0 reaches $a$. It is not hard to argue (c.f. the urn model of [2] sec. 4) that the limits

$$L_a = \lim_{s_0 \to \infty} \frac{1}{s_0} \int_0^{s_0} Z_s^a ds \qquad (61)$$

exist a.s. From the construction of Brownian motion $\mathbf{X}$ on $\mathcal{S}$ in section 5.2 we see that the process $(L_a, a \geq 0)$ is the "inverse local time at the root" process for $\mathbf{X}$. (This construction should be regarded as "unconditional on $\mathcal{S}$"). Of course, conditionally on $\mathcal{S}$ the process $(L_a)$ is a subordinator, and $E(L_a|\mathcal{S}) = a$. Getting explicit distributional information about $(L_a)$ seems the most important potentially solvable open question in this area.

Now consider the SSCRT. As in section 2.7 (relation 2) we may regard the SSCRT as a semi-infinite baseline with "bushes" attached, the bushes being rescaled copies of the CCRT. More precisely, let $(\mathcal{S}_c, \mu_c)$ denote the CCRT $(\mathcal{S}, \mu)$ after scaling by multiplying lengths by $c$ and making the total measure $= c^2$. Then mark the baseline $[0, \infty)$ according to a marked Poisson process, marks in $[c, c + dc]$ appearing at rate $\sqrt{\frac{2}{\pi}} c^{-2} dc$. Wherever a mark $c$ appears on the baseline, we attach a copy of $(\mathcal{S}_c, \mu_c)$.

Now consider Brownian motion $\mathbf{X}$ on the SSCRT, run until the time $T_1$ it first travels unit distance along the baseline. This has an occupation density (w.r.t. $\nu$) process $(\hat{Z}(x), 0 \leq x \leq 1)$ on the baseline, which is just occupation density for 1-dimensional Brownian motion on the half-line started at 0 and run until first hitting 1. Specifically, $(\hat{Z}(1 - x), 0 \leq x \leq 1)$ is the other Ray-Knight diffusion with drift rate $\mu(z) = z$ and variance rate $\sigma^2(z) = 2z$.

We are interested in the occupation density (over all $\mathcal{R}$) for $\mathbf{X}$ at time $T_1$. What happens on a bush depends only on the occupation density $\hat{Z}(x)$ at the point $x$ where the bush attaches to the baseline. A scaling argument shows that inverse local time $L_a^c$ for the Brownian motion on $(\mathcal{S}_c, \mu_c)$ scales as

$$(L_a^c, a \geq 0) \stackrel{d}{=} (c^3 L_{a/c}, a \geq 0).$$

Thus the amount of time spent in a bush with scaling factor $c$ attached at $x$ is distributed as $c^3 L_{\hat{Z}(x)/c}$. Adding over bushes gives

$$T_1 \stackrel{d}{=} \sum_{(x_i, c_i) \in \text{POIS}} c_i^3 L_{\hat{Z}(x_i)/c_i}^{(i)} \qquad (62)$$

where POIS is the Poisson process of points $(x, c) \in (0, 1) \times (0, \infty)$ with rate $\sqrt{\frac{2}{\pi}} c^{-2}$ and where $L_a^{(i)}$ is distributed as at (61), independent as $i$ varies. This is the advertized connection between SSCRT first passage times and CCRT inverse local time.

Changing topics, another quantity which (surprisingly) can be calculated explicitly is related to the *cover time*

$$C = \inf\{t : \mathcal{S} = \cup_{0 \le s \le t} \mathbf{X}_s\}$$

for Brownian motion on $\mathcal{S}$, i.e. the first time at which the sample path has hit *every* point of $\mathcal{S}$. Consider the related cover-and-return time

$$C^+ = \inf\{t \ge C : \mathbf{X}_t = \text{root}\}.$$

Clearly $C^+$ is at least the time to visit the furthest leaf and return, which by (59) has mean $2\sqrt{2\pi}$. In fact I assert

$$EC^+ = 6\sqrt{2\pi}. \tag{63}$$

I do not know any simple symmetry argument for the factor of 3, though it is tempting to seek some overlooked symmetry. Pedestrianly, one can study random walk on unconditioned Galton-Watson trees and set up a recursion for the analogous cover-and-return time. Write $C_n^+$ for this time, conditioned on the size of the tree $= n$. In the case of Poisson offspring (i.e. the uniform unordered random labelled tree) it is proved in [6] that

$$\text{if } EC_n^+ \sim cn^{3/2} \text{ then } c = 6\sqrt{2\pi}. \tag{64}$$

This is an intuitively convincing argument for (63). At the rigorous level, a major issue is that, even granted weak convergence of rescaled random walks on discrete trees to the limit Brownian motion on $\mathcal{S}$, this does *not* imply convergence of cover times.

## 6  SUPERPROCESSES

This section is directed at readers already familiar with the subject of superprocesses. I have no technical knowledge of the subject, but am merely aiming to set out in conversational style some remarks about what is intuitively obvious, given a "random tree" background.

There is an underlying nice continuous-time Markov process $(X_t; t \ge 0)$ taking values in a space $\Sigma$. The associated superprocess takes values in the set $M(\Sigma)$ of non-negative measures on $\Sigma$, starting at time 0 with (say) the unit mass at a point $x_0$. It can be constructed as a weak limit of finite-population branching Markov processes, or directly via martingale characterization. Recently attention has been given to the associated "historical process" which gives the family tree of the limit population: see e.g. Dawson and Perkins [15], Dynkin [18, 19], Le Gall [23]. My viewpoint is to take this as the starting place: think about tree-indexed $\Sigma$-valued processes rather than time-indexed $M(\Sigma)$-valued processes.

Given a rooted tree $t$ with a finite number of edges of positive length, let t be the point-set of all points of the tree, i.e. the points in the edges as well as the branchpoints and endpoints. Given a starting position $x_0$ for the underlying Markov process, there is an obvious construction of a tree-indexed Markov process $(X_s; s \in t)$, as follows. Put $X_{\text{root}} = x_0$ and then define $X$ on one edge at a time, working away from the root. For an edge $[[a, b]]$ for which $X_a = x_a$ has already been defined, we define $(X_s, s \in [[a, b]])$ to be distributed as the underlying process started at $x_a$ and run for time $d(a, b)$, independently of the previously-defined parts of the tree-indexed process.

Now consider a general continuum tree $(\mathcal{S}_0, \mu_0)$ (e.g. constructed from some function $f$ as in section 2.2). By the Kolmogorov extension theorem we can define a tree-indexed process $(\mathbf{X}_s; s \in \mathcal{S}_0)$ such that for each finite set of points $v_1, \ldots, v_k$ in $\mathcal{S}_0$ the process restricted to the subtree t spanned by $(v_i)$ is distributed as specified above. Of course such constructions by extension are unsatisfactory because different "versions" may have different sample path properties. To get a cleaner construction, suppose the underlying process has cadlag paths. Then we can construct $(\mathbf{X}_s; s \in \text{skeleton}(\mathcal{S}_0))$ such that every realization is cadlag at each point in the skeleton. For leaves $x \in \mathcal{S}_0$ one can seek to define by continuity:

$$\mathbf{X}_x = \lim\{\mathbf{X}_s : s \in [[0, x]], \ s \to x\}$$

In general the limit will exist outside a $\mu_0$-null set of leaves, on which we must let $\mathbf{X}$ be undefined.

With each $s \in \mathcal{S}_0$ we associate the "intrinsic time" $t(s)$ which is just the distance from the root to $s$:

$$t(s) = d(\text{root}, s).$$

We can then define a time-indexed $M(\Sigma)$-valued process

$$\Theta_t(\cdot) = \int_{\mathcal{S}_0} 1_{(\mathbf{X}_s \in \cdot)} 1_{(t(s) \leq t)} \mu_0(ds). \tag{65}$$

If sufficiently smooth in $t$ we can differentiate to get

$$\theta_t(\cdot) = \frac{d}{dt}\Theta_t(\cdot). \tag{66}$$

Some notation: for a positive measure $\theta$, write $||\theta||$ for its total mass.

Now consider this construction of a tree-indexed process $\mathbf{X}$ in the two special cases where the indexing tree is the CCRT $(\mathcal{S}, \mu)$ or the SSCRT $(\mathcal{R}, \nu)$. I claim these are the same as the usual superprocess, but with different conditionings. In the second case (the <u>immortal superprocess</u>) we are conditioning on the process being created (with infinitesimal mass) at point $x_0$ at time 0, and on the process surviving for ever. In the first case (the <u>superprocess excursion</u>) we are conditioning on the process being created (with infinitesimal mass) at point $x_0$ at time 0, and on the total (i.e. integrated over time up to extinction) population size being equal to 1. Arguably these conditionings are more natural in many biological applications than the usual "start with unit mass" model. To fit this usual model into our set-up, we use as index the continuum random tree pictured on the bottom in figure 3, i.e. the part of the SSCRT which branches off the first unit of the baseline, measuring "time" as "distance from baseline". In all these examples, what is usually called the superprocess is the $M(\Sigma)$-valued process $(\theta_t)$ at (66).

The assertions above are intuitively obvious from the interpretation of $\mathcal{S}$ and $\mathcal{R}$ as limits of family trees in critical branching processes. As with the development of historical processes, the point is to "decouple" the family tree structure from the Markov motion in the space $\Sigma$. In the next section I make some observations about superprocesses based on this "continuum random tree" viewpoint.

### 6.1   Five Observations

1. *Computer simulation.* Suppose you want to estimate some distribution associated with some explicit superprocess by computer simulation. The naive way would be to simulate a discrete-time critical branching process. Starting with 50, say, individuals and running until extinction would require order $50^2 = 2,500$ calls to the random number generator just to simulate the "family tree". It is much more efficient to use the global constructions in section 2.3 and simulate the first 50, say, branches of the CCRT, which requires only $2 \times 50 = 100$ calls to the generator (and then finally simulate the Markov process along the edges).

2. *Different models for random trees.* Consider the IMST(n) model of section 4. This is a random tree on $n$ vertices, where we pay no attention to the order in which edges were added in the construction. Now imagine one vertex placed at $0 \in R^d$ and the edges having length $n^{-1/2}$ and independent uniform random directions in $R^d$. Let $\Phi^k$ be the empirical distribution of

the vertex positions. Granted the conjecture that this model also rescales to $S$, it is intuitively obvious that $\Phi^k$ converges in distribution to $\Theta_\infty$, the total occupation density (65) associated with the superprocess excursion built over $d$-dimensional Brownian motion. The point is that the discrete IMST(n) model has no notion of "Markovian branching" or even of "time", but still leads to a superprocess.

3. *A connection between superprocesses and Brownian motion on continuum trees.* One quantity of interest in the latter context was $L_a$ at (61), the time at which Brownian motion on the CCRT has accumulated local time $a$ at the root. It is intuitively clear that

$$L_a \overset{d}{=} \int_S \mathbf{X}_s^{(a)} \, \mu(ds)$$

where $\mathbf{X}^{(a)}$ is the $S$-indexed process (i.e. superprocess excursion) built over the Ray-Knight diffusion (60).

4. *Symmetry properties.* The symmetry properties of $S$ and $\mathcal{R}$ we have discussed lead to symmetry properties of superprocesses. Here is one example. Suppose the underlying Markov process is ergodic with stationary distribution $\pi$. Choose $X_0$ from $\pi$, then run the superprocess excursion starting at $X_0$. Let $\Theta_\infty$ be the total occupation measure (65) and pick $X^*$ according to $\Theta_\infty$. If the underlying Markov process is reversible then

$$(X_0, \Theta_\infty, X^*) \overset{d}{=} (X^*, \Theta_\infty, X_0).$$

In words, the distribution of $\Theta_\infty$ relative to a random individual in the population is the same as its distribution relative to the progenitor: so the progenitor is not special. This property is intuitively clear from the "random re-rooting" property (20) of the CCRT.

5. *The immortal superprocess.* This has recently been studied rigorously by Evans and Perkins [21, 20]. From our viewpoint of the superprocess as the Markov process $\mathbf{X}$ indexed by the SSCRT, some elementary properties are obvious.

(a) The "population size at time $t$" $|\theta_t|$ has Gamma$(2, 2/t)$ distribution, by Corollary 6 and (18).

(b) If the underlying Markov process has absorbing states then (because the SSCRT consists of bounded bushes attached to an infinite baseline) the superprocess gets absorbed with the same probabilities as for the underlying Markov process ([20]).

(c) If the underlying Markov process converges from any start to a unique stationary distribution $\pi$ then

$$\theta_t/|\theta_t| \xrightarrow{P} \pi. \tag{67}$$

For using a standard exchangeability fact (that for exchangeable sequences, pairwise independence implies independence) to prove (67) it suffices to prove

$$\mathrm{dist}(\mathbf{X}_{V_1(t)}, \mathbf{X}_{V_2(t)}) \to \pi \times \pi \tag{68}$$

where $V_1(t)$ and $V_2(t)$ are picked uniformly from the population at time $t$. The individuals $V_1(t)$ and $V_2(t)$ had last common ancestor at time $G_t$, say, and

$$P(\mathbf{X}_{V_1(t)} \in A, \mathbf{X}_{V_2(t)} \in B | G_t = g, \mathbf{X}_{G_t} = x) = P_x(\mathbf{X}_{t-g} \in A)P_x(\mathbf{X}_{t-g} \in B).$$

By the self-similarity property of the SSCRT we have $G_t \overset{d}{=} tG_1$, and then (68) follows easily from the convergence assumption on the underlying Markov chain [21].

In fact, the exact distribution of $G_1$ was calculated in Proposition 12, as was the distribution $C_1$ of the time of the last common ancestor of the entire time-1 population.

## References

[1] D.J. Aldous. Asymptotic fringe distributions for general families of random trees. 1991. To appear in Annals of Applied Probability, May 1991.

[2] D.J. Aldous. The continuum random tree I. *Ann. Probab.*, 19:1–28, 1991.

[3] D.J. Aldous. The continuum random tree III. In preparation.

[4] D.J. Aldous. A random tree model associated with random graphs. *Random Structures and Algorithms*, 1:383–402, 1990.

[5] D.J. Aldous. The random walk construction of uniform spanning trees and uniform labelled trees. *SIAM J. Discrete Math.*, 3:450–465, 1990.

[6] D.J. Aldous. Random walk covering of some special trees. *J. Math. Analysis Appl.*, 157:xxx, 1991. To appear.

[7] D.J. Aldous and J.M. Steele. Asymptotics for Euclidean minimal spanning trees on random points. 1991. Unpublished.

[8] M.T. Barlow and E.A. Perkins. Brownian motion on the Sierpinski gasket. *Probab. Th. Rel. Fields*, 79:543–624, 1988.

[9] R.N. Bhattacharaya and E.C. Waymire. *Stochastic Processes with Applications*. Wiley, 1990.

[10] P. Biane. Relations entre pont et excursion du mouvement Brownien reel. *Ann. Inst. Henri Poincare*, 22:1–7, 1986.

[11] P. Biane and M. Yor. Valeurs principales associees aux temps locaux Browniens. *Bull. Sci. Math. (2)*, 111:23–101, 1987.

[12] K.L. Chung. Excursions in Brownian motion. *Arkiv fur Matematik*, 14:155–177, 1976.

[13] E. Csaki and G. Mohanty. Excursion and meander in random walk. *Canad. J. Statist.*, 9:57–70, 1981.

[14] D.A. Darling. On the supremum of a certain Gaussian process. *Ann. Probab.*, 11:803–806, 1983.

[15] D.A. Dawson and E. Perkins. Historical processes. 1991. Mem. Amer. Math. Soc., to appear.

[16] R. Durrett, D.L. Iglehart, and D.R. Miller. Weak convergence to Brownian meander and Brownian excursion. *Ann. Probab.*, 5:117–129, 1977.

[17] R. Durrett, H. Kesten, and E. Waymire. On weighted heights of random trees. *J. Theoretical Probab.*, 4:223–237, 1991.

[18] E.B. Dynkin. *Branching Particle Systems and Superprocesses*. Technical Report, Cornell University, 1989.

[19] E.B. Dynkin. *Path Processes and Historical Processes*. Technical Report, Cornell University, 1989.

[20] S.N. Evans. *The Entrance Space of a Measure-Valued Markov Branching Process Conditioned on Non-extinction*. Technical Report, U.C.Berkeley, 1991. Canad. Math. Bull., to appear.

[21] S.N. Evans and E. Perkins. Measure-valued Markov branching processes conditioned on non-extinction. *Israel J. Math.*, 71:329–337, 1990.

[22] P. Flajolet and A. Odlyzko. The average height of binary trees and other simple trees. *J. Comput. System Sci.*, 25:171–213, 1982.

[23] J.-F. Le Gall. *Brownian Excursions, Trees and Measure-Valued Branching Processes.* Technical Report, Universite P. et M. Curie, Paris, 1989.

[24] G. R. Grimmett. Random labelled trees and their branching networks. *J. Austral. Math. Soc. (Ser. A)*, 30:229–237, 1980.

[25] V.J. Gupta, O.J. Mesa, and E. Waymire. Tree dependent extreme values: the exponential case. *J. Appl. Probab.*, 27:124–133, 1990.

[26] W. Gutjahr and G. Ch. Pflug. *The Asymptotic Contour Process of a Binary Tree is Brownian Excursion.* Technical Report, University of Vienna, 1991. To appear in Stochastic Proc. Appl.

[27] J.-P. Imhof. On Brownian bridge and excursion. *Studia Sci. Math. Hungar.*, 20:1–10, 1985.

[28] K. Ito and H.P. McKean. *Diffusion Processes and their Sample Paths.* Springer, 1965.

[29] D.P. Kennedy. The distribution of the maximum Brownian excursion. *J. Appl. Prob.*, 13:371–376, 1976.

[30] D.P. Kennedy. The Galton-Watson process conditioned on the total progeny. *J. Appl. Prob.*, 12:800–806, 1975.

[31] H. Kesten. Subdiffusive behavior of random walk on a random cluster. *Ann. Inst. H. Poincare Sect. B*, 22:425–487, 1987.

[32] F. B. Knight. On the excursion process of Brownian motion. *Trans. Amer. Math. Soc.*, 258:77–86, 1980.

[33] V.F. Kolchin. *Random Mappings.* Optimization Software, New York, 1986. (Translation of Russian original).

[34] T. Lindstrom. Brownian motion on nested fractals. *Memoirs of the A.M.S.*, 420, 1989.

[35] A. Meir and J.W. Moon. On the altitude of nodes in random trees. *Canad. J. Math.*, 30:997–1015, 1978.

[36] J.W. Moon. Random walks on random trees. *J. Austral. Math. Soc.*, 15:42–53, 1973.

[37] J. Neveu and J. Pitman. The branching process in a Brownian excursion. In *Seminaire de Probabilites XXIII*, pages 248–257, Springer, 1989. Lecture Notes in Math. 1372.

[38]  A.M. Odlyzko and H.S. Wilf. Bandwidths and profiles of trees. *J. Combin. Theory Ser. B*, 42:348–370, 1987.

[39]  R. Pemantle. *Choosing a Spanning Tree for the Integer Lattice Uniformly.* Technical Report, Cornell University, 1989. To appear in Ann. Probability.

[40]  J.W. Pitman and M. Yor. Bessel processes and infinitely divisible laws. In *Stochastic Integrals*, pages 285–370, Springer, 1981. Lecture Notes in Math. 851.

[41]  J.W. Pitman and M. Yor. A decomposition of Bessel bridges. *Z. Wahrsch. Verw. Gebiete*, 59:425–457, 1982.

[42]  A. Renyi and G. Szekeres. On the height of trees. *J. Austral. Math. Soc.*, 7:497–507, 1967.

[43]  L.C.G. Rogers and D. Williams. *Diffusions, Markov Processes and Martingales.* Wiley, 1987.

[44]  P. Salminen. Brownian excursions revisited. In *Seminar on Stochastic Processes 1983*, pages 161–187, Birkhauser Boston, 1984.

[45]  J.M. Steele. Gibb's measures on combinatorial objects and the central limit theorem for an exponential family of random trees. *Prob. Engineering Inf. Sci.*, 1:47–60, 1987.

[46]  G. Szekeres. Distribution of labelled trees by diameter. In *Combinatorial Mathematics X*, pages 392–397, Springer-Verlag, 1982. Lecture Notes in Math. 1036.

[47]  L. Takacs. Limit theorems for random trees. 1991. Unpublished.

[48]  L. Takacs. On the total height of a random planted trivalent plane tree. 1991. Unpublished.

[49]  L. Takacs. Queues, random graphs and branching processes. *J. Appl. Math. and Simulation*, 1:223–243, 1988.

# Harmonic Morphisms and the Resurrection of Markov Processes

## P. J. Fitzsimmons*

Department of Mathematics
University of California, San Diego
La Jolla, CA 92093-0112 USA

## 1 INTRODUCTION

Our purpose in this paper is to carry out the program, outlined at the end of §5 in Csink, Fitzsimmons & Øksendal [**CFØ**], of extending the stochastic characterization of harmonic morphisms given in that paper to the case where the constant functions are not necessarily harmonic. Various consequences of this characterization were established in [**CFØ**]; they remain valid, with only minor changes, in the present context. The reader can also consult [**CFØ**] for references to the earlier literature on harmonic morphisms.

We shall now describe the main result in detail. Let $(E, \mathcal{U})$ be a $\mathfrak{P}$-harmonic space in the sense of Constantinescu & Cornea [**CC72**]. In particular, $E$ is a locally compact, second countable Hausdorff space, and $\mathcal{U} : G \to \mathcal{U}(G)$ ($G$ open in $E$) is the sheaf of *positive $\mathcal{U}$-hyperharmonic functions*. Equivalently, $(E, \mathcal{U})$ is a *balayage space* in the sense of Bliedtner & Hansen [**BH**] which satisfies the local truncation property [**BH**, p.125], and which has no absorbing points.

### 1.1 Definition
Given two $\mathfrak{P}$-harmonic spaces $(E, \mathcal{U})$ and $(F, \mathcal{V})$, a continuous function $\varphi : E \to F$ is a *harmonic morphism* provided $u \circ \varphi \in \mathcal{U}(\varphi^{-1}(G))$ whenever $u \in \mathcal{V}(G)$ and $G \subset F$ is open.

*Research supported in part by NSF grant DMS 8721347

The notion of harmonic morphism seems to be due to Constantinescu and Cornea [**CC65**]. Actually our definition is somewhat broader than that used by previous authors, since the *positive* hyperharmonic functions have been substituted for the superharmonic functions. Clearly if the pullback $u \mapsto u \circ \varphi$ preserves (local) superharmonicity then it also preserves (local) harmonicity, since $u$ is harmonic if and only if $u$ and $-u$ are superharmonic.

Roughly stated, our main result asserts that if $X$ and $Y$ are Markov processes associated with $(E, \mathcal{U})$ and $(F, \mathcal{V})$ respectively, then a continuous mapping $\varphi : E \to F$ is a harmonic morphism if and only if $\varphi$ carries the sample paths of (a time change of) $X$ into those of $Y$.

To associate a Markov process with $(E, \mathcal{U})$ we must assume that the constant function 1 is in $\mathcal{U}(E)$. Then there is a bounded continuous, strict potential $p > 0$ in $\mathcal{U}(E)$; see [**CC72**, Prop.7.2.1]. Let $X = (\Omega, \mathcal{F}, \mathcal{F}_t, \theta_t, X_t, P^x)$ be the Hunt process associated with $(E, \mathcal{U})$ and $p$. The state space of $X$ is $E \cup \{\Delta\}$, where $\Delta$ is the cemetery point adjoined to $E$ as the point at infinity if $E$ is noncompact, and as an isolated point otherwise. Writing $(P_t)$ for the semigroup of $X$ and $U = \int_0^\infty P_t \, dt$ for the associated potential kernel, we have

$$p(x) = U1(x) = \int_0^\infty P_t 1(x) \, dt = P^x(\zeta),$$

where $\zeta = \inf\{t : X_t = \Delta\}$ is the lifetime of $X$. A different choice of the strict potential $p$ would result in a Hunt process related to $X$ by time change. The kernel $U$ is strong Feller (i.e. $Uf$ is continuous for any bounded Borel function $f$) and consequently $X$ has a reference measure; cf. Blumenthal & Getoor [**BG**, p.197]. The paths of $X$ are continuous on $[0, \zeta[$, and evidently $P^x(\zeta < \infty) = 1$ for all $x$.

If $G \subset E$ is open, then the cone $\mathcal{U}(G)$ of positive hyperharmonic functions on $G$ is precisely the class of excessive functions of the subprocess $X^G$ obtained by killing $X$ at the exit time $\tau_G = \tau(G) = \inf\{t \geq 0 : X_t \notin G\}$. That is,

$$X_t^G = \begin{cases} X_t, & 0 \leq t < \tau_G; \\ \Delta, & t \geq \tau_G. \end{cases}$$

An alternative description of $\mathcal{U}$ is provided by the *harmonic measures* $H_G(x, \cdot)$ defined for open $G \subset E$ by

$$H_G(x, f) = H_G f(x) = P^x(f(X_{\tau_G}); \tau_G < \zeta) = P^x(f(X_{\tau_G})).$$

(Here and elsewhere we use the convention that a function $f : E \to \mathbb{R}$ is extended to $E \cup \{\Delta\}$ by setting $f(\Delta) = 0$.) A function $u \colon G \to [0, \infty]$ lies in $\mathcal{U}(G)$ if and only if it is lower semi-continuous and $H_W u \le u$ on $G$ for all open $W$ with $\bar{W} \subset G$. Moreover a bounded continuous function $h : G \to \mathbb{R}$ is *harmonic on* $G$ if and only if $H_W h = h$ on $G$ for all $W$ as before. The path-continuity of $X$ is reflected in the fact that the measure $H_G(x, \cdot)$ is carried by the boundary of $G$ if $x \in G$.

Let

$$1 = h + p_0 \tag{1}$$

be the Riesz decomposition of the constant function 1 into harmonic and potential components. We assume without loss of generality that

$$p_0 = Uk, \tag{2}$$

where $k$ is Borel measurable and $0 \le k \le 1$. (This can always be arranged by substituting $p + p_0$ for $p$, which amounts to performing a time-change on $X$, but has no effect on the $\mathfrak{P}$-harmonic space $(E, \mathcal{U})$.)

Let $(F, \mathcal{V})$ be a second $\mathfrak{P}$-harmonic space with $1 \in \mathcal{V}(F)$, and let $q$ be a bounded continuous strict potential $q$. Let $Y = (Y_t, Q^x)$ denote the associated Hunt process. We use $V$ to denote the potential kernel for $Y$, and we assume that the analog of (2) is valid:

$$1 = \hat{h} + V\ell, \tag{3}$$

where $\hat{h}$ is $\mathcal{V}$-harmonic and $0 \le \ell \le 1$.

Recall that a continuous additive functional (CAF) of $X$ is an $(\mathcal{F}_t)$-adapted real-valued process $A = (A_t)$ such that $A_0 = 0$, $t \mapsto A_t$ is continuous and increasing, and

$$A_{t+s}(\omega) = A_t(\omega) + A_s(\theta_t \omega), \qquad s, t \ge 0, \omega \in \Omega. \tag{4}$$

That (4) holds with no exceptional $\omega$-set is a consequence of a "perfection" theorem found in the appendix of Getoor [G]. By the same theorem we can (and do) assume that any CAF is adapted to the filtration $(\mathcal{F}_{t+}^*)$, where $\mathcal{F}_t^*$ is the universal completion of $\sigma\{X_s; 0 \le s \le t\}$.

Let $\varphi: E \to F$ be continuous, let $A = (A_t)$ be a CAF of $X$ with inverse

$$B(t) = \inf\{s : A_s > t\},$$

and let $\pi$ be a Borel function on $E$ with $0 \le \pi \le 1$. Fix an open set $G \subset E$ and put $\tau = \tau_G$. Define $C \subset \Omega \times [0, 1]$ by

$$C = \{(\omega, u) : \tau(\omega) = \zeta(\omega) \text{ and } \pi(X_{\zeta-}(\omega)) \ge u, \text{ or } \tau(\omega) < \zeta(\omega)\}.$$

Consider the stochastic process

$$Z_t(\omega, \hat{\omega}, u) = \begin{cases} \varphi(X_{B(t)}(\omega)), & 0 \le t < A_\tau(\omega); \\ Y_{t-A_\tau(\omega)}(\hat{\omega}), & t \ge A_\tau(\omega), \ (\omega, u) \in C; \\ \hat{\Delta}, & t \ge A_\tau(\omega), \ (\omega, u) \notin C; \end{cases}$$

under the law $\tilde{P}^x$ defined on $\Omega \times \hat{\Omega} \times [0, 1]$ by

$$\tilde{P}^x(d\omega, d\hat{\omega}, du) = P^x(d\omega)Q^{\varphi(X_\tau - (\omega))}(d\hat{\omega})\, du,$$

where $\hat{\Omega}$ is the path space and $\hat{\Delta}$ the cemetery for $Y$. Roughly speaking, $Z$ is constructed by continuing $\varphi(X_{B(t)})$ beyond time $A_\tau$ by tacking on a copy of $Y$ started at $y = \varphi(X_\tau)$, provided $X_\tau \in E$. (Aside from the initial condition $Y_0 = y$, $Y$ is independent of $X$.) Also, if $X_\tau = \Delta$ but $X_{\tau-} \in E$, then we toss a coin whose probability of heads is $\pi(X_{\tau-})$ and continue (now with $y = \varphi(X_{\tau-})$) if a head is tossed. This "randomized continuation" accounts for any differences in the "killing rates" $k$ and $\ell$ of $X$ and $Y$. We call $(Z_t, \tilde{P}^x)$ the $\tau_G$-splicing of $\varphi(X_{B(t)})$ and $Y$ with continuation probability $\pi$.

## 1.2 Definition

A continuous mapping $\varphi: E \to F$ is *X-Y path preserving with time change B and continuation probability $\pi$* provided the $\tau_G$-splicing of $\varphi(X_{B(t)})$ and $Y$ with continuation probability $\pi$ has the same law (under $\tilde{P}^x$) as $(Y_t, P^{\varphi(x)})$, for all $x \in G$ and all precompact open sets $G \subset E$.

Here is our main result.

## 1.3 Theorem

*Let $(E, \mathcal{U})$ and $(F, \mathcal{V})$ be $\mathfrak{P}$-harmonic spaces, and let $p \in \mathcal{U}(E)$ and $q \in \mathcal{V}(F)$ be bounded continuous strict potentials such that (2) and (3) hold. Let $X$ and $Y$ be the associated Hunt processes. If $\varphi: E \to F$ is continuous, then the following three assertions are equivalent:*

(i) *$\varphi$ is a harmonic morphism (of $(E, \mathcal{U})$ into $(F, \mathcal{V})$);*

(ii) *there is a CAF $A = (A_t)$ of $X$ and a Borel function $\pi$ on $E$ with $0 \le \pi \le 1$ such that*

$$V f(\varphi(x)) = P^x \left( \int_0^{\tau(G)} f(\varphi(X_t)) \, dA_t \right) + H_G(Vf \circ \varphi)(x)$$

$$+ P^x \left( \int_0^{\tau(G)} V f \circ \varphi(X_t) \pi(X_t) k(X_t) \, dt \right) \tag{5}$$

*for all bounded positive continuous functions $f$ on $E$, $x \in G$, and precompact open sets $G \subset E$;*

(iii) *there is a CAF $(A_t)$ of $X$ with inverse $B$ and a Borel function $\pi$ on $E$ with $0 \le \pi \le 1$ such that $\varphi$ is $X$-$Y$ path preserving with time change $B$ and continuation probability $\pi$.*

## 1.4 Remark

It should be noted that the CAF $A$ of Theorem 1.3 need not be strictly increasing. In fact, one consequence of the theorem is that $t \mapsto \varphi(X_t)$ is constant on any interval of constancy of $t \mapsto A_t$; cf. [CFØ, Prop. 3.1]. This is consistent with the continuity of $t \mapsto \varphi(X_{B(t)})$. See Remark 2.5 for a reasonably explicit characterization of the ingredients $A$ and $\pi$ in the decomposition (5).

## 1.5 Remark

For a better understanding of (5) one should note that

$$P^x(g(X_{\tau(G)-})) = H_G g(x) + P^x \left( \int_0^{\tau(G)} g(X_t) k(X_t) \, dt \right). \tag{6}$$

This identity, which follows from results of section 2, makes the implication (iii)$\Rightarrow$(ii) in Theorem 1.5 quite transparent.

## 1.6 Example

The need for the "splicing" in Theorem 1.3 is easily illustrated. Let $F$ be the complex plane and $E$ the cut plane $F\backslash ]-\infty, 0]$; let $Y$ be complex Brownian motion, and let $X$ be $Y$ killed at the first hitting time of $]-\infty, 0]$. The mapping $\varphi : E \to F$ defined by $\varphi(z) = z^2$ is analytic, hence a harmonic morphism . Indeed, by a famous theorem of Lévy, if we put $A_t = \int_0^t |\varphi'(X_s)|^2\, ds$, and $B(t) = A^{-1}(t)$, then $\tilde{Y}_t = X^2_{B(t)}$, $0 \le t < A_\zeta$, is equal in law to $Y$ killed at a certain random time $\sigma$. The catch is that $\sigma$ is not a stopping time of $Y$ and $\tilde{Y}$ is not itself a Markov process. Only after continuing $\tilde{Y}$ beyond $\sigma$ with an independent copy of planar Brownian motion do we obtain a Markov process.

If 1 is $\mathcal{U}$-harmonic then we can take $k \equiv 0$; also, $P^x(\tau_G < \zeta) \equiv 1$ in this case so the function $\pi$ is irrelevant. This special case of Theorem 1.3 is the main result (Theorem 1.1) of [**CFØ**]. Our proof of Theorem 1.3 amounts to showing that the present situation can be reduced to that of [**CFØ**] by a process of "resurrection."

For if 1 is not $\mathcal{U}$-harmonic, then $P^x(X_{\zeta-} \in E, \zeta < \infty) > 0$ for some $x \in E$; see Lemma 2.1. One might say (with tongue in cheek) that on $\{X_{\zeta-} \in E, \zeta < \infty\}$, $X$ has died an unnatural death. We can remedy this situation as follows. Consider the $\Omega$-valued Markov chain $(X^n; n = 0, 1, 2, \ldots)$ with initial law $P^x$ and transition kernel

$$\Pi(\omega, \omega') = P^{x(\omega)}(d\omega'),$$

where $x(\omega) = X_{\zeta-}(\omega)$. Note that $P^\Delta$ is the unit mass at the constant path $t \mapsto \Delta$. Let $\zeta^n$ denote the lifetime of $X^n$, and define a process $\bar{X}$ by

$$\bar{X}_t = X^n_t, \quad \text{if } \zeta^0 + \cdots + \zeta^{n-1} \le t < \zeta^0 + \cdots + \zeta^n, \quad n = 0, 1, \ldots.$$

Note that if $X^n_{\zeta^n-} = \Delta$ then $\bar{X}_t = \Delta$ for all $t \ge \zeta^0 + \cdots + \zeta^n$. Thus, at the time of the "$n$th death" we resuscitate if possible (i.e. if $X^n_{\zeta^n-} \in E$); otherwise the killing is final. It is intuitively clear that $\bar{X}$ is a Markov process whose lifetime is predictable. Therefore 1 is harmonic for $\bar{X}$. Moreover, it turns out that if $\varphi$ is a harmonic morphism between $X$ and $Y$, then it is also a harmonic morphism between $\bar{X}$ and $\bar{Y}$ (the resurrection of $Y$). Thus the main result of [CFØ] applies to $\varphi$, $\bar{X}$ and $\bar{Y}$, and Theorem 1.3 obtains after we unravel the connection between $X$, $Y$, $\bar{X}$ and $\bar{Y}$.

The construction of $\bar{X}$ outlined above is a special case of a general resurrection procedure studied by Ikeda, Nagasawa & Watanabe [INW], Meyer [M], and others. The present case is simpler than the general case since $X$ and $\bar{X}$ are related by a multiplicative functional. In fact, as the reader can easily prove from the above description, the semigroup $(\bar{P}_t)$ of $\bar{X}$ is the (minimal) solution of

$$\bar{P}_t f(x) = P_t f(x) + \int_E \int_{]0,t]} \bar{P}_{t-s} f(y) \, \Gamma^x(ds, dy), \tag{7}$$

where $\Gamma^x(ds, dy) = P^x(\zeta \in ds, X_{\zeta-} \in dy)$. In view of (6), (7) can be cast into a more useful form:

$$\bar{P}_t f(x) = P_t f(x) + P^x \left( \int_0^t \bar{P}_{t-s} f(X_s) k(X_s) \, ds \right),$$

which suggests that

$$\bar{P}_t f(x) = P^x (f(X_t) \exp(\int_0^t k(X_s) \, ds)). \tag{8}$$

In fact, in the next section we shall see that (8) is appropriate as a definition of the resurrected process $\bar{X}$.

We close this section with a bit of notation. Recall [BG, III] that a (positive) multiplicative functional (MF) of $X$ is a $[0, \infty]$-valued process $M = (M_t)$ adapted to $(\mathcal{F}_t)$ such that $M_{t+s} = M_t + M_s \circ \theta_t$ for all $s, t \geq 0$. If also $M$ is a $P^x$-supermartingale for all $x \in E$ (equivalently, if $P^x(M_t) \leq 1$ for all $x$) then the formula

$$P_t^* f(x) = P^x(f(X_t) M_t)$$

defines a subMarkov semigroup $(P_t^*)$. One can associate with $(P_t^*)$ a Markov process $X^*$. For example, if $X$ is a right Markov process then so is $X^*$. We refer the reader to [BG] and Sharpe [S] for details on the transformation of processes by multiplicative functionals. In the sequel it will be convenient to indicate the dependence of $X^*$ on $M$ by refering to $X^*$ as $(X, M)$.

Both of the processes $X$ and $\bar{X}$ will be realized as the coordinate process on the canonical sample space $\Omega$ of paths $\omega : [0, \infty[ \to E$ that are continuous on $[0, \zeta(\omega)[$ and absorbed in $\Delta$ at time $\zeta(\omega)$. The two processes are distinguished by their respective laws $\{P^x; x \in E\}$ and $\{\bar{P}^x; x \in E\}$. In particular, a (perfect) CAF of $\bar{X}$ can be viewed as a CAF of $X$. Likewise, at times a CAF of $\bar{X}^G$ will be thought of as a CAF of $X^G$.

## 2 PROOF OF THEOREM 1.3

We begin this section with the rigorous definition of the resurrected process $\bar{X}$. Since $X$ is a Hunt process, hence quasi-left continuous, the accessible part of the lifetime $\zeta$ is predictable, and the totally inaccessible part of $\zeta$ is given by

$$\zeta_i = \begin{cases} \zeta, & X_{\zeta-} \in E, \, \zeta < \infty; \\ \infty, & \text{otherwise.} \end{cases} \tag{9}$$

In particular, if $G \subset E$ is open, then $\{\tau_G = \zeta < \infty\} = \{\tau_G = \zeta_i < \infty\}$. Also, since $U1 = p$ is bounded we have

$$\infty > U(x, 1_G) = P^x \left( \int_0^\infty 1_G(X_t) \, dt \right),$$

so $P^x(\tau_G < \infty) = 1$ for all $x \in E$.

### 2.1 Lemma
*Define functions $h_0(x) = P^x(\zeta_i = \infty)$ and $u(x) = 1 - h_0(x)$ for $x \in E$. Then $h_0$ is $\mathcal{U}$-harmonic and $u \in \mathcal{U}(E)$. Moreover, the only non-negative $\mathcal{U}$-harmonic minorant of $u$ is 0 (i.e. $u$ is a potential). In particular, $u = Uk$.*

*Proof.* Using (9), the identity $\zeta_i \circ \theta_{\tau(G)} = (\zeta_i - \tau_G)^+$ and the strong Markov property we compute

$$H_G h(x) = P^x(\zeta_i \circ \theta_{\tau(G)} = \infty, \tau_G < \zeta)$$
$$= P^x(\zeta_i = \infty, \tau_G < \zeta)$$
$$= P^x(\zeta_i = \infty) = h(x).$$

Thus $h$ is $\mathcal{U}$-harmonic. Since $1 \in \mathcal{U}(E)$ and $h \leq 1$, we have $u = 1 - h \in \mathcal{U}(E)$. If $(G_n)$ is a sequence of precompact open sets with union $E$, then clearly

$\tau_{G_n} \uparrow \zeta$. Therefore

$$\lim_n H_{G_n} u(x) = \lim_n P^x(\zeta_i < \infty, \tau_{G_n} < \zeta) = \lim_n P^x(\tau_{G_n} < \zeta_i < \infty)$$
$$= P^x(\tau_{G_n} < \zeta_i < \infty, \forall n) = 0,$$

since $\zeta_i$ is totally inaccessible. The last assertion of the lemma now follows immediately. $\square$

Let $K = (K_t)$ denote the dual predictable projection of the additive functional $t \mapsto 1_{\{\zeta_i \leq t\}}$. Thus $K$ is a CAF of $X$ and

$$J_t := K_t - 1_{\{\zeta_i \leq t\}} \text{ is a } P^x\text{-martingale}, \quad \forall x \in E.$$

Recalling (2) and Lemma 2.1 we see that

$$P^x(K_\zeta) = P^x(K_\infty) = Uk(x), \quad \forall x \in E. \tag{10}$$

It follows from (10) and [**BG**, V(2.8)] that

$$K_t = \int_0^t k(X_s)\,ds, \quad t \geq 0.$$

Consider now the *Doléans equation*:

$$dL_t = L_{t-}\,dJ_t, \quad t \geq 0, \qquad L_0 = 1. \tag{11}$$

As is well known, (11) has a unique (semimartingale) solution $(L_t)$. Since $J$ is a martingale additive functional of $X$, $L$ is a positive local martingale *multiplicative functional* of $X$. In particular $L$ is a supermartingale, so if $T$ is any stopping time,

$$P^x(L_T) \leq 1, \qquad \forall x \in E.$$

Moreover we have the explicit formula (cf. [**S**, A4])

$$L_t = \exp(K_t)\left(1_{\{t < \zeta_i < \infty\}} + 1_{\{\zeta_i = \infty\}}\right). \tag{12}$$

We now introduce the resurrected process $\bar{X} = (X, L)$. That is, $\bar{X}$ is the strong Markov process on $E$ with semigroup

$$\bar{P}_t f(x) = P^x(f(X_t)L_t).$$

The process $\bar{X}$ is realized as the coordinate process on the same path space $\Omega$ as $X$ under the laws $\{\bar{P}^x, x \in E\}$ determined by

$$\bar{P}^x(\Phi; T < \bar{\zeta}) = P^x(\Phi \cdot L_T; T < \zeta), \qquad (13)$$

where $T$ is any $(\mathcal{F}^{\circ}_{t+})$-stopping time, $\mathcal{F}^{\circ}_t = \sigma\{X_s, 0 \le s \le t\}$, and $\Phi$ is bounded and $\mathcal{F}^{\circ}_{T+}$-measurable; cf. [S, §62]. (We write $\bar{\zeta}$ rather than $\zeta$ on the left side of (3) only for emphasis; actually $\bar{\zeta}(\omega) = \zeta(\omega)$ for all $\omega \in \Omega$.) Using the fact that $L_t = \exp(K_t) > 0$ if $0 \le t < \zeta$, it is easy to check that $\bar{X}$ is a standard process. In fact, much more is true:

## 2.2 Proposition
$\bar{X}$ is a Hunt process with continuous paths. The lifetime of $\bar{X}$ is predictable, so the constants are $\bar{X}$-harmonic. If $G \subset E$ is a precompact open set such that $x \mapsto \bar{P}^x(\tau_G)$ is bounded, then the harmonic space associated with the killed process $\bar{X}^G$ is a $\mathfrak{P}$-harmonic space.

So as not to interrupt the main argument, we defer the proof of Proposition 2.2 to section 3.

The same construction yields a process $\bar{Y}$, the resurrection of $Y$. In the sequel the overbar will be used to signal objects defined relative to $\bar{X}$ or $\bar{Y}$. (We also use $\bar{N}$ to denote the closure of the set $N$, but no confusion should arise.)

We now prove that the harmonic morphism property of the mapping $\varphi$ is preserved by the resurrection procedure. For this we require a lemma, which is stated in terms of $X$, but applies just as well to $Y$, $\bar{X}$, or $\bar{Y}$. In what follows we say that a set $G \subset E$ is an *exit set* (for $X$) provided $G$ is precompact and open, and $x \mapsto P^x(\tau_G)$ is bounded. It is known [BG, p.240] that there is a countable cover of $E$ consisting of exit sets.

## 2.3 Lemma
Let $u \ge 0$ be bounded and continuous, and such that

$$\liminf_{W \downarrow \{x\}} \frac{u(x) - H_W u(x)}{P^x(\tau_W)} \ge 0, \quad (W \text{ open}) \quad \forall x \in E.$$

*Then $u \in \mathcal{U}(E)$.*

*Proof.* For $\epsilon > 0$ define $u_\epsilon = u + \epsilon U1$. Then

$$u_\epsilon - H_W u_\epsilon = u - H_W u + \epsilon P^{\cdot}(\tau_W),$$

hence

$$\liminf_{W \downarrow \{x\}} \frac{u_\epsilon(x) - H_W u_\epsilon(x)}{P^x(\tau_W)} \geq \epsilon > 0.$$

Thus for each $x \in E$ there is a neighborhood $W_x$ of $x$ such that

$$H_W u_\epsilon(x) \leq u_\epsilon(x), \quad \forall \text{ open } W \subset W_x.$$

As is well known [**BH**, III.4.4], this together with the continuity of $u$ implies that $u_\epsilon \in \mathcal{U}(E)$. The lemma follows upon letting $\epsilon \downarrow 0$. $\square$

## 2.4 Proposition

*Suppose that $\varphi$ is a harmonic morphism between $(E,\mathcal{U})$ and $(F,\mathcal{V})$. Then it is also a harmonic morphism between $(E, \bar{\mathcal{U}})$ and $(F, \bar{\mathcal{V}})$.*

*Proof.* Let $N$ be an open subset of $F$. We must show that if $v \in \bar{\mathcal{V}}(N)$, then $u := v \circ \varphi \in \bar{\mathcal{U}}(\varphi^{-1}(N))$. Fix $x \in \varphi^{-1}(N)$ and $\epsilon > 0$. Let $N' \subset N$ be an exit set (for $\bar{Y}$) containing $\varphi(x)$. Then as a consequence of Proposition 2.2 there is a sequence $(v_n)$ of bounded continuous $\bar{Y}^{N'}$-potentials increasing to $v$ on $N'$; once we show that $v_n \circ \varphi \in \mathcal{U}(\varphi^{-1}(N))$, the analogous inclusion will obtain for $v$ by a passage to the limit. Thus, there is no loss of generality in assuming that $v$ is bounded and continuous. Let $G \subset N'$ be an open neighborhood of $\varphi(x)$ such that $G \subset \{|v - v(\varphi(x))| < \epsilon\}$. Let $c = u(x) - \epsilon$ and note that $v - c \in \bar{\mathcal{V}}(G)$ since the constants are $\bar{\mathcal{V}}$-harmonic. Since $\bar{\mathcal{V}}(G) \subset \mathcal{V}(G)$, it follows that $u - c \in \mathcal{U}(\varphi^{-1}(G))$. Now let $W \subset \varphi^{-1}(G)$ be any exit set (for $\bar{X}$) with $x \in W$, and put $\tau = \tau_W$. Then by (13)

$$\begin{aligned}
u(x) - c &\geq H_W(u - c)(x) = P^x(u(X_\tau) - c; \tau < \zeta) \\
&= P^x(e^{K_\tau}(u(X_\tau) - c); \tau < \zeta) - P^x((e^{K_\tau} - 1)(u(X_\tau) - c); \tau < \zeta) \\
&= \bar{H}_W(u - c) - P^x((e^{K_\tau} - 1)(u(X_\tau) - c); \tau < \zeta).
\end{aligned}$$

But

$$\begin{aligned}
P^x((e^{K_\tau} - 1)(u(X_\tau) - c); \tau < \zeta) &\leq \epsilon P^x((e^{K_\tau} - 1); \tau < \zeta) \\
&= \epsilon(\bar{H}_W 1 - P^x(\tau < \zeta)) = \epsilon P^x(\tau = \zeta) \\
&= \epsilon P^x(K_\tau) \leq \epsilon P^x(\tau) \leq \epsilon \bar{P}^x(\tau).
\end{aligned}$$

Thus, since $\bar{H}_W c(x) = c$,

$$u(x) - \bar{H}_W u(x) \geq -\epsilon \bar{P}^x(\tau)$$

and the proposition follows from the lemma. $\square$

*Proof of Theorem 1.3.* The implications (ii)$\Rightarrow$(iii) and (iii)$\Rightarrow$(i) can be proved by straightforward modifications of the arguments used in [**CFØ**]. The only new ingredient is the identity

$$P^x(f(X_{\tau(G)-}); \tau_G = \zeta) = P^x(f(X_{\zeta_i-}); \zeta_i \leq \tau_G)$$
$$= P^x \left( \int_0^{\tau(G)} f(X_t) \, dK_t \right),$$

which is an immediate consequence of the definition of $K$.

Passing to the proof of (i)$\Rightarrow$(ii), let us assume that $\varphi$ is a harmonic morphism between $X$ and $Y$. We begin by noting that it suffices to prove that there is a CAF $A$ of $X$ such that

$$V f(\varphi(x)) = P^x \left( \int_0^\tau [f \circ \varphi(X_t) - (\ell \cdot Vf) \circ \varphi(X_t)] \, dA_t \right) \tag{14}$$
$$+ P^x \left( \int_0^\tau V f \circ \varphi(X_t) \, dK_t \right) + H_G(V f \circ \varphi)(x),$$

where $G$ is any precompact open set, $\tau = \tau_G$, and $x \in G$. Indeed suppose (14) has been proved, and recall the Riesz decomposition $1 = \hat{h} + V\ell$ where $\hat{h}$ is $Y$-harmonic. Taking $f = \ell$ in (14), and then discarding the $X^G$-harmonic terms, using the fact that $1 - Uk$ is $X$-harmonic, we obtain

$$P^{\cdot} \left( \int_0^\tau \hat{h} \circ \varphi(X_t) \, dK_t \right) = P^{\cdot} \left( \int_0^\tau (\ell \hat{h}) \circ \varphi(X_t) \, dA_t \right) + p^* \quad \text{on } G, \tag{15}$$

where $p^*$ is the $X^G$-potential part of $\hat{h} \circ \varphi$. Now by Lemma 2.1 and (13)

$$\hat{h}(y) = Q^y(\tau_{\hat{G}} < \zeta(\bar{Y})) = \bar{Q}^y \left( \exp\left( -\int_0^{\tau(\hat{G})} \ell(Y_t) \, dt \right); \tau_{\hat{G}} < \zeta(\bar{Y}) \right) > 0,$$

so by the uniqueness theorem [**BG**, IV(2.13)] for potentials of CAF's, applied here to $X^G$, it follows that

$$K_t = \int_0^t \ell \circ \varphi(X_s) \, dA_s + C_t, \quad t < \tau,$$

for some CAF $C$. Consequently [**BG**, V(2.8)] there is a Borel function $\nu$ with $0 \leq \nu \leq 1$ such that

$$\int_0^t \ell{\circ}\varphi(X_s)\,dA_s = \int_0^t \nu(X_s)\,dK_s, \quad t < \tau. \tag{16}$$

Substituting (16) into (14) and writing $\pi = 1 - \nu$, we finally arrive at

$$Vf(\varphi(x)) = P^x\left(\int_0^\tau f{\circ}\varphi(X_t)\,dA_t\right)$$
$$+ P^x\left(\int_0^\tau Vf{\circ}\varphi(X_t)\pi(X_t)\,dK_t\right) + H_G(Vf{\circ}\varphi)(x) \qquad \text{on } G,$$

which is formula (5) in condition (ii) of Theorem 1.3, as desired.

For later use we note that the above argument reveals that $P^x(A_{\tau(G)})$ is bounded above by $q{\circ}\varphi$ independently of $G$.

Formula (14) will be proved in three steps. The first of these is the main argument and it amounts to proving a localized version of (14). In the final two steps the localization is lifted. Recall that $G \subset E$ is an *exit set* for $X$ if $G$ is precompact and open, and $x \mapsto P^x(\tau_G)$ is bounded.

$1°$ Let $G$ be an exit set for $\bar{X}$, and assume that $\hat{G} \subset F$ is an exit set for $\bar{Y}$ such that $G \subset \varphi^{-1}(\hat{G})$. Let $\tau$ (resp. $\hat{\tau}$) denote the first exit time of $X$ (resp. $Y$) from $G$ (resp. $\hat{G}$). Let $v(x) = Q^x(\int_0^{\hat{\tau}} f(Y_t)\,dt)$, where $f$ is a bounded positive Borel function on $F$. We are going to show that there is a CAF $A = A^G$ of $X^G$ such that

$$v(\varphi(x)) = P^x\left(\int_0^\tau [f{\circ}\varphi(X_t) - (\ell v){\circ}\varphi(X_t)]\,dA_t\right)$$
$$+ P^x\left(\int_0^\tau v{\circ}\varphi(X_t)\,dK_t\right) + H_G(v{\circ}\varphi)(x), \tag{17}$$

for all $x \in G$.

In view of Proposition 2.4, $\varphi$ is a harmonic morphism between $\bar{X}^G$ and $\bar{Y}^{\hat{G}}$; because $G$ and $\hat{G}$ are exit sets we can apply Theorem 1.1 of [**CFØ**]: there is a (perfect) CAF $A$ of $\bar{X}^G$ such that

$$\bar{V}^* f(\varphi(x)) = \bar{P}^x \int_0^\tau f{\circ}\varphi(\bar{X}_t)\,dA_t + \bar{H}_G(\bar{V}^* f{\circ}\varphi)(x), \quad x \in G. \tag{18}$$

Here $\bar{V}^*$ is the potential kernel for the process $\bar{Y}^{\hat{G}}$, and $\bar{H}_G$ is the harmonic measure for $\bar{X}$. Similarly, we write $V^*$ for the potential kernel of $Y$, and if $B$ is a CAF

$$\bar{U}_B^* g(x) := \bar{P}^x \int_0^\tau g(\bar{X}_t)\, dB_t, \qquad U_B^* g(x) := P^x \int_0^\tau g(X_t)\, dB_t.$$

The following variants of the resolvent equation are simple consequences of (13):

$$\bar{V}^* f = V^* f + \bar{V}^*(\ell \cdot V^* f) = V^* f + V^*(\ell \cdot \bar{V}^* f); \qquad (19)$$

$$\bar{U}_A^* g = U_A^* g + U^*(k \cdot \bar{U}_A^* g) = U_A^* g + \bar{U}^*(k \cdot U_A^* g); \qquad (20)$$

$$\bar{H}_G f = H_G f + \bar{U}^*(k \cdot H_G f) = H_G f + U^*(k \cdot \bar{H}_G f). \qquad (21)$$

Now replace $f$ in (18) by $\ell \cdot V^* f$ and subtract the resulting identity from (18). Using (19)–(21) we obtain

$$\begin{aligned}
V^* f(\varphi(x)) &= \bar{U}_A^*(f\circ\varphi - (\ell \cdot V^* f)\circ\varphi)(x) + \bar{H}_G(V^* f\circ\varphi)(x) \\
&= U_A^*(f\circ\varphi - (\ell \cdot V^* f)\circ\varphi)(x) + H_G(V^* f\circ\varphi)(x) \\
&\quad + \left[ U_K^*(\bar{U}_A^*(f\circ\varphi - (\ell \cdot V^* f)\circ\varphi) + \bar{H}_G(V^* f\circ\varphi))(x) \right].
\end{aligned}$$

But the term in square brackets is evidently equal to $U_K^*(V^* f\circ\varphi)$, so (17) follows. A variant of the above argument shows that if $N$ is an open subset of $G$, then (17) remains valid (on $N$) if $\tau = \tau_G$ (resp. $G$) is replaced by $\tau_N$ (resp. $N$).

$2°$ We use the notation of $1°$. In addition we assume that there is an open set $D \subset \hat{G}$ such that $\bar{G} \subset \varphi^{-1}(D)$. Suppose $\tilde{h} \in \mathcal{V}(\hat{G})$ is bounded and $Y$-harmonic on $\hat{G}$. Then on $G$ we have

$$\tilde{h}\circ\varphi = U_K^*(\tilde{h}\circ\varphi) - U_A^*((\ell\tilde{h})\circ\varphi) + H_G(\tilde{h}\circ\varphi). \qquad (22)$$

To see this let $N$ be an open set with $N \subset \bar{N} \subset G$, so that $\varphi(\bar{N})$ is a compact subset of $\hat{G}$. Since $\tilde{h} \in \mathcal{V}(\hat{G})$, there is a sequence $(V^* f_n)$ of $Y^{\hat{G}}$-potentials with $V^* f_n \uparrow \tilde{h}$; since $\tilde{h}$ is $Y$-harmonic on $\hat{G}$ we can assume that

$f_n = 0$ on $\varphi(\bar{N})$ for each $n$. By $1°$,

$$V^* f_n(\varphi(x)) = P^x \left( \int_0^{\tau(N)} V^* f_n \circ \varphi(X_t) \, dK_t \right)$$

$$- P^x \left( \int_0^{\tau(N)} (\ell \cdot V^* f_n) \circ \varphi(X_t) \, dA_t \right) + H_N(V^* f_n \circ \varphi)(x),$$

Letting $n \to \infty$ and then $N \uparrow G$ through a sequence $(N_j)$ we obtain (22) since $\tau_{N_j} \uparrow \tau_G$, $\{\tau_{N_j} < \tau_G, \forall j\} \subset \{\tau_G < \zeta\}$, and since $\tilde{h} \circ \varphi$ is bounded and continuous on $\varphi^{-1}(D) \supset \bar{G}$.

Now fix a bounded positive Borel function $f$ on $E$. Then $\tilde{h} := Vf - V^* f$ is $Y$-harmonic on $\hat{G}$. Using (22) and (17), we see that on $G$

$$\begin{aligned}
Vf \circ \varphi &= \tilde{h} \circ \varphi + V^* f \circ \varphi \\
&= U_K^*((Vf - V^* f) \circ \varphi) - U_A^*((\ell(Vf - V^* f)) \circ \varphi) \\
&\quad + H_G((Vf - V^* f) \circ \varphi) + V^* f \circ \varphi \\
&= U_K^*(Vf \circ \varphi) - U_A^*((\ell \cdot Vf) \circ \varphi) + H_G(Vf \circ \varphi) \\
&\quad - U_K^*(V^* f \circ \varphi) + U_A^*((\ell \cdot V^* f) \circ \varphi) \\
&\quad - H_G(V^* f \circ \varphi) + V^* f \circ \varphi \\
&= U_A^*(f \circ \varphi) + U_K^*(Vf \circ \varphi) - U_A^*((\ell \cdot Vf) \circ \varphi) + H_G(Vf \circ \varphi).
\end{aligned}$$

Thus

$$Vf \circ \varphi = U_A^*(f \circ \varphi) + U_K^*(Vf \circ \varphi) - U_A^*((\ell \cdot Vf) \circ \varphi) + H_G(Vf \circ \varphi). \tag{23}$$

$3°$ By the obvious adaptation of [BG, p.240] there is a sequence $(W_n)$ of exit sets (for $\bar{X}$) with $\cup_n W_n = E$, and we can certainly assume that for each $x \in E$ and each $\epsilon > 0$ there are infinitely many indices $n$ such that $W_n$ is of diameter $< \epsilon$ and contains $x$. It is then easy to see that the (increasing) sequence of iterated exit times defined by $T_0 = 0$ and

$$T_{n+1} = T_n + \tau_{W_{n+1}} \circ \theta_{T_n} \tag{24}$$

satisfies $T_n \uparrow \zeta$ almost surely. Likewise there is a cover $(\hat{W}_n)$ of $F$ by $\bar{Y}$-exit sets. A compactness argument shows that we can arrange that $W_n \subset$

$\varphi^{-1}(\hat{W}_n)$ for all $n$, and that for each $n$ there is an open set $D_n \subset \hat{W}_n$ with $\bar{W}_n \subset \varphi^{-1}(D_n)$. Thus the considerations of steps $1°$ and $2°$ apply. We now call to the reader's attention the fact that the CAF produced in those steps depended implicitly on $G$. But if $G_1$ and $G_2$ are two exit sets (for $\bar{X}$) then it follows easily from (23) that

$$P^x(A^{G_1}_{\tau(G_1 \cap G_2)}) = P^x(A^{G_2}_{\tau(G_1 \cap G_2)}), \quad x \in G_1 \cap G_2, \qquad (25)$$

where we have written $A^{G_i}$ to emphasize the dependence on $G_i$. The compatibility condition (25) allows one to use the "piecing together" argument of [**BG**, V,§5] to produce a single CAF $A$ of $X$ such that

$$A_{t \wedge \tau(G)} = A^G_{t \wedge \tau(G)}, \quad \forall t \geq 0, \text{ a.s. } P^x, \quad \forall x \in G.$$

Thus (23) holds with this CAF $A$ for all (sufficiently small) $\bar{X}$-exit sets $G$. Moreover, as noted following (16), $x \mapsto P^x(A_{\tau(G)})$ is bounded by $\|q\|_\infty < \infty$ for any such exit set $G$.

Let $G$ be a precompact open subset of $E$. It follows that if we define stoppings time $S_n$ in the same way as the $T_n$ but with $W_n$ replaced by $G_n := W_n \cap G$, then

$$S_n \uparrow \tau_G \quad \text{a.s.} \qquad (26)$$

Now fix a bounded positive Borel function $f$ on $F$. Write $\tau = \tau_G$, $\tau_n = \tau_{G_n}$, and

$$C_t = \int_0^t [f \circ \varphi(X_s) - (\ell \cdot Vf) \circ \varphi(X_s)] \, dA_s + \int_0^t Vf \circ \varphi(X_s) \, dK_s.$$

Then $C$ is a (signed) CAF and clearly $x \mapsto P^x(|C_\tau|)$ is bounded. By (24) and (26),

$$\begin{aligned}
P^x(C_\tau) &= \sum_{n \geq 1} P^x(C_{S_n} - C_{S_{n-1}}) \\
&= \sum_{n \geq 1} P^x(C_{\tau_n} \circ \theta_{S_{n-1}}) \\
&= \sum_{n \geq 1} H_{G_1} \cdots H_{G_{n-1}}(x, P^\cdot(C_{\tau_n})),
\end{aligned} \qquad (27)$$

and because of (23) the $n^{\text{th}}$ term of this series is equal to

$$H_{G_1} \cdots H_{G_{n-1}}(Vf \circ \varphi - H_{G_n} Vf \circ \varphi)$$

so the last sum in (27) telescopes to yield

$$P^x(C_\tau) = Vf\circ\varphi(x) - \lim_n H_{G_1}\cdots H_{G_n}(Vf\circ\varphi)(x)$$
$$= Vf\circ\varphi(x) - \lim_n P^x(Vf\circ\varphi(X_{S_n}))(x) \tag{28}$$
$$= Vf\circ\varphi(x) - H_G(Vf\circ\varphi)(x),$$

since $S_n \uparrow \tau_G$ and $\{S_n < \tau_G, \forall n\} \subset \{\tau_G < \zeta\}$, and since $Vf\circ\varphi$ is continuous on $E$. Clearly (28) implies (14), so the proof of Theorem 1.3 is complete, modulo Proposition 2.2.

## 2.5 Remark

We can extract from the above proof the following characterizations of $A$ and $\pi$. If $\bar{V}1$ is bounded, then the (bounded) potential of $A$ is simply $U_A1 = u - U(ku)$, where $u$ is the $\bar{X}$-potential part of $\bar{V}1\circ\varphi$. In the general case let $G$ and $\hat{G}$ be exit sets for $\bar{X}$ and $\bar{Y}$ respectively such that $G \subset \varphi^{-1}(\hat{G})$. Put

$$v_G(y) = \bar{Q}^y(\tau(\hat{G})) = Q^y\left(\int_0^{\tau(\hat{G})} \exp(\int_0^t \ell(Y_s)\,ds)\,dt\right),$$

and let $u_G$ be the $\bar{X}^G$-potential part of $v_G\circ\varphi$. Then

$$a_G(x) := P^x(A_{\tau(G)}) = u_G(x) - P^x\left(\int_0^{\tau(G)} (ku_G)(X_t)\,dt\right).$$

(It is easy to verify directly that the third term in the above display does not depend on the choice of $\hat{G}$.) Now let $(W_n)$ be a sequence of exit sets covering $E$ and such that the stopping times defined by (24) satisfy $T_n \uparrow \zeta$. Then the (bounded) potential of $A$ is given by

$$U_A1 = P^\cdot(A_\zeta) = \sum_{n\geq 1} H_{W_1}\cdots H_{W_{n-1}} a_{W_n}.$$

Finally note that once $A$ is known, $\pi$ is determined up to a set of potential zero by the relation

$$Uk = U_A(\ell\circ\varphi) + U(k\pi).$$

## 3 PROOF OF PROPOSITION 2.2

### 3.1 Lemma

*The lifetime of $\bar{X}$ is predictable.*

*Proof.* Since $K_t \leq t$ for all $t \geq 0$, the local martingale $L$ is reduced by the sequence of constant stopping times $R_n = n$; that is, $(L_{\cdot \wedge n})$ is a uniformly integrable martingale (relative to each law $P^x$) for each integer $n$. Fix $x \in E$. Let $\zeta_p$ the predictable part of $\zeta$, and choose an increasing sequence of $(\mathcal{F}^\circ_{t+})$-stopping times such that $S_n < \zeta_p$ for all $n$ and $\lim_n S_n = \zeta_p$ a.s. $P^x$. Set $T_n = S_n \wedge n$, and $T :=\uparrow \lim_n T_n$. Then since $L_{T_n} = 0$ on $\{\zeta \leq T_n < \zeta_p\} = \{\zeta_i \leq T_n < \zeta_p\}$,

$$
\begin{aligned}
1 &= P^x(L_{T_n}) \\
&= P^x(L_{T_n}; T_n < \zeta) + P^x(L_{T_n}; \zeta_p \leq T_n) \\
&= P^x(L_{T_n}; T_n < \zeta) = \bar{P}^x(T_n < \zeta),
\end{aligned}
$$

because of (13). Moreover, by (12)

$$
\begin{aligned}
\bar{P}^x(T < \bar{\zeta}) &= P^x(L_T; T < \zeta) = P^x(e^{K_T}; T < \zeta_i, T < \zeta) \\
&= P^x(e^{K_T}; T < \zeta_i, T < \zeta, T = \zeta_p) = 0.
\end{aligned}
$$

It follows that $\bar{\zeta}$ is announced by the sequence $(T_n)$ under $\bar{P}^x$, so $\bar{\zeta}$ is predictable. $\square$

We shall now prove that $\bar{X}$ is a Hunt process, and that for each $\alpha > 0$ the $\alpha$-potential kernel $\bar{U}^\alpha = \int_0^\infty e^{-\alpha t} \bar{P}_t \, dt$ is a strong Feller kernel.

### 3.2 Lemma

*For each $\alpha > 0$, $\bar{U}^\alpha f$ is continuous whenever $f$ is a bounded Borel function on $E$.*

*Proof.* In view of the resolvent equation, it is enough to prove the assertion for *one* $\alpha > 0$, say $\alpha = 2$. But we can view $\bar{U}^2$ as the potential kernel of the process $(X, \exp(- \int_0^t (2 - k(X_s)) \, ds))$. By the perturbation theorem (V.2.7) of [**BH**], the strong Feller property of $U$ is inherited by $\bar{U}^2$. $\square$

Lemma 3.1 and the next result imply the assertions in the first sentence of Proposition 2.2.

### 3.3 Lemma

For $\alpha > 0$ and $G$ an open subset of $E$, let $\bar{\mathcal{U}}^\alpha(G)$ denote the class of $\alpha$-excessive functions of $\bar{X}^G$. Then $(E, \bar{\mathcal{U}}^\alpha)$ is a $\mathfrak{P}$-harmonic space. In particular $\bar{X}$ is a Hunt process with continuous paths.

*Proof.* We first need to check that the cone of excessive functions of the process $(\bar{X}, e^{-\alpha t})$ satisfies the axioms (B1)–(B4) on p.57 of [**BH**]. Axioms (B1)–(B3) follow from the fact that $\bar{X}$ is a right process with a reference measure. Moreover (B4) follows from the perturbation argument used in proving the previous lemma in case $\alpha > 1$. To prove (B4) for $\alpha \in ]0, 1]$ we must show that for each bounded positive Borel function $f$ on $E$ there exists a continuous $\alpha$-$\bar{X}$-excessive function $v > 0$ such $\bar{U}^\alpha f / v \in C_0(E)$. See [**BH**, II.4.6]. But $\bar{U}^\alpha f \in \bar{\mathcal{U}}^\alpha(E) \subset \bar{\mathcal{U}}^2(E)$, so there exists a continuous $\tilde{v} \in \bar{\mathcal{U}}^2(E)$ with $\tilde{v} > 0$ and $\bar{U}^\alpha f / \tilde{v} \in C_0(E)$. Define $v = \tilde{v} + (2 - \alpha)\bar{U}^\alpha \tilde{v}$. Then $v \geq \tilde{v}$, and because of the resolvent equation $v \in \bar{\mathcal{U}}^\alpha(E)$. In view of Lemma 3.2

$$0 \leq \bar{U}^\alpha f / v \leq \bar{U}^\alpha f / \tilde{v} \in C_0(E),$$

so (B4) follows. The Hunt property for $\bar{X}$ is now a consequence of the general theory of balayage spaces; see [**BH**, IV.7.6]. It is clear that $\bar{X}$ has continuous paths on $[0, \zeta[$ because of (13) and the analogous property of $X$. The quasi-left-continuity of $\bar{X}$ and the predictability of $\bar{\zeta}$ (Lemma 3.1) imply that $\bar{X}_{\zeta-} = \Delta$ on $\{\bar{\zeta} < \infty\}$, so $t \mapsto \bar{X}_t$ is continuous on all of $[0, \infty[$. It now follows that $(E, \bar{\mathcal{U}}^\alpha)$ is a $\mathfrak{P}$-harmonic space; see [**BH**, III,§8]. $\square$

If $G$ is an exit set for $\bar{X}$, then the argument used in the proof of Lemma 3.3 is easily modified to show that the sheaf of positive hyperharmonic functions associated with the killed process $\bar{X}^G$ forms a $\mathfrak{P}$-harmonic space. This proves the final assertion of Proposition 2.2. $\square$

### References

[BG] Blumenthal, R.M. and Getoor, R.K. (1968). *Markov Processes and Potential Theory*. New York: Academic Press.

[BH] Bliedtner, J. and Hansen, W. (1986). *Potential Theory*. Universitext, Berlin: Springer.

[CC65] Constantinescu, C. and Cornea, A. (1965). 'Compactifications of harmonic spaces', *Nagoya Math. J.* **25** 1–57.

[CC72] Constantinescu, C. and Cornea, A. (1972). *Potential Theory on Harmonic Spaces*. Berlin: Springer.

[CFØ] Csink, L., Fitzsimmons, P.J. and Øksendal, B. (1990). 'A stochastic characterization of harmonic morphisms'. *Math Annalen* **287** 1–18.

[G] Getoor, R.K. (1990). *Excessive Measures*. Birkhäuser, Boston.

[INW] Ikeda, N., Nagasawa, M. and Watanabe, S. (1966). 'A construction of Markov processes by piecing out'. *Proc. Japan Acad.* **42** 370–375.

[M] Meyer, P.-A. (1975). 'Renaissance, recollements, mélanges, ralentissement de processus de Markov'. *Ann. Inst. Fourier Grenoble* **25** 465–497.

[S] Sharpe, M.J. (1988). *General Theory of Markov Processes*. New York: Academic Press.

# Statistics of Local Time and Excursions for the Ornstein-Uhlenbeck Process

## J. HAWKES and A. TRUMAN

Department of Mathematics and Computer Science
University College of Swansea
SWANSEA    SA2 8PP
Wales, U.K.

**ABSTRACT**

The Poisson-Lévy excursion measures for the Ornstein-Uhlenbeck process are calculated explicitly by utilising some simple connections with the quantum mechanical harmonic oscillator.

## 1   INTRODUCTION

In this paper we discuss the local time and excursion theory for the Ornstein-Uhlenbeck process $X$ ($\in \mathbb{R}$), satisfying the Langevin equation

$$dX(t) = -X(t)dt + dB(t) , \tag{1}$$

B being a BM($\mathbb{R}$) process, with $\mathbb{E}(B(t)B(s)) = \min(s,t)$ , $\mathbb{E}$ being the expectation with respect to Wiener measure $\mu$ .

Nelson showed that the Ornstein-Uhlenbeck process is associated with the stochastic mechanics of the ground state of a quantum mechanical harmonic oscillator with Hamiltonian $H = 2^{-1}\left(\dfrac{-d^2}{dx^2} + x^2\right)$ (Refs.(4), (5)). Here, utilizing some results of Truman and Williams (Ref.(8)) and bringing together some quantum mechanical results for the harmonic oscillator Hamiltonian, we obtain detailed information for the statistics of the local time and excursions for this Ornstein-Uhlenbeck process.

The local time at a upto time t, $L^a(t)$ , is defined by

$$L^a(t) = \lim_{h \downarrow 0} h^{-1} \operatorname{Leb}\left\{ s \in (0,t) : X(s) \in \left(a - \frac{h}{2}, a + \frac{h}{2}\right) \right\} , \tag{2}$$

or formally $L^a(t) = \int_0^t \delta(X(s) - a)\, ds$ . Defining the corresponding subordinator $\gamma^a$ by

$$\gamma^a(t) = \inf\{s > 0 : L^a(s) > t\}, \qquad t > 0, \tag{3}$$

we shall show that for $a = 0$

$$\mathbb{E}_0\left\{e^{-\lambda\gamma_0(t)}\right\} = \exp\left\{-t \int_0^\infty (1 - e^{-\lambda s})\, dv_0(s)\right\}, \tag{4}$$

for each $\lambda > 0$, where

$$\frac{dv_0(s)}{ds} = \frac{2}{\sqrt{\pi}}\, e^{-s}\, (1 - e^{-2s})^{-\frac{3}{2}}. \tag{5}$$

The map $s \mapsto X(s)$ is $\mu$ a.s. continuous so for each $a \in \mathbb{R}$ $\{s > 0 : X(s) \gtrless a\}$ is an open subset of the real line which can be decomposed into its component intervals - the excursion intervals. Define $\#_0(s,t)$ by

$\#_0(s,t)$ = the number of excursions away from 0

of duration $s$ to $(s + ds)$ upto the local time at 0 equals $t$.

Then the last result is equivalent to

$$\mathbb{P}(\#_0(s,t) = N) = e^{-tdv_0(s)} \frac{(tdv_0(s))^N}{N!}, \text{ for } N = 0, 1, 2, ..., \tag{6}$$

$v_0$ being given by Eq.(5).

In this paper we establish more general results for excursions above or below any point $a$, namely: if $\#_a^\pm(s,t)$ denotes the number of excursions $^{above}_{below}$ $a$, respectively, then

$$\mathbb{P}(\#_a^\pm(s,t) = N) = e^{-tdv_a^\pm(s)} \frac{(tdv_a^\pm(s))^N}{N!}, \text{ for } N = 0, 1, 2, ..., \tag{7}$$

where

$$\frac{dv_a^{\pm}(s)}{ds} = \sum_{\alpha_n^{\pm}} (\alpha_n^{\pm})^2 \, e^{s\alpha_n^{\pm} + a^2} \frac{\left\{ \int\limits_{y \gtrless a} D_{-\alpha_n^{\pm}}(\pm\sqrt{2}y) \, e^{-y^2/2} \, dy \right\}^2}{\int\limits_{y \gtrless a} \left\{ D_{-\alpha_n^{\pm}}(\pm\sqrt{2}y) \right\}^2 dy} ,$$

(8)

D being Weber's parabolic cylinder function, $\alpha = \alpha_n^{\pm}$ being the successive values of $\alpha$ for which

$$D_{-\alpha}(\pm\sqrt{2}a) = 0 .$$

(9)

As we shall see, Eq.(5) is a special case of Eq.(8), corresponding to setting $a = 0$, with $v_a = v_a^+ + v_a^-$ . The above $v_a$ is called the Poisson-Lévy excursion measure.

Defining $L^{\pm}(t) = \text{Leb} \{s \in (0,t) : X(s) \gtrless a\}$ , $L^{\pm}(\gamma^a(t))$ are independent random variables, with

$$L^+(\gamma^a(t)) + L^-(\gamma^a(t)) = \gamma^a(t) ,$$

(10)

and the above result is equivalent to

$$\mathbb{E}_a \left\{ e^{-\lambda L^{\pm}(\gamma^a(t))} \right\} = \exp \left\{ -t \int_0^{\infty} (1 - e^{-\lambda s}) \, dv_a^{\pm}(s) \right\} ,$$

(11)

for each $\lambda > 0$, $v_a^{\pm}$ being given by Eq.(8). We establish the above results by finding the distribution of $\tau_x(a) = \inf \{s > 0 : X(s) = a \mid X(0) = x\}$ in terms of the quantum mechanical Hamiltonian H.

## 2  LOCAL TIME AND EXCURSIONS FROM THE ORIGIN FOR O-U PROCESSES

We begin with a more or less standard result. (See Ref.(1)).

### Proposition

The quantum mechanical harmonic oscillator Hamiltonian $H = 2^{-1}(\frac{-d^2}{dx^2} + x^2)$ is limit point at $\pm\infty$ , the unique $L^2$ eigenfunction near $\pm\infty$, corresponding to eigenvalue $\lambda$ being $D_{\lambda-\frac{1}{2}}(\pm\sqrt{2}x)$, respectively, $D$ being Weber's parabolic cylinder function.

**Proof**

The well-known sufficient condition for limit point behaviour of $\left(-2^{-1}\dfrac{d^2}{dx^2} + V(x)\right)$ near $\infty$ is that a positive differentiable function $M(x)$ and positive constants $k_1$, $k_2$ and a can be found such that (i) $V(x) > -k_1 M(x)$ for $x > a$ ,

(ii) $|M'(x)| (M(x))^{\frac{3}{2}} < k_2$ for $x > a$ and (iii) $\int_a^\infty (M(x))^{\frac{1}{2}} dx$ diverges (See Ref.(6)). Taking e.g. $M(x) = x^2$ proves that the harmonic oscillator Hamiltonian is limit point at $\infty$ , and $-\infty$ can be treated similarly. Laplace's method of solving the o.d.e. gives $D_{\lambda-\frac{1}{2}}(\pm\sqrt{2}x)$ as the unique $L^2$ eigenfunctions of $\left(-2^{-1}\dfrac{d^2}{dx^2} + 2^{-1}x^2\right)$ near $\pm\infty$ , with

$$D_{\lambda-\frac{1}{2}}(x) = \frac{e^{-\frac{1}{4}x^2}\Gamma(\frac{1}{2}+\lambda)}{2\pi i} \int_\gamma e^{xs}\, e^{-\frac{1}{2}s^2}\, s^{-\lambda-\frac{1}{2}}\, ds \ , \qquad (12)$$

$\gamma$ being the contour starting just below the cut on the negative reals at $-\infty$, looping once around the origin and finishing just above the cut at $-\infty$. //

**Corollary**
For the O-U process

$$\mathbb{E}_a\left\{e^{-\lambda\gamma^a(t)}\right\} = \exp\left\{\frac{-t}{\tilde{p}_\lambda(a,a)}\right\}$$

$$= \exp\left\{\frac{-t\sqrt{\pi}}{\Gamma(\lambda)\, D_{-\lambda}(\sqrt{2}a)\, D_{-\lambda}(-\sqrt{2}a)}\right\} \ , \qquad (13)$$

$\tilde{p}_\lambda(x,y) = \int_0^\infty e^{-\lambda s}\, p_s(x,y)\, ds$ being the Laplace transform of $p_s(x,y)$, the transition density, $D$ Weber's parabolic cylinder function.

**Proof**

The first part of the above identity follows because $X$ is a stationary Markov process with transition density (explicitly) given by

$$p_t(x,y) = \left(\pi(1 - e^{-2t})\right)^{-\frac{1}{2}} \exp\left\{\frac{-(y - xe^{-t})^2}{(1 - e^{-2t})}\right\} , \qquad (14)$$

whose Laplace transform $\tilde{p}_\lambda(x,y)$ (resolvent kernel) is bounded and continuous in $(x,y)$. Because of the operator identity:

$$-\left(2^{-1}\frac{d^2}{dx^2} - x\frac{d}{dx}\right) = e^{x^2/2}\left(-\frac{1}{2}\frac{d^2}{dx^2} + \frac{x^2}{2} - \frac{1}{2}\right) e^{-x^2/2}$$

$$= f_E^{-1}(x)\,(H-E)\,f_E(x) , \qquad (15)$$

$f_E$ being the ground state of the Hamiltonian $H$, $E$ the ground state eigenvalue $\left(f_E(x) = \pi^{-\frac{1}{4}} e^{-x^2/2}, E = \frac{1}{2} \text{ in our case}\right)$, we obtain

$$\tilde{p}_\lambda(a,a) = (H + \lambda - \frac{1}{2})^{-1}(a,a) = \frac{2\,D_{-\lambda}(\sqrt{2}a)\,D_{-\lambda}(-\sqrt{2}a)}{W} , \qquad (16)$$

$W$ being the Wronskian of $D_{-\lambda}(\sqrt{2}x)$ and $D_{-\lambda}(-\sqrt{2}x)$. $W$ is a constant and so it can be evaluated at the origin. By using Hankel's formula[1], we obtain

$$D_{-\lambda}(0) = \frac{\sqrt{\pi}}{2^{\frac{\lambda}{2}}\,\Gamma(\frac{1}{2} + \frac{\lambda}{2})} \qquad (17)$$

and

$$D'_{-\lambda}(0) = \frac{-\sqrt{\pi}}{2^{\frac{\lambda}{2} - \frac{1}{2}}\,\Gamma(\frac{\lambda}{2})} , \qquad (18)$$

so by the duplication formula (see Ref.(1))

$$W = \frac{2\sqrt{\pi}}{\Gamma(\lambda)} , \qquad (19)$$

giving the desired result. //

---

1   Hankel's formula states: $\Gamma^{-1}(z) = (2\pi i)^{-1}\int_\gamma e^t\,t^{-z}\,dt$, each $z \in C$, $\gamma$ being the contour above, $t^{-z}$ being defined as $e^{-z\,\ln t}$, ln being principal value of the logarithm.

We now come to the main result in this section - an explicit formula for the Poisson-Lévy excursion measure $v_0$.

**Proposition**

For the O-U process, for $\lambda, t > 0$,

$$\mathbb{E}_0\left\{e^{-\lambda \gamma^0(t)}\right\} = \exp\left\{-t\int_0^\infty (1 - e^{-\lambda s})\, dv_0(s)\right\}, \qquad (20)$$

where

$$\frac{dv_0(s)}{ds} = \frac{2}{\sqrt{\pi}}\, e^{2s}\, (e^{2s}-1)^{-\frac{3}{2}}. \qquad (21)$$

**Proof**

Using the above results in Eqs. (17) and (18) and the duplication formula gives

$$\mathbb{E}_0\left\{e^{-\lambda \gamma^0(t)}\right\} = \exp\left\{\frac{-2t\Gamma(\frac{1}{2}+\frac{\lambda}{2})}{\Gamma(\frac{\lambda}{2})}\right\}. \qquad (22)$$

Therefore, setting $H(u) = v_0\{[u,\infty)\}$, we wish to find $H(u)$ such that

$$\frac{\Gamma(\frac{1}{2}(\lambda+1))}{\frac{\lambda}{2}\Gamma(\frac{\lambda}{2})} = \int_0^\infty e^{-\lambda u}\, H(u)\, du, \qquad (23)$$

or

$$\frac{\Gamma(\frac{1}{2}(\lambda+1))}{\Gamma(\frac{1}{2}\lambda+1)} = \int_0^\infty e^{-\lambda u}\, H(u)\, du, \qquad (24)$$

or

$$B(\tfrac{1}{2}(\lambda+1), \tfrac{1}{2}) = \sqrt{\pi}\int_0^\infty e^{-\lambda u}\, H(u)\, du. \qquad (25)$$

Therefore

$$2\int_0^\infty e^{-\lambda x}\, e^{-x}\, (1 - e^{-2x})^{\frac{1}{2}}\, dx = \sqrt{\pi}\int_0^\infty e^{-\lambda x}\, H(x)\, dx. \qquad (26)$$

Finally

$$H(x) = \frac{2}{\sqrt{\pi}} (e^{2x} - 1)^{-\frac{1}{2}} \text{ and } dv_0(x) = -H'(x) dx \ . \qquad //$$

We shall generalise this result in the next section, where we find a connection between the Poisson-Lévy excursion measure $v$ and the quantum mechanical Hamiltonian H.

## 3  GENERAL EXCURSIONS FOR O-U PROCESSES

We need an elementary consequence of the Feynman-Kac formula (see section (25), Ref.(7)):

**Proposition**

Let $H = 2^{-1}\left(\frac{-d^2}{dx^2} + x^2\right)$ , with corresponding ground-state eigenfunction

$f_E(x) = \pi^{-\frac{1}{4}} e^{-x^2/2}$ and corresponding eigenvalue $E = \frac{1}{2}$ . Define

$$H_\pm = \lim_{\lambda \uparrow \infty} (H + \lambda \chi_\mp) , \qquad (27)$$

$\chi_\pm$ being the characteristic function of $\{x : x \gtrless a\}$ , so $H_\pm$ is the Dirichlet Hamiltonian with a Dirichlet boundary condition at $x = a$ . Then for a.e. $x \gtrless a$ , respectively,

$$\mathbb{P}(\tau_x(a) > t) = f_E^{-1}(x) (\exp \{-t (H_\pm - E)\} f_E(x)) . \qquad (28)$$

**Proof**

Set $\tau(x,N) = \tau_x(N) \wedge \tau_x(-N)$ . Then, because $\int^y f_E^{-2}(u) du$ diverges as $y \to \pm\infty$, $\tau(x,N) \uparrow \infty \ \mu$ a.s. as $N \uparrow \infty$ (see e.g. Ref. (2) p.159). Thus, for each $\lambda > 0$,

$$\mathbb{E}\left\{\exp(-\lambda \int_0^t \chi_\mp (X_x(s)) ds)\right\} = \lim_{N \uparrow \infty} \mathbb{E}\left\{\chi_{\{\tau(x,N) > t\}}\right.$$

$$\left. \exp(-\lambda \int_0^t \chi_\mp (X_x(s)) ds)\right\}, \quad (29)$$

$\chi_{\{\tau(x,N) > t\}}$ being 1, if $\tau(x,N) > t$ , and 0 , otherwise.

By the Girsanov-Cameron-Martin theorem, for $B_x(\cdot) = x + B(\cdot)$, Brownian motion starting at $x$, with $\tau^B(x,N) = \inf\{s > 0 : |B_x(s)| = N\}$, using Itô's formula (see Ref.(8) and (9)),

$$\text{r.h.s.} = \lim_{N \uparrow \infty} \mathbb{E}\Big\{ \chi_{\{\tau^B(x,N) > t\}} \exp\Big\{(-\int_0^t (V_0 + \lambda \chi_{\mp} - E)(B_x(s)) ds\Big\}$$

$$f_E(B_x(t)) \, f_E^{-1}(x) \Big\} \, , \tag{30}$$

for $V_0(x) = 2^{-1} x^2$. Using the local integrability and positivity of $(V_0 + \lambda \chi_{\mp})$, the Feynman-Kac formula gives via the dominated convergence theorem, for a.e.x.,

$$\text{r.h.s.} = f_E^{-1}(x) \left( \exp\{-t(H + \lambda \chi_{\mp} - E) \, f_E(x) \right) \tag{31}$$

(see section 25 in Ref.(7)). Since for $x \gtrless a$, respectively,

$$\mathbb{P}(\tau_x(a) > t) = \lim_{\lambda \uparrow \infty} \mathbb{E}\Big\{ \exp(-\lambda \int_0^t \chi_{\mp} (X_x(s)) \, ds) \Big\} \, , \tag{32}$$

the desired result follows. //

For our one-dimensional time-homogeneous process

$$\mathbb{E}\{e^{-\lambda \tau_x(a)}\} = \frac{\tilde{p}_\lambda(x,a)}{\tilde{p}_\lambda(a,a)} = \frac{f_E^{-1}(x)(H + \lambda - E)^{-1}(x,a) f_E(a)}{(H + \lambda - E)^{-1}(a,a)} \, , \tag{33}$$

where $\tilde{p}_\lambda(x,a) = \int_0^\infty e^{-\lambda s} p_s(x,a) \, ds$. Multiplying both sides of the last equation by $f_E^2(x)$ and integrating by Fubini's theorem, we obtain for the Poisson-Lévy excursion measure $\nu_a$

$$\lambda^{-1}\{(H + \lambda - E)^{-1}(a,a)\}^{-1} = \int_0^\infty e^{-\lambda u} \nu_a\{[u, \infty)\} \, du$$

$$= f_E^{-2}(a) \int_{-\infty}^\infty f_E^2(x) \, \mathbb{E}\{e^{-\lambda \tau_x(a)}\} dx \, , \tag{34}$$

for each $\lambda > 0$. Combining this result with the above proposition we obtain:

**Proposition**

For $H = 2^{-1}\left(\dfrac{-d^2}{dx^2} + x^2\right)$, $E = \dfrac{1}{2}$, $f_E(x) = \pi^{-\frac{1}{4}} e^{-x^2/2}$ and $H_{\pm} = \lim_{\lambda \uparrow \infty} (H + \lambda\chi_{\mp})$,

$\chi_{\mp}$ being the characteristic function of $\{x : x \gtrless a\}$, respectively,

$$\left\{(H + \lambda - E)^{-1} (a,a)\right\}^{-1} = \int_0^{\infty} (1 - e^{-\lambda s}) \, dv_a(s) , \qquad (35)$$

where

$$v_a\{[s, \infty)\} = f_E^{-2}(a)\left\{\left(f_E, (H_+ - E) e^{-s(H_+ - E)} f_E\right)_{L^2}\right.$$

$$\left. + \left(f_E, (H_- - E) e^{-s(H_- - E)} f_E\right)_{L^2}\right\} . \qquad (36)$$

We can now establish the main result of this section:

**Proposition**

Let $\alpha = \alpha_n^{\pm}$ be the successive values of $\alpha$ for which $D_{-\alpha}(\pm\sqrt{2}a) = 0$,

respectively. Then

$$\frac{dv_a(s)}{ds} = \frac{dv_a^+(s)}{ds} + \frac{dv_a^-(s)}{ds} \qquad (37)$$

with

$$\frac{dv_a^{\pm}(s)}{ds} = \sum_{\alpha_n^{\pm}} (\alpha_n^{\pm})^2 \, e^{(a^2 + s\alpha_n^{\pm})} \frac{\left\{\displaystyle\int_{y \gtrless a} D_{-\alpha_n^{\pm}}(\pm\sqrt{2}y) \, e^{-y^2/2} \, dy\right\}^2}{\displaystyle\int_{y \gtrless a} \left\{D_{-\alpha_n^{\pm}}(\pm\sqrt{2}y)\right\}^2 dy} , \qquad (38)$$

respectively.

**Proof**

Because $V_0(x) \uparrow \infty$ as $|x| \uparrow \infty$, the Hamiltonian $H_{\pm}$ has pure point spectrum $\left\{\frac{1}{2} - \alpha_n^{\pm}\right\}_{n=0,1,2,\ldots}$ , $H_{\pm}$ having corresponding eigenfunctions $D_{-\alpha_n^{\pm}}(\pm\sqrt{2}x)$ (before normalisation), respectively. This is a consequence of singular boundary value theory (see Chapter 10 in Ref.(3), in particular p.524). The resulting spectral theorem for $H_\pm$ gives the reality of the spectrum of $H_\pm$ and the above expressions for $v_a^\pm$. //

We state without proof here that $v_a^\pm$ give the excursions $\begin{smallmatrix}\text{above}\\\text{below}\end{smallmatrix}$ a, respectively (see Ref.(8)).

For a = 0 evidently $\alpha_n^+ = \alpha_n^-$ as expected. These $\alpha$ values can be found from the condition $D_{-\alpha}(0) = 0$. Using Eq.(17), $\alpha$ must satisfy:

$$\Gamma^{-1}\left(\tfrac{1}{2} + \tfrac{\alpha}{2}\right) = 0 \tag{39}$$

$$\therefore \; \alpha_n^+ = \alpha_n^- = -1 - 2n \;, \quad \text{for } n = 0, 1, 2, \ldots \tag{40}$$

The corresponding eigenvalues of $H_\pm$ are $\lambda$, where

$$\lambda - \tfrac{1}{2} = 2n + 1 \;, \quad \text{for } n = 0, 1, 2, \ldots \tag{41}$$

i.e.

$$\lambda = \tfrac{1}{2} + (2n + 1) \;, \quad \text{for } n = 0, 1, 2, \ldots \tag{42}$$

The corresponding eigenfunctions are $H_{2n+1}(x)\, e^{\frac{1}{2}x^2}$, $H_{2n+1}$ being the odd Hermite polynomials, which are zero at the origin, so satisfying the Dirichlet boundary condition. $\left(H_n(x) = e^{x^2}\left(-\frac{d}{dx}\right)^n e^{-x^2}\right)$. It only remains to evaluate the inner product in the numerator of Eq.(38) to confirm that Eq.(5) is a special case of Eq.(38) with a = 0. The key to this is the observation that for n = 1, 3, 5, ...

$$\int_0^\infty e^{-x^2} H_n(x)\, dx = -\frac{d^{n-1}}{dx^{n-1}}\, e^{-x^2}\Big|_{x=0} \tag{43}$$

and r.h.s. can be evaluated from Maclaurin's series for $e^{-x^2}$.

## References

(1)    M. Abramowitz and I.A. Stegun, Handbook of Mathematical Functions, Dover, New York, 1970.

(2)    I.T. Gihman and A.V. Skorohod, Stochastic Differential Equations, Springer, Berlin, 1972.

(3)    E. Hille, Lectures on Ordinary Differential Equations, Addison Wesley, Reading (Mass), 1969.

(4)    E. Nelson, Dynamical Theories of Brownian Motion, Princeton University Press, Princeton, 1967.

(5)    E. Nelson, Quantum Fluctuations, Princeton University Press, Princeton, 1984.

(6)    M. Reed and B. Simon, Mathematical Physics, Volume 2, Fourier Analysis and Self-adjointness, Academic, New York, 1975.

(7)    B. Simon, Functional Integration and Quantum Physics, Academic, New York, 1979.

(8)    A. Truman and D. Williams, A Generalised arc-sine law and Nelson's stochastic mechanics of one-dimensional time-homogeneous diffusions, in Diffusion Processes and Related Problems in Analysis, Birkhauser (to be published).

(9)    D. Williams and L.C.G. Rogers, Diffusions, Markov Processes and Martingales, Volume 2, Itô Calculus, Wiley, 1987.

# $L^p$–CHEN FORMS ON LOOP SPACES

## J.D.S. JONES AND R. LEANDRE

University of Warwick and Université de Strasbourg

## INTRODUCTION

Let $M$ be a Riemannian manifold. Our aim is to study differential forms on the following infinite dimensional manifolds:

(1) the path space $PM$ consisting of paths $w : [0,1] \to M$,
(2) the loop space $LM$ consisting of paths $w$ such that $w(0) = w(1)$,
(3) the based loop space $L_x M$ consisting of loops $w$ such that $w(0) = w(1) = x$ where $x$ is a chosen base point in $M$.

One consequence of the fact that these manifolds are infinite dimensional is that there are infinite sequences $\alpha_n$ of forms with each $\alpha_n$ homogeneous of degree $n$. These infinite sequences are very important in the geometrical applications of loop spaces; for example they are essential in the theory of equivariant cohomology in infinite dimensions as is made quite clear in [25].

In [11] Chen describes the theory of "iterated integrals"; this is a method of constructing differential forms on these infinite dimensional manifolds. We will refer to forms constructed by this means as Chen forms. The purpose of this paper is to study some of the analytical properties of Chen forms; in particular to make estimates for suitable $L^p$-norms and to consider various decay conditions which one might put on the terms in an infinite sequence of the kind mentioned in the previous paragraph.

The motivation for studying forms on the loop space comes from the relation between the $S^1$-equivariant cohomology of the loop space of $M$ and the Atiyah-Singer index theorem [1], [7]. Here the circle acts on the loop space by rotating loops. In [17] and [16] this link between index theory and loop spaces is studied using Chen forms (called iterated integrals in [17] and [16]) and it is shown how the cohomology and the equivariant cohomology of the loop space $LM$ can be computed using Chen forms. The result is best expressed using Hochschild homology, for ordinary cohomology, and Connes's theory of cyclic homology, for equivariant cohomology. Thus there is a rather remarkable interaction between loop spaces, index theory, and cyclic homology. More information and further references can be found in [1], [7], and [17]. These ideas are summarised in [24].

Because of this connection with index theory we are particularly interested in $LM$. However there is no advantage to be gained by restricting attention to this case alone. Indeed at several points it is simplest to argue first with the full path space $PM$ and then deduce the corresponding result for the loop spaces $LM$ and $L_xM$. Thus we take care to develop the theory in all three cases.

In [17] the loop space means the smooth loop space $L^\infty M$ considered as an infinite dimensional manifold modelled on the Fréchet space $C^\infty(S^1, \mathbb{R}^d)$ where $d$ is the dimension of $M$. Of course the natural measures coming from stochastic analysis do not make sense on the smooth loop space and so to do $L^p$-estimates it is necessary to work with the full loop space $LM$ of continuous loops. This provides us with our first problem: give a precise definition of what we mean by differential forms on $LM$ and extend Chen forms to differential forms on $LM$. This, inevitably, requires stochastic analysis. It leads to the theory of stochastic Chen forms which is described in detail in §1. Stochastic Chen forms are also described in [16], where they are called stochastic iterated integrals. Our approach, in keeping with our aims, is based on very down-to-earth estimates and is very different from that of [16]. The essential estimates, on which our work is based, are mostly carried out in §4 and §5.

In §2 and §3 we introduce natural $L^p$-norms on the space of forms on $LM$ and study the completions of the space of Chen forms in these $L^p$-norms. These completed spaces are the spaces of $L^p$-Chen forms. In forming these completions we complete the direct sum of the homogeneous terms, not just each of the homogeneous terms separately. So the spaces of $L^p$-Chen forms contain infinite sums $\sum \alpha_n$ where each of the terms $\alpha_n$ is homogeneous of degree $n$; we will refer to such sums as forms of infinite degree. The condition that such a sum has finite $L^p$-norm imposes decay conditions on the terms $\alpha_n$. We show that these spaces of $L^p$-Chen forms have the minimum algebraic properties we might hope for: the intersection of all the $L^p$-spaces is closed under exterior products of forms and interior products of forms with vector fields.

One way to think of these constructions is to use the following analogy. If we take the Hilbert space $H = L^2[0, 1]$ we can consider two, isomorphic, versions of the associated symmetric Fock space. The first, which is essentially algebraic, is defined as a completion of the sum of all the symmetric powers of $H$. Thus we allow infinite sums $\sum \beta_n$, where $\beta_n$ is in the $n$-th symmetric power of $H$, which satisfy certain decay conditions. The relation between these two versions of the symmetric Fock space is given by iterated integrals and Wiener chaos. In our context the algebraic space is constructed as follows. Let $\Omega$ be the space of forms on $M$ and $\bar\Omega$ the space of forms of strictly positive degree. Then we take $\mathcal{C} = \sum \Omega \otimes \bar\Omega^{\otimes n}$ suitably completed. The link between the algebraic space and the space of forms on the loop space is provided by stochastic Chen forms.

The similarity with the theory of stochastic differential equations appears very naturally. In [9], the convergence, in $L^p$, of the Picard series of a linear equation is studied; the same problem is studied in [35] for the general equation. In both cases the components of the equation are scalars. Exponential forms on loop spaces are, formally, solutions of a linear equation whose components are differential forms and they give forms of infinite degree. Bismut's equivariant Chern character [7] is an example of such a form. We choose our $L^p$-norms so that the path ordered exponential series is absolutely convergent in all the $L^p$. These exponential forms, which are allowed to take their values in a matrix algebra, could be called the coherent forms on the loop space, in analogy with quantum physics.

In the theory of cyclic cohomology, Connes's entire cyclic cohomology, see [13], [18], and [26], is designed for "infinite dimensional cyclic cocycles": this means we allow infinite sums $\sum \varphi_n$ where each term is homogeneous of degree $n$, but impose decay conditions. We make the link between our spaces of $L^p$-Chen forms and Connes's entire cyclic cohomology and so give a simple condition under which a form of infinite degree lies in the intersection of the spaces of $L^p$-Chen forms. One of our main theorems is that if an infinite sum $\sum \omega_n \in \mathcal{C}$ satisfies an algebraic condition, then the corresponding stochastic Chen form on the loop space belongs to all the spaces of $L^p$-Chen forms. This condition shows that Bismut's equivariant Chern character, see [7] and [17], belongs to all the spaces of $L^p$-Chen forms. The method of proving this is to use the construction of the equivariant Chern character given in [17] in terms of Chen forms and then to use estimates to show that this infinite sum satisfies the appropriate algebraic condition.

The subject is a mixture of homological algebra, geometry of loop spaces, and stochastic analysis. Many of the constructions have a natural interpretation in terms of homological algebra, compare [17]. We have tried to point out these links but, in order to keep the paper to a reasonable length, we have not tried to describe them in much detail.

We would like D. Elworthy and M. Emery for many helpful discussions and comments on the contents of this paper. The first author would also like to thank Ezra Getzler for many invaluable discussions on Chen's iterated integrals which have been a major influence on some of the ideas behind this paper.

## §1 STOCHASTIC CHEN FORMS

Let $M$ be a compact Riemannian manifold of dimension $d$ and let $PM$, **the path space** of $M$, be the Banach manifold of all continuous functions $w : [0,1] \to M$. The **loop space** $LM$ is the submanifold consisting of those

$w$ such that $w(0) = w(1)$. The **based loop space**, $L_x M$ is the submanifold of loops $w$ such that $w(0) = w(1) = x$ where $x \in M$ is a chosen base point. Note that

$$L_x M \subset LM \subset PM$$

and each manifold has codimension $d$ in the next.

Let $dP_{x,y}$ be the usual Brownian bridge measure on the space of paths $P_{x,y} M$ which start at $x$ and end at $y$; we use the notation $E^{x,y}$ for expectations computed with respect to this measure. We use the measure

$$p_1(x, x)\, dx\, dP_x$$

on $LM$, where $p_t(x, y)$ is the heat kernel of $M$. This is the unique measure of the form

$$f(x)\, dx\, dP_x$$

which is invariant under rotations of loops, see [7] and [15]. We use the notation $E_L$ for expectations computed using this measure on the loop space. We use the measure

$$dP = p_1(x, y) dx\, dy\, dP_{x,y}$$

on $PM$, this is the equilibrium measure for Brownian paths, and $E$ for the corresponding expectation.

Let $T_x = T_x M$ be the tangent space to $x \in M$ and, given a path $w$ in $M$, let

$$\tau_s(w) : T_{w_0} \to T_{w_s}$$

be the (stochastic) parallel transport along $w$. This operator is a linear isometry and it is defined for all continuous paths $w$, except for a set of measure zero. Our discussion is based on two simple observations:

(1) The tangent space to the Banach manifold $PM$ at a continuous path $w$ is the space of continuous vector fields along $w$.
(2) Parallel transport defines an isomorphism from vector fields along $w$ to a space of paths in $T_{w_0}$.

We begin by defining suitable Hilbert spaces of paths in a $d$-dimensional inner product space $T$. Let $\mathcal{H}(T)$ be the space of paths $X : [0, 1] \to T$, which are absolutely continuous with square integrable derivative. We equip $\mathcal{H}(T)$ with the norm

$$\|X\|^2 = \int_0^1 \|X(t)\|^2 dt + \int_0^1 \left\| \frac{dX(t)}{dt} \right\|^2 dt.$$

It is straightforward to check that $\mathcal{H}(T)$ is a Hilbert space. We define $\mathcal{H}^0(T)$ to be the codimension $2d$ subspace consisting of those $X$ with $X(0) = X(1) = 0$.

Now let $w$ be a path in $M$ and let $X$ be a path in $\mathcal{H}(T_{w_0})$ and assume that parallel transport $\tau_s(w)$ is defined. Then define $\tau(w)X$ by

$$(\tau(w)X)_s = \tau_s(w)X_s.$$

Thus $\tau(w)X$ is a vector field along $w$. Now define $\mathcal{H}_w$ to be the space of all vector fields along the path $w$ of the form $\tau(w)X$ where $X \in \mathcal{H}(T_{w_0})$. There is a linear isomorphism $\tau(w) : \mathcal{H}(T_{w_0}) \to \mathcal{H}_w$ so we make $\mathcal{H}_w$ into a Hilbert space using the inner product inherited from $\mathcal{H}(T_{w_0})$. The spaces $\mathcal{H}_w$ form a measurable field of Hilbert spaces tangent to $PM$. This means that $\mathcal{H}_w$ is defined for all $w$ except for a set of measure zero and each of the spaces $\mathcal{H}_w$ is a linear subspace of the tangent space of the Banach manifold $PM$ at the path $w$.

Suppose now that $w$ is a loop based at $x \in M$. Then define $\mathcal{H}_w^0$ to be the space of vector fields along $w$ of the form $\tau(w)X$ where $X \in \mathcal{H}^0(T_{w_0})$. Since $\tau_1(w)(0) = 0$ it follows that the vector field $\tau(w)X$ along the loop $w$ starts and ends at $0 \in T_x M$. In this case we get an isomorphism

$$\tau(w) : \mathcal{H}^0(T_{w_0}) \to \mathcal{H}_w^0$$

and the spaces $\mathcal{H}_w^0$ form a measurable field of Hilbert spaces tangent to $L_x M$.

To define the corresponding field of Hilbert spaces on $LM$ let $h_w = \tau_1(w) : T_{w_0} \to T_{w_0}$ be the holonomy around the loop $w$. Then the path $\tau(w)X$ is a vector field along the loop $w$ if

$$h_w(X_1) = X_0.$$

We define $\mathcal{K}_w$ to be the subspace of $\mathcal{H}(T_{w_0})$ consisting of those paths which satisfy this holonomy condition and then define $\mathcal{T}_w$ to be the subspace of $\mathcal{H}_w$ given by $\tau(w)X$ where $x \in \mathcal{K}_w$. In this case

$$\tau(w) : \mathcal{K}_w \to \mathcal{T}_w$$

is an isomorphism and the spaces $\mathcal{T}_w$ form a measurable field of Hilbert spaces tangent to $LM$. If $w$ is a loop in $M$

$$\mathcal{H}_w^0 \subset \mathcal{T}_w \subset \mathcal{H}_w$$

and each space has codimension $d$ in the next. The space $\mathcal{H}_w$ consists of paths in the tangent manifold $TM$ which cover the loop $w$, $\mathcal{T}_w$ consists of loops in $TM$ which cover $w$, and $\mathcal{H}_w^0$ consists of loops in $TM$ which cover $w$ and start and end at $0 \in T_{w_0} M$.

Next we construct a basis for the Hilbert spaces $\mathcal{H}_w$. First we consider the space $\mathcal{H} = \mathcal{H}(\mathbb{R})$. We start with the linearly independent set

$$a_n(s) = \begin{cases} \sin(2\pi n s), & \text{if } n > 0 \\ \cos(2\pi n s), & \text{if } n \leq 0. \end{cases}$$

This set is an orthogonal basis for $\mathcal{H}$ and, for a suitable constant $C$,

$$\|a_n\|^2 = 1 + Cn^2.$$

We define

(1.1)
$$b_n = \frac{a_n}{\|a_n\|}$$

so $b_n$ is a basis for $\mathcal{H}_w$ and there is a constant $C$ such that

(1.2)
$$\sup_{s \in [0,1]} |b_n(s)| \leq \frac{C}{|n|}, \qquad \text{for } n \neq 0.$$

Now let $w$ be a path in $M$ and let $e_1, \ldots, e_d$ be an orthonormal basis for $T_{w_0}$. For $i = 1, \ldots, d$ and $n \in \mathbb{Z}$ we define

(1.3)
$$\begin{aligned} e_{n,i} &\in \mathcal{H}(T_{w_0}), & e_{n,i}(s) &= b_n(s)e_i, \\ \varepsilon_{n,i} &\in \mathcal{H}_w, & \varepsilon_{n,i}(s) &= \tau_s(w)(e_{n,i}(s)). \end{aligned}$$

The $e_{n,i}$ are an orthonormal basis for the Hilbert space $\mathcal{H}(T_{w_0})$ and the $\varepsilon_{n,i}$ an orthonormal basis for $\mathcal{H}_w$.

It is possible to obtain a basis for $\mathcal{H}_w^0$ and $\mathcal{T}_w$ by a similar, but more intricate, construction. However we shall avoid doing this and refer all our computations to the bases (1.3) for $\mathcal{H}_w$ and $\mathcal{H}(T_{w_0})$. The reason for doing this is that the subspace $\mathcal{T}_w$ is an "anticipating" function of $w$ in the sense that it involves the holonomy around the loop $w$. In particular any natural construction of a basis for $\mathcal{T}_w$ will involve anticipating terms and lead to anticipating stochastic integrals.

Choose an orthonormal basis $e_1(x), \ldots, e_d(x)$ for $T_x$ such that each $e_i$ is a measurable function of $x$. Such a basis always exists as can be verified by a straightforward argument. Let $\varepsilon_{n,i}(w)$ be the orthonormal basis for $\mathcal{H}_w$ defined by $e_1(w_0), \ldots, e_d(w_0)$. Now if $b$ is any section of the field of Hilbert spaces $\mathcal{H}_w$ then we say that $b$ is **measurable** if each of the functions

$$w \mapsto \langle b(w), \varepsilon_{n,i}(w) \rangle$$

is a measurable function on $PM$. It is straightforward to check that this notion of measurability does not depend on the choice of basis $e_i(x)$. The

$L^p$-**norm** of a measurable vector field is defined to be the $L^p$-norm of the function $\|b(w)\|$ on $PM$. If $b$ is a section of the field of Hilbert spaces $T_w$ then $b$ is measurable if the functions $w \mapsto \langle b(w), \varepsilon_{n,i}(w) \rangle$ are measurable functions on $LM$ and the $L^p$-norm of $b$ is the $L^p$-norm of the function $w \mapsto \|b(w)\|$ defined on $LM$. The definition of measurability for sections of $\mathcal{H}_w^0$ is identical. We shall refer to measurable sections of these fields of Hilbert spaces as **measurable vector fields**. In the case of the loop space $LM$ it can be checked that these notions are invariant, in the natural sense, under the action of the circle given by rotation of loops.

We shall now give the definition of measurable forms. The main reason for this is to be explicit about our conventions for the constants which appear in the theory of differential forms. Often these conventions do not matter but since one of our main purposes is to keep track of decay conditions we need to be precise.

We use the notation $|J|$ for the number of elements in a finite set $J$. Let $b_i$, $i \in I$, be any set of measurable vector fields on $PM$ indexed by a totally ordered set $I$. If $J$ is a finite subset of $I$ we write $b_J$ for the ordered $|J|$-tuple given by the $b_j$ where $j \in J$. A **measurable $r$-form** $\sigma$ on $PM$ is defined to be a continuous, multilinear, alternating function

$$\sigma_w : \mathcal{H}_w \times \cdots \times \mathcal{H}_w \to \mathbb{R},$$

defined for almost all paths $w$, with the property that given any $r$ measurable vector fields $b_1, \ldots, b_r$ the function

$$w \mapsto \sigma_w(b_1(w), \ldots, b_r(w))$$

is a measurable function on $PM$. The definition of measurable forms on $LM$ and $L_x M$ is identical.

To define the exterior product, and also for later use, we recall the definition of shuffles. Let $I = \{1, \ldots, r\}$; then a $(r_1, \ldots, r_n)$-**shuffle** of $I$ is a partition of $I$ into subsets $I_1, \ldots, I_n$ with $|I_k| = r_k$. Let $\Phi(r_1, \ldots, r_n)$ be the set of $(r_1, \ldots, r_n)$-shuffles of $I$; there are

$$\frac{r!}{r_1! \cdots r_n!}$$

such shuffles. A shuffle defines a permutation of $I$ by re-ordering $I$ as

$$I_1 \cup I_2 \cup \cdots \cup I_n.$$

and $\text{sign}(I_1, \ldots, I_n)$ is defined to be the sign of this permutation.

Let $\sigma$ be a measurable $r$-form and $\tau$ a measurable $s$-form on $PM$. The **exterior product** $\sigma \wedge \tau$ is defined by the formula

$$(\sigma \wedge \tau)(b_1, \ldots, b_{r+s}) = \sum_{(I,J) \in \Phi(r,s)} \text{sign}(I,J) \sigma(b_I) \tau(b_J).$$

The **interior product** of a measurable $r$-form $\sigma$ with a measurable vector field $b$ is defined by

$$i_b(\sigma)(b_1, \ldots, b_{r-1}) = \sigma(b, b_1, \ldots, b_{r-1}).$$

Let $\varepsilon_{n,i}$ be the orthonormal basis for $\mathcal{H}_w$ defined by a measurable orthonormal basis $e_1(x), \ldots, e_d(x)$ for $T_x$. This basis is indexed by $\mathbb{Z} \times \{1, \ldots, d\}$ ordered lexicographically. Let $\beta_{n,i}(w)$ denote the dual basis of $\mathcal{H}_w^*$. Now let

$$I = \{(n_1, i_1), \ldots, (n_r, i_r)\}$$

be a finite ordered subset of $\mathbb{Z} \times \{1, \ldots, d\}$ and define $\beta_I(w)$ by the formula

(1.4) $$\beta_I = \beta_{n_1, i_1} \wedge \cdots \wedge \beta_{n_r, i_r}.$$

With our conventions for exterior products $\beta_I$ is the unique $r$-form such that

$$\beta_I(\varepsilon_{n_1, i_1}, \ldots, \varepsilon_{n_r, i_r})_w = 1$$

and $\beta_I$ vanishes on any $r$-tuple of the $\varepsilon_{n,i}$ indexed by a set other than $I$. The $\beta_I$ are a basis, over the space of measurable functions on $PM$, for the space of measurable forms on $PM$.

Now we introduce **Chen forms** as a means of constructing differential forms on $PM$. The original definition is due to Chen [11], and the general theory of Chen forms on the smooth loop space is extensively studied in [17]. Here we study stochastic Chen forms in the Stratonovitch sense. Let $\Omega$ be the space of all smooth forms on $M$ and let $\bar{\Omega}$ be the subspace of all forms of degree $r$ with $r \geq 1$. Let

$$\omega : [0,1] \to \bar{\Omega}^{\otimes n}, \qquad s \mapsto \omega_{1,s} \otimes \cdots \otimes \omega_{n,s}$$

be a map which is piecewise constant. We suppose that each $\omega_i$ is homogeneous and define

$$r_i = \deg \omega_i - 1.$$

Let $\Delta^n$ be the simplex

$$\Delta^n = \{(s_1, \ldots, s_n) \in \mathbb{R}^n : 0 \leq s_1 \leq \cdots \leq s_n \leq 1\}.$$

Now define

(1.5)
$$\sigma(\omega) = \int_{\Delta^n} \omega_1(dw, -)_{s_1} \wedge \cdots \wedge \omega_n(dw, -)_{s_n}.$$

We must explain how $\sigma(\omega)$ is a measurable form on $PM$ of degree

$$r = r_1 + \cdots + r_n.$$

Let $b_1(w), \ldots, b_r(w)$ be measurable vector fields on $PM$, then

(1.6) $\sigma(\omega)(b_1, \ldots, b_r)_w =$

$$\sum_{\Phi(r_1,\ldots,r_n)} \text{sign}(I_1, \ldots, I_n) \int_{\Delta^n} \omega_1(dw, b_{I_1})_{s_1,w} \cdots \omega_n(dw, b_{I_n})_{s_n,w}.$$

We must make sense of the stochastic integral in (1.6).

The first step is to express the $b_j$ in terms of the basis $\varepsilon_{n,i}$

(1.7)
$$b_j(w) = \sum_{n,i} \lambda_{n,i}^{(j)}(w)\varepsilon_{n,i}(w),$$

where the $\lambda_{n,i}^{(j)}(w)$ are measurable functions of $w$. Now (1.6) becomes

(1.8)
$$\sigma(\omega)(b_1, \ldots, b_r) = \sum \lambda_{n_1,i_1}^{(1)} \cdots \lambda_{n_r,i_r}^{(r)} \sigma(\omega)(\varepsilon_{n_1,i_1}, \ldots, \varepsilon_{n_r,i_r})$$

where the sum is taken over

$$(n_1, \ldots, n_r) \in \mathbf{Z}^r, \qquad 1 \le i_1, \ldots, i_r \le d.$$

In view the definition of the $\varepsilon_{n,i}$, given in (1.3), there is no difficulty in defining the stochastic integrals $\sigma(\omega)(\varepsilon_{n_1,i_1}, \ldots, \varepsilon_{n_r,i_r})_w$, for almost all $w$. It remains to show that the series (1.8) converges almost everywhere and this requires an estimate for $\sigma(\omega)(\varepsilon_{n_1,i_1}, \ldots, \varepsilon_{n_r,i_r})_w$.

Fix $x \in M$ and let $f_1, \ldots, f_r : [0,1] \to T_x M$ be continuous functions which have bounded variation. We need some hypotheses on the $f_i$ since we will use Stratonvitch integrals and these hypotheses are adequate for our applications. Presumably, it is possible to get away with much weaker conditions on the $f_i$. Let

$$f = (f_1, \ldots, f_r)$$
$$I_i = \{r_1 + \cdots + r_i + 1, \ldots, r_1 + \cdots + r_i + r_{i+1}\}.$$

and define

$$(1.9) \qquad A(f;\omega) = \int_{\Delta^n} \omega_1 \left(dw, \tau(f_{I_1})\right)_{s_1} \cdots \omega_n \left(dw, \tau(f_{I_n})\right)_{s_n}.$$

Choose a constant $C(f)$ such that

$$\|f_i\|_\infty \leq C(f), \qquad 1 \leq i \leq r.$$

Recall that if $\omega_s$ is a form on $M$ which depends on $s$

$$\|\omega\|_{k,\infty} = \sup_s \left(\|\omega_s\|_\infty + \|\nabla \omega_s\|_\infty + \cdots + \|\nabla^k \omega_s\|_\infty\right).$$

Let $N$ be the frame bundle of $M$ and let $\pi : N \to M$ be the projection. Let $\tilde{w}_s(u)$, $s \geq 0$, $u \in N$, be the horizontal lift of the diffusion $w_s(x)$, $s \geq 0$, where $x = \pi(u)$. The diffusion $\tilde{w}_s(u)$ is **hypoelliptic** if it satisfies the following condition: There is a sub-bundle $P$ of the frame bundle $N$ with structure group $G \subset O(d)$ such that the diffusion $\tilde{w}_s(u)$ takes place in $P$ and the components of the vector fields over $P$ which occur in the stochastic differential equation of $\tilde{w}_s(u)$ satisfy the strong Hörmander hypothesis, [30]. We refer the reader to [30] for a more detailed discussion of the hypoelliptic condition. If the dimension of $G$ is zero, for example if $M = \mathbb{R}^d$, there is no extra condition and the diffusion is hypoelliptic. If, at each point in $M$, the components of the curvature of $M$ span the Lie algebra of $G$, then once more the diffusion is hypoelliptic. Furthermore the diffusion $\tilde{w}_s(u)$ is hypoelliptic for a generic metric on $M$.

**Lemma 1.10.** (1) *For all $x \in M$ we have the estimate*

$$(1.11) \qquad E^{x,x}\left[\,|A(f;\omega)|^p\,\right]^{1/p} \leq \frac{C(f)^r C(p)^n}{\sqrt{n!}} \|\omega_1\|_{2d+3,\infty} \cdots \|\omega_n\|_{2d+3,\infty}.$$

*where $d$ is the dimension of $M$.*

(2) *Suppose that for all $u$ the diffusion $\tilde{w}_s(u)$ is hypoelliptic. Then, for all $x \in M$, we have the estimate*

$$(1.12) \qquad E^{x,x}\left[\,|A(f;\omega)|^p\,\right]^{1/p} \leq \frac{C(f)^r C(p)^n}{\sqrt{n!}} \|\omega_1\|_{1,\infty} \cdots \|\omega_n\|_{1,\infty}.$$

(3) *If each of the $\omega_i$ is a 1-form then (1.12) holds even if the diffusion $\tilde{w}_s(u)$ is not hypoelliptic.*

This lemma is the most important technical ingredient in the proofs of our main results. It is proved in §4 and §5. Note that the constant $C(p)$ occuring in (1.11) and (1.12) depends on the Riemannian manifold $M$.

Furthermore the term $C(f)^r$ occuring in (1.11) and (1.12) can be replaced by $\|f_1\|_\infty \cdots \|f_r\|_\infty$ but it is more convenient, at several points in writing the proof, to choose the constant $C(f)$.

If we take open Brownian motion, that is we consider the expectation $E\left[|A(f;\omega)|^p\right]^{1/p}$ without any conditioning, the estimate (1.12) is valid without the hypoellipticity hypothesis; this follows from the method of Schwartz [35]. It seems reasonable to conjecture that in fact (1.12) is valid in general without the hypoellipticity hypothesis.

By combining Lemma (1.10) with (1.2) and (1.3) we immediately get the following estimate.

**Lemma 1.13.** *For all $p > 1$ there is a constant $C(p,\omega)$, depending only on $p$ and $\|\omega_i\|_{2d+3}$, $1 \le i \le n$, such that*

$$(1.14) \qquad \|\sigma(\omega)(\varepsilon_{n_1,i_1},\ldots,\varepsilon_{n_r,i_r})\|_{L^p} \le \frac{C(p,\omega)}{(|n_1|+1)\ldots(|n_r|+1)}$$

*where $\| - \|_{L^p}$ is the $L^p$-norm on $PM$.* $\square$

The terms in the denominator are $|n_i|+1$ rather than $|n_i|$ simply to deal with the possibility $n_i = 0$. This is, of course, a very weak application of Lemma (1.10). To prove that the series in (1.8) is absolutely convergent almost everywhere we first prove the following lemma.

**Lemma 1.15.** *Suppose that*

$$\sum_{n,i} |\lambda_{n,i}^{(j)}|^2 \in L^p$$

*for all $p$. Then the series (1.8) is absolutely convergent almost everywhere.*

*Proof.* It follows that

$$\sum |\lambda_{n_1,i_1}^{(1)}(w)\ldots\lambda_{n_r,i_r}^{(r)}(w)|^2 < \infty$$

for almost all $w$ and that

$$E\left[\left(\sum |\lambda_{n_1,i_1}^{(1)}\ldots\lambda_{n_r,i_r}^{(r)}|^2\right)\right] < \infty$$

where each of the sums is taken over

$$(n_1,\ldots,n_r) \in \mathbf{Z}^r, \qquad (i_1,\ldots,i_r) \in \{1,\ldots,d\}^r.$$

Now the Schwartz inequality shows that, almost always,

$$(1.16) \quad \sum |\lambda_{n_1,i_1}^{(1)}(w)\ldots\lambda_{n_r,i_r}^{(r)}(w)| \, |\sigma(\omega)(\varepsilon_{n_1,i_1},\ldots,\varepsilon_{n_r,i_r})w| \le$$

$$\left(\sum |\lambda_{n_1,i_1}^{(1)}(w)\ldots\lambda_{n_r,i_r}^{(r)}(w)|^2\right)^{1/2} \left(\sum |\sigma(\omega)(\varepsilon_{n_1,i_1},\ldots,\varepsilon_{n_r,i_r})w|^2\right)^{1/2}.$$

Taking expectations and applying the Schwartz inequality again we get

$$(1.17) \quad E\left[\sum |\lambda^{(1)}_{n_1,i_1} \ldots \lambda^{(r)}_{n_r,i_r}| \, |\sigma(\omega)(\varepsilon_{n_1,i_1}), \ldots, \varepsilon_{n_r,i_r})| \right] \leq$$

$$E\left[\left(\sum |\lambda^{(1)}_{n_1,i_1} \ldots \lambda^{(r)}_{n_r,i_r}|^2\right)\right]^{1/2} E\left[\left(\sum |\sigma(\omega)(\varepsilon_{n_1,i_1}, \ldots, \varepsilon_{n_r,i_r})|^2\right)\right]^{1/2}$$

The first expectation on the right hand side of (1.17) is finite and, from Lemma (1.13), the second is less than

$$\left(\sum \frac{C}{(|n_1|+1)^2 \ldots (|n_r|+1)^2}\right)^{1/2}.$$

Thus the right hand side of (1.17) is finite and this proves that the series in (1.8) is almost always absolutely convergent. □

We now use a truncation procedure to show that, in general, we can reduce the proof of (1.8) to the case covered by Lemma (1.15). Since each of the $b_j$ occuring in (1.7) is a measurable section of the field of Hilbert spaces $\mathcal{H}_w$, it follows that

$$\sum_{n,i} |\lambda^{(j)}_{n,i}(w)|^2 < \infty$$

for almost all paths $w$. Define

$$\bar{\lambda}^{(j)}_{n,i}(w) = \begin{cases} \lambda^{(j)}_{n,i}(w) & \text{if } \sum |\lambda^{(j)}_{n,i}(w)|^2 \leq K \\ 0, & \text{otherwise} \end{cases}$$

so that

$$\sum_{n,i} |\bar{\lambda}^{(j)}_{n,i}|^2 \in L^p.$$

Now let $K \to \infty$ to complete the proof of (1.8).

We have now given a precise meaning to the formulas (1.5) and (1.6) and therefore we have defined

$$\sigma(\omega)_w : \mathcal{H}_w \times \cdots \times \mathcal{H}_w \to \mathbb{R}.$$

It is not difficult to get the estimates needed to prove that this function $\sigma(\omega)_w$ is continuous. In §3 we carry out several such estimates; for example (3.11). It is clear that $\sigma(\omega)_w$ is alternating and so we see that $\sigma(\omega)$ is indeed a measurable form on $PM$.

We now explain how to "restrict" $\sigma(\omega)$ to the loop space $LM$ to get a measurable form on $LM$. The point to take care of is that

$$\sigma(\omega)_w : \mathcal{H}_w \times \cdots \times \mathcal{H}_w \to \mathbb{R}$$

is only defined for almost all $w \in P(M)$ and the loop space has measure zero in $PM$. So it could happen that $\sigma(\omega)_w$ it is never defined for $w \in LM$. To see that this is not the case we simply repeat the procedure used to define $\sigma(\omega)$. So we first note that the stochastic integrals

$$\sigma(\omega)(\varepsilon_{n_1,i_1}, \ldots, \varepsilon_{n_r,i_r})_w$$

are defined for almost all loops $w$. Then we express the general integral

$$\sigma(\omega)(b_1, \ldots, b_r)_w$$

in terms of the previous integrals as in (1.7) and (1.8), where the $\lambda_{n,i}$ are now functions on $LM$. Finally we argue, as above, computing expectations using the measure on $LM$, that the series (1.8), where the $\lambda_{n,i}$ are functions on $LM$, converges for almost all $w \in LM$. This shows that

$$\sigma(\omega)_w : \mathcal{H}_w \times \cdots \times \mathcal{H}_w \to \mathbb{R}$$

is defined for almost all loops $w \in LM$. Finally we restrict to the subspaces $\mathcal{T}_w$

$$\sigma(\omega)_w : \mathcal{T}_w \times \cdots \times \mathcal{T}_w \to \mathbb{R}$$

to get a measurable form on $LM$. An exactly analogous argument shows how to restrict $\sigma(\omega)$ to the based loop space to get a measurable form on $L_x M$.

A Chen form on $LM$ is defined on $\mathcal{T}_w$ for almost all loops $w$. However, from the definition, it has a canonical extension to $\mathcal{H}_w$, for almost all loops $w$. For the purposes of calculation it is often convenient to use this extension and we sometimes use this trick, and its analogue for $L_x M$, without making it explicit.

We can construct a larger class of forms on $PM$ and $LM$ using the $\sigma(\omega)$ as follows. Let $\theta$ be an $r$-form on $M$ and for $t \in [0,1]$ define a measurable form $\theta_t$ as follows

$$\theta_t : \mathcal{H}_w \times \cdots \times \mathcal{H}_w \to \mathbb{R}, \qquad \theta_t(b_1, \ldots, b_r) = \theta(b_1(t), \ldots, b_r(t)).$$

Now let $\omega$ be as above and suppose we are given forms $\theta$ and $\varphi$ on $M$, then we define

$$(1.18) \qquad \sigma(\theta; \omega; \varphi) = \theta_0 \wedge \sigma(\omega) \wedge \varphi_1.$$

This is a measurable form on $PM$. On the loop space $LM$ we need a form $\theta$ on $M$ and then we define

$$(1.19) \qquad \sigma(\theta; \omega) = \theta_0 \wedge \sigma(\omega).$$

**Definition 1.20.** The spaces of **Chen forms** $C(PM)$, $C(LM)$, and $C(L_xM)$ on $PM$, $LM$, and $L_xM$ are defined to be the linear span of the forms

$$\sigma(\theta; \omega; \varphi), \qquad \sigma(\theta; \omega), \qquad \sigma(\omega)$$

respectively. We write $C^r$ for the space of Chen forms of degree $r$.

Notice how the relation between the spaces of Chen forms mimics, at the level of differential forms, the inclusions

$$L_xM \subset LM \subset PM.$$

One reason for the importance of Chen forms is that the cohomology of $PM$, $LM$, and $L_xM$, and also the equivariant cohomology of $LM$ can be computed, algebraically, from the spaces $C(PM)$, $C(LM)$, and $C(L_xM)$, see [17]. Indeed in [17, §3] the relation between these spaces of Chen forms and various constructions from homological algebra, in particular the bar complex, is described in detail.

Recall the multiplicative formula for Chen forms [17, §4]. Given $\psi = \psi_1 \otimes \ldots \otimes \psi_n$ and $\omega = \omega_1 \otimes \ldots \otimes \omega_m$ then the shuffle product $S(\psi, \omega)$ is defined to be

$$(1.21) \qquad S(\psi, \omega) = \sum S_\pi(\psi_1 \otimes \ldots \otimes \psi_n \otimes \omega_1 \otimes \ldots \otimes \omega_m).$$

Here the sum is taken over all $(n, m)$ shuffles of $\{1, \ldots, n + m\}$; $S_\pi$ is, up to a sign, the permutation $\pi$ of the components. The sign is determined by the following convention: if we interchange $\alpha$ and $\beta$ then we introduce the sign $(-1)^{(\deg \alpha - 1)(\deg \beta - 1)}$, the exponent involves $\deg \alpha - 1$ to take account of the interior product in the definition of Chen forms. It follows that

$$(1.22) \qquad \sigma(S(\psi, \omega)) = \sigma(\psi)\sigma(\omega);$$

the proof is identical to that given in [17] since we are using Stratonovitch integrals.

Note that if $\omega = \omega_1 \otimes \ldots \otimes \omega_n$ where $\deg \omega_i = 1$ and $\theta$, $\varphi$ are functions on $M$ then $\sigma(\theta; \omega; \varphi)$ is a function on $PM$; indeed $C^0(PM)$ is the linear span of elements of this form. Similarly we can look at the spaces of functions $C^0(LM)$ and $C^0(L_xM)$.

**Theorem 1.23.** The spaces $C^0(PM)$, $C^0(LM)$, and $C^0(L_xM)$ are dense in all the spaces of $L^p$ functions on $PM$, $LM$, and $L_xM$ respectively.

*Proof.* We give the proof for $PM$. The proofs for $LM$ and $L_xM$ are almost identical. By Itô's formula in the Stratonovitch context we see that if $g$ is a smooth function on $M$ then

$$g(w_t) = g(w_0) + \int_0^t \langle dg(w_s), dw_s \rangle$$
$$= g(w_0) + \sigma(\alpha_s)$$

with $\alpha_s = dg$ if $s < t$ and $\alpha_s = 0$ if $s > t$. Thus, if $0 < t < 1$, the function $w \mapsto g(w_t)$ is a linear combination of Chen forms. Now (1.22) shows that all finite products $g_1(w_{t_1}) \ldots g_k(w_{t_k})$, where $g_i$ is smooth and $0 < t_i < 1$, can be expressed as finite linear combinations of Chen forms. Therefore, for any smooth functions $g_0, \ldots, g_{k+1}$ on $M$,

$$g_0(w_0)g_1(w_{t_1}) \ldots g_k(w_{t_k})g_{k+1}(w_1) \in C^0.$$

and since the space spanned by these products is dense in all the $L^p$ spaces this proves the lemma.  □

We finish this section by showing that a Chen form is completely determined by its restriction to the space of paths, or loops, with finite energy. We use the notation $\sigma^{(1,2)}(\omega)$ for the restriction to the space of paths or loops with finite energy.

**Theorem 1.24.**
$$\sigma^{(1,2)}(\omega) = 0 \iff \sigma(\omega) = 0$$

*Proof.* We write out $\sigma(\omega)$ in terms of the basis $\beta_I$ of forms defined in (1.4)

$$\sigma(\omega) = \sum \sigma_I(\omega)\beta_I$$

where the $\sigma_I(\omega)$ are functions. From (1.6)

$$\sigma_I(\omega)_\varphi = \sigma(\omega)(\varepsilon_I)_\varphi$$

is defined for all paths $\varphi$ of finite energy. We will show that $\sigma_I(\omega)_\varphi = 0$ for all paths $\varphi$ with finite energy if and only if $\sigma_I(\omega)_w = 0$ for almost all $w$. To prove this we consider the measure $\mu$ on $M$ defined by

$$\mu(f) = E^x \left[ \sigma_I(\omega)_w^2 f(w_1) \right]$$

where the expectation $E^x$ is taken over all paths starting at $x$. This measure has a density

$$q(y) = E^{x,y} \left[ \sigma_I(\omega)_w^2 \right] p_1(x,y)$$

and since $p_1(x,y) > 0$ the result for the path space follows from [21, Theorem 8.1] and for the loop space and the based loop space it follows from [5, Theorem II.1].  □

## §2 $L^{H,p,\alpha}$-CHEN FORMS

Recall that if $\mathcal{H}$ is a Hilbert space then the natural inner product on the exterior power $\Lambda^r \mathcal{H}$ is given by the formula

$$\langle x_1 \wedge \cdots \wedge x_r, y_1 \wedge \cdots \wedge y_r \rangle = \det(\langle x_i, y_j \rangle).$$

This makes $\Lambda^r(\mathcal{H})$ into a Hilbert space and we get a Hilbert space norm $\| - \|_H$ on $\Lambda^r \mathcal{H}$. In our context this inner product defines a pointwise norm $\|\sigma\|_{H,w}$ on measurable forms $\sigma$ on $PM$, $LM$, and $L_x M$. We want to allow infinite sums of homogeneous forms with some prescribed decay conditions and to do this we weight the pointwise norm $\|\sigma\|_{H,w}$ by a function $\alpha(r,p)$ where $r = \deg \sigma$. The properties we require of this weighting function are:

(2.1) $$\alpha(r,p) \leq \alpha(r,q), \qquad \text{if } p < q,$$

(2.2) $$r\alpha(r-1,p)^2 \leq C_1(p)\alpha(r,2p)^2,$$

(2.3) $$r!\,\alpha(r,p) \leq C_2(p)^r,$$

(2.4) $$2^{r+s}\alpha(r+s,p)^2 \leq C_3(p)\alpha(r,2p)^2\alpha(s,2p)^2.$$

Examples of suitable weighting functions are given by

$$\alpha(r,p) = \frac{2^{rp}}{(r-ap)!}$$

where $a$ is an integer. Now define the **random $(H,p,\alpha)$-norm** of the homogeneous measurable $r$-form $\sigma$ by

(2.5) $$\|\sigma\|_{H,p,\alpha,w} = \alpha(r,p)\|\sigma\|_{H,w}.$$

If $\sigma$ is not homogeneous then we define the inner product and the norm by making the different homogeneous components orthogonal. Thus the random $(H,p,\alpha)$-norm of an inhomogeneous form $\sigma$ is defined by

(2.6) $$\|\sigma\|_{H,p,\alpha,w}^2 = \sum_r \alpha(r,p)^2\|\sigma_r\|_{H,w}^2.$$

where $\sigma_r$ is the homogeneous component of $\sigma$ with degree $r$. The $L^{H,p,\alpha}$-**norm** of $\sigma$ is defined by

(2.7) $$\|\sigma\|_{L^{H,p,\alpha}} = \left( E\left[ \|\sigma\|_{H,p,\alpha}^p \right] \right)^{1/p}.$$

This expectation $E$ is taken over $PM$, $LM$, or $L_x M$, according to the case, using the measures described at the beginning of §1. There are infinite sums of homogeneous terms with finite $L^{H,p,\alpha}$-norm. We are particularly interested in such infinite sums so from now on we will assume that a measurable form $\sigma$ on $PM$, $LM$, or $L_x M$ is not necessarily homogeneous unless explicitly stated otherwise.

The $H$ in the notation for these norms is there to make it clear that we are dealing with the random $(p,\alpha)$-norm and the $L^{p,\alpha}$-norm for measurable forms based on the Hilbert norm $\| - \|_H$. This will be implicitly assumed for the rest of this section so it will cause no confusion if we drop the $H$ from the notation. So if $\sigma$ is a measurable form, then, for the rest of this section, $\|\sigma\|$ will mean $\|\sigma\|_H$, $\|\sigma\|_{p,\alpha}$ will mean $\|\sigma\|_{H,p,\alpha}$, and $\|\sigma\|_{L^{p,\alpha}}$ will mean $\|\sigma\|_{L^{H,p,\alpha}}$.

If $p < q$, it follows from (2.1) that

$$\|\sigma\|_{p,\alpha} \leq \|\sigma\|_{q,\alpha}, \qquad \|\sigma\|_{L^{p,\alpha}} \leq \|\sigma\|_{L^{q,\alpha}}.$$

Next we estimate the $L^{p,\alpha}$-norm of a wedge product.

**Lemma 2.8.** *Let $\sigma$, $\tau$ be measurable forms on $PM$, $LM$, or $L_x M$, then:*

$$\|\sigma \wedge \tau\|_w \leq \|\sigma\|_w \|\tau\|_w$$
$$\|\sigma \wedge \tau\|_{L^{p,\alpha}} \leq C_3(p)\|\sigma\|_{L^{2p,\alpha}} \|\tau\|_{L^{2p,\alpha}}.$$

*Proof.* We give the proof for $PM$, the other cases are almost identical. Pick a countable orthonormal basis $b_i$ for $\mathcal{H}_w$ and let $\beta_I$ be the corresponding orthonormal basis for alternating multilinear functions on $\mathcal{H}_w$. With respect to this basis we write

$$\sigma_w = \sum_I \sigma_{I,w} \beta_I, \qquad \tau_w = \sum_J \tau_{J,w} \beta_J$$

so that

$$\|\sigma \wedge \tau\|_w = \left\| \sum \sigma_{I,w} \tau_{J,w} \beta_I \wedge \beta_J \right\|$$
$$\leq \sum |\sigma_{I,w}| \, |\tau_{J,w}| = \|\sigma\|_w \|\tau\|_w.$$

This proves the first inequality.

To prove the second inequality we start from

$$(\sigma \wedge \tau)_w = \sum \sigma_{I,w} \tau_{J,w} \beta_I \wedge \beta_J$$

and express the right hand side of this equation in terms of its homogeneous components. Let $K = I \cup J$ and then we get

$$(\sigma \wedge \tau)_w = \sum_K \sum_{I \cap J = \varnothing} \sigma_{I,w} \tau_{J,w} \beta_K.$$

Now we estimate the random $(p, \alpha)$-norms:

$$\|\sigma \wedge \tau\|_{p,\alpha,w}^2 = \left\| \sum \sigma_{I,w} \tau_{J,w} \beta_I \wedge \beta_J \right\|_{p,\alpha,w}^2$$

$$= \sum_K \alpha(|K|, p)^2 \left| \sum_{I \cap J = \varnothing} \sigma_{I,w} \tau_{J,w} \right|^2$$

$$\leq \sum_K \alpha(|K|, p)^2 \, 2^{|K|} \sum_{I \cap J = \varnothing} \sigma_{I,w}^2 \tau_{I,w}^2.$$

To complete the proof of the second inequality we use (2.4) and the Hölder inequality. $\square$

Now we estimate the $L^{p,\alpha}$-norm of $i_b(\omega)$.

**Lemma 2.9.** *Let $b$ be a measurable vector field and $\sigma$ a measurable form on $PM$, $LM$, or $L_xM$; then*

$$\|i_b(\sigma)\|_{L^{p,\alpha}} \leq C_1(p) \|b\|_{L^{2p}} \|\sigma\|_{L^{2p,\alpha}}.$$

*Proof.* Once more we only give the proof for $PM$. We work with the basis $\varepsilon_{n,i}$ for measurable vector fields on $PM$ given in (1.3) and the basis $\beta_I$ for measurable forms given by (1.4). In terms of these bases

$$b = \sum \lambda_{n,i} \varepsilon_{n,i}, \qquad \sigma = \sum \sigma(I) \beta_I.$$

Then, since

$$i_{\varepsilon_{n,i}}(\beta_J) = \pm \beta_{J \setminus (n,i)}, \qquad \text{if } (n,i) \in J$$
$$= 0, \qquad \text{if } (n,i) \notin J$$

it follows that the coefficient of $\beta_I$ in the formal expression for $i_b(\sigma)$ is

(2.10) $$\mu_I = \sum_{(n,i) \notin I} \pm \lambda_{n,i} \sigma(I_{n,i})$$

where $I_{n,i} = I \cup \{(n,i)\}$. We must prove that the sum in (2.10) converges almost everywhere:

$$E\left[|\mu_I|^p\right]^{1/p} \leq E\left[\left(\sum_{(n,i) \notin I} |\lambda_{n,i}| \, |\sigma(I_{n,i})|\right)^p\right]^{1/p}$$

$$\leq E\left[\left(\sum_{(n,i) \notin I} |\lambda_{n,i}|^2\right)^{p/2} \left(\sum_{(n,i) \notin I} |\sigma(I_{n,i})|^2\right)^{p/2}\right]^{1/p}$$

$$\leq E\left[\left(\sum_{(n,i) \notin I} |\lambda_{n,i}|^2\right)^p\right]^{1/2p} E\left[\left(\sum_{(n,i) \notin I} |\sigma(I_{n,i})|^2\right)^p\right]^{1/2p}.$$

If $\sigma$ has finite $L^{2p,\alpha}$ norm and $b$ has finite $L^{2p}$ norm then the right hand side of this inequality is finite and so (2.10) converges almost everywhere.

We now compute the $L^{p,\alpha}$ norm of $i_b(\sigma)$. Since

$$|\mu_I|^2 \leq \left( \sum_{(n,i) \notin I} |\lambda_{n,i}|^2 \right) \left( \sum_{(n,i) \notin I} |\sigma(I_{n,i})|^2 \right)$$

it follows that

$$\|i_b(\sigma)\|_{p,\alpha}^2 \leq \left( \sum_I \alpha(|I|,p)^2 \sum_{(n,i) \notin I} \sigma(I_{n,i})^2 \right) \left( \sum_{n,i} \lambda_{n,i}^2 \right)$$

Now we replace $I_{n,i}$ by $J$ to get

$$\|i_b(\sigma)\|_{p,\alpha}^2 \leq \left( \sum_J \alpha(|J|-1,p)^2 \sum_{(n,i) \in J} |\sigma(J)|^2 \right) \left( \sum_{n,i} \lambda_{n,i}^2 \right)$$

$$\leq \left( \sum_J \alpha(|J|-1,p)^2 \, |J| \, |\sigma(J)|^2 \right) \left( \sum_{n,i} \lambda_{n,i}^2 \right)$$

Now using (2.2) we conclude that

$$\|i_b(\sigma)\|_{p,\alpha}^2 \leq C_1(p)\|b\|_2^2 \|\sigma\|_{2p,\alpha}.$$

Finally the Cauchy-Schwartz inequality shows that

$$\|i_b(\sigma)\|_{L^{p,\alpha}} \leq C_1 \|b\|_{L^{2p}} \|\sigma\|_{L^{2p,\alpha}}. \quad \square$$

Now suppose that $\sigma(\omega)$ is a homogeneous Chen form of degree $r$ as in (1.6) and write out $\sigma(\omega)$, in terms of the basis $\beta_I$ for forms given by (1.5), as

$$\sigma(\omega) = \sum \sigma_I(\omega)\beta_I.$$

We show that $\sigma(\omega)$ has finite $L^{p,\alpha}$-norm for all $p$ by an argument very similar to the proof that (1.9) is almost always absolutely convergent. In fact the Hölder inequality for $p \geq 2$ shows that

$$(2.12) \qquad E\left[ \left( \sum |\sigma_I(\omega)|^2 \right)^{p/2} \right] \leq \left( \sum E[\, |\sigma_I(\omega)|^p \,]^{2/p} \right)^{1/2}.$$

Then, as in the previous section, if $I = \{(n_1, i_1), \ldots, (n_r, i_r)\}$ we get the following inequality

$$E\left[\sigma_I(\omega)^p\right]^{1/p} \leq \frac{C}{(|n_1| + 1) \ldots (|n_r| + 1)}$$

and so the right hand side of (2.12) is less than

$$\left(\sum \frac{C^2}{(|n_1| + 1)^2 \ldots (|n_r| + 1)^2}\right)^{1/2}$$

and, since this term is finite, this shows that $\sigma(\omega)$ has finite $L^{p,\alpha}$-norm. Now if $\sigma(\theta; \omega; \varphi)$ is a general Chen form as in (1.18) a simple argument using Lemma (2.8) shows that $\sigma(\theta; \omega; \varphi)$ has finite $L^{p,\alpha}$-norm.

**Definition 2.13.** The spaces of $L^{p,\alpha}$-**Chen forms** $\mathcal{C}_{p,\alpha}(PM)$, $\mathcal{C}_{p,\alpha}(LM)$, and $\mathcal{C}_{p,\alpha}(L_x M)$ on $PM$, $LM$, and $L_x M$ are defined to be the the closure of the spaces $\mathcal{C}(PM)$, $\mathcal{C}(LM)$, and $\mathcal{C}(L_x M)$ in the $L^{p,\alpha}$-norm.

Note that in forming $\mathcal{C}_{p,\alpha}$ we complete the whole space $\mathcal{C} = \bigoplus \mathcal{C}^r$, not just its homogeneous subspaces $\mathcal{C}^r$, so we allow certain infinite sums of homogeneous elements. In terms of homological algebra, compare [17], the spaces $\mathcal{C}_{p,\alpha}$ are completions of the various Chen normalised bar complexes.

From now on we write $\mathcal{C}_{p,\alpha}$ for any of the spaces of $L^{p,\alpha}$-Chen forms and only specify which of the spaces $PM$, $LM$, or $L_x M$ we are working on if it is absolutely necessary. In particular any statement about $\mathcal{C}_{p,\alpha}$ is valid for $PM$, $LM$, and $L_x M$ unless specified otherwise.

It follows that

$$\mathcal{C}_{q,\alpha} \subset \mathcal{C}_{p,\alpha}, \qquad \text{if } q > p$$

and we define

$$\mathcal{C}_{\infty,\alpha} = \bigcap_p \mathcal{C}_{p,\alpha}.$$

The following lemma follows from a standard diagonal argument.

**Lemma 2.14.** If $\sigma$ is in $\mathcal{C}_{\infty,\alpha}$ there is a sequence $\sigma_n \in \mathcal{C}$ such that $\sigma_n$ converges to $\sigma$ in each of the $L^{p,\alpha}$-norms. $\square$

Now we examine the wedge product.

**Theorem 2.15.** If $\sigma, \tau \in \mathcal{C}_{\infty,\alpha}$ then the exterior product $\sigma \wedge \tau$ is also in $\mathcal{C}_{\infty,\alpha}$.

*Proof.* Using (2.14) pick sequences $\sigma_n$ and $\tau_n$ in $\mathcal{C}$ which converge to $\sigma$ and $\tau$, respectively, in all the $L^{p,\alpha}$-norms. Then $\sigma_n \wedge \tau_n \in \mathcal{C}$ and using Lemma (2.8) it follows that $\sigma_n \wedge \tau_n$ converges in all the $L^{p,\alpha}$-norms. This proves the theorem. $\square$

**Theorem 2.16.** *Suppose that $\sigma$ is in $\mathcal{C}_{\infty,\alpha}$ and that $b$ is a vector field which has finite $L^p$-norm for all $p$; then*

$$i_b(\sigma) \in \mathcal{C}_{\infty,\alpha}.$$

*Proof.* Since $\sigma$ is in all the $L^{p,\alpha}$-spaces and $b$ has finite $L^p$-norm for all $p$ (2.9) shows that $i_b(\sigma)$ is in all the $L^{p,\alpha}$-spaces. It only remains to prove that that if $\sigma$ is a Chen form then so is $i_b(\sigma)$. As in the proof of (2.8) we write

$$\sigma = \sum \sigma(I)\beta_I, \qquad b = \sum \lambda_{n,i}\varepsilon_{n,i}.$$

Since each $\lambda_{n,i}$ is measurable, Lemma (1.23) shows that it is enough to prove that if $\sigma$ is a homogeneous Chen form then each $i_{\varepsilon_{n,i}}(\sigma)$ is in $\mathcal{C}_{H,\infty,\alpha}$. We prove this for the forms $\sigma(\omega)$ defined in (1.5). The more general case of forms $\sigma(\theta;\omega;\varphi)$ or $\sigma(\theta;\omega)$, defined in (1.18) and (1.19), follows using the fact that $i_b$ is a derivation and Theorem (2.15).

By using a partition of unity we can reduce to the case where there is an orthonormal basis $e_1(k)(x), \ldots, e_d(k)(x)$ for $T_x$ which is smooth on the support of $\omega_k$ and we choose such bases. If $X$ is a vector field defined on the support of $\omega_k$ then we write $X^j(k)$, where $1 \leq j \leq d$, for the components of $X$ defined using the basis $e_j(k)$. Define $e_{n,i}(k)$ and $\varepsilon_{n,i}(k)$ as in (1.3); then it is enough to prove that the form

$$\int_{\Delta^m} \omega_1(dw, -)_{s_1} \wedge \cdots \wedge \omega_j(dw, \varepsilon_{n,i}^j(k), -)_{s_j} \wedge \cdots \wedge \omega_m(dw, -)_{s_m}$$

is in $\mathcal{C}_{\infty,\alpha}$ for all $j$, $k$. We use the approximation of a loop by a piecewise geodesic path and the corresponding approximation for parallel transport [8]. Then, compare [8], the above form becomes the limit, as $l \to \infty$, of a sequence of forms

$$(2.17) \quad \int_{\Delta^m} \omega_1(dw, -)_{s_1} \wedge \cdots \wedge \omega_j(dw, \varphi_l e_{n,i}^j(k), -)_{s_j} \wedge \cdots \wedge \omega_m(dw, -)_{s_m}$$

where the function $\varphi_l(s, w)$ is of the following type: If $s_j \in [l/N, (l+1)/N]$, then

$$\varphi_l(s_j, w) = \psi_{l,k}(w_{1/N}, w_{2/N}, \ldots, w_{(l-1)/N}, w_{s_j})$$

where

$$\psi_{l,k} : M^l \to \mathbb{R}$$

is a suitable smooth function. By the Stone-Weierstrass theorem we can assume that each $\psi_{l,k}$ is a sum of products of functions on $M$. The form (2.17) now becomes a sum of forms of the type

$$\int_{\Delta^m} \omega_1(dw, -)_{s_1} \wedge \cdots \wedge \tilde{\omega}_j(dw, -)_{s_j} \cdots \wedge \cdots \wedge \omega(dw, -)_{s_m}$$

and is thus a Chen form.  $\square$

Now we come to our main results, motivated by [12], [18], and [26]. We will state the theorem for the loop space $LM$, but there are obvious analogues for the path space and the based loop space.

**Theorem 2.18.** *Let $\theta_n$, and $\omega_{i,n}$, where $1 \leq i \leq n$, be differential forms on $M$ and define $\omega_n = \omega_{1,n} \otimes \cdots \otimes \omega_{n,n}$.*

(1) *Suppose that the power series*

$$\sum_{n=1}^{\infty} \left( \|\theta_n\|_{k,\infty} \prod_{i=1}^{n} \|\omega_{i,n}\|_{k,\infty} \frac{z^n}{\sqrt{n!}} \right)$$

*has infinite radius of convergence for all $k \leq 2d + 3$ where $d$ is the dimension of $M$; then*

$$\sum_{n=1}^{\infty} \sigma(\theta_n; \omega_n) \in \mathcal{C}_{\infty,\alpha}(LM).$$

(2) *Suppose that for all $u$ the diffusion $\tilde{w}_s(u)$ is hypoelliptic and that the power series*

$$\sum_{n=1}^{\infty} \left( \|\theta_n\|_{1,\infty} \prod_{i=1}^{n} \|\omega_{i,n}\|_{1,\infty} \frac{z^n}{\sqrt{n!}} \right)$$

*has infinite radius of convergence; then*

$$\sum_{n=1}^{\infty} \sigma(\theta_n; \omega_n) \in \mathcal{C}_{\infty,\alpha}(LM).$$

*Remarks.*

(1) We believe that part (2) is true without the hypoelliptic assumption but we have not yet been able to prove it.
(2) There is an exactly analogous theorem giving a condition which ensures that

$$\sum_{n=1}^{\infty} \sigma(\theta_n; \omega_n; \varphi_n) \in \mathcal{C}_{\infty,\alpha}(PM).$$

The condition is given by multiplying the coefficient of $z^n$ in the above power series by $\|\varphi_n\|_{k,\infty}$ in part (1) and by $\|\varphi_n\|_{1,\infty}$ in part (2). In this case however, part (2) is always valid since, as

we remarked after the statement of Lemma (1.10), the basic estimate (1.12) is valid without any hypoellipticity hypothesis. The
analogous condition for the based loop space is given by removing
the term $\theta_n$. The proofs are obvious modifications of the proof we
give below.

(3) The importance of this theorem is that it makes a link between
Connes's entire cyclic theory and the spaces $\mathcal{C}_{\infty,\alpha}$. More precisely,
in terms of homological algebra, it shows the iterated integral map
$\sigma$, described in [17], maps Connes's entire cyclic complex, see [13],
for the algebra $\Omega$ into the space of $L^{\infty,\alpha}$-Chen forms on $LM$.

*Proof of 2.18.* We must prove that

$$\sum_n \|\sigma(\theta_n; \omega_n)\|_{L^{p,\alpha}} < \infty$$

for all $p$. It is enough to deal with the case where $\theta_n = 1$ for all $n$ and each
of the forms $\omega_{i,n}$ is homogeneous; the general case follows easily from this
case and Lemma (2.8). Set $r_{i,n} = \deg \omega_{i,n} - 1$ and $r_n = r_{1,n} + \cdots + r_{n,n}$.
In terms of the orthonormal basis $\beta_I$ for forms defined in (1.4)

$$\sigma(\omega_n) = \sum \sigma_I(\omega_n)\beta_I$$

and we compute $L^{p,\alpha}$-norms using the expectation $E_L$ on the loop space
described at the beginning of §1:

$$\|\sigma(\omega_n)\|_{L^{p,\alpha}} = \alpha(r_n, p) E_L \left[ \|\sigma(\omega_n)\|^p \right]^{1/p}$$

$$(2.19) \qquad = \alpha(r_n, p) E_L \left[ \left( \sum |\sigma_I(\omega_n)|^2 \right)^{p/2} \right]^{1/p}$$

$$\leq \alpha(r_n, p) \left( \sum E_L \left[ |\sigma_I(\omega_n)|^p \right]^{2/p} \right)^{1/2}.$$

Using (1.6) we can express $\sigma_I(\omega_n)$ as a sum of at most $r_n!$ terms of the form

$$A(f; \omega) = \int_{\Delta^n} \omega_1 \left( dw, \tau(f_{I_1}) \right)_{s_1} \cdots \omega_n \left( dw, \tau(f_{I_n}) \right)_{s_n}$$

where $f_1, \ldots, f_{r_n}$ are taken from the basis $e_{n,i}$ for $\mathcal{H}(T_{w_0})$ defined in (1.3).
Now we can use part (1) of Lemma (1.10) to estimate the expectation of
each of these integrals. We get

$$E^{x,x} \left[ |\sigma_I(\omega_n)|^p \right]^{1/p} \leq \frac{r_n! \, C(p)^n}{\sqrt{n!} \, (|m_1| + 1) \cdots (|m_{r_n}| + 1)} \|\omega_n\|_{k,\infty}$$

where

$$I = \{(m_1, i_1), \ldots, (m_{r_n}, i_{r_n})\}$$
$$\|\omega_n\|_{k,\infty} = \|\omega_{1,n}\|_{k,\infty} \cdots \|\omega_{n,n}\|_{k,\infty}$$
$$k \geq 2d + 3.$$

From this it follows that

$$E_L \left[|\sigma_I(\omega_n)|^p\right]^{1/p} = \left(\int_M p_1(x,x) E^{x,x} \left[|\sigma_I(\omega_n)|^p\right] dx\right)^{1/p}$$
$$\leq \frac{r_n! \, C(p)^n}{\sqrt{n!}(|m_1|+1)\ldots(|m_{r_n}|+1)} \, \|\omega_n\|_{k,\infty}$$

and from (2.19) it follows that

(2.20) $\quad \|\sigma(\omega_n)\|_{L^{p,\alpha}} \leq$

$$\frac{\alpha(r_n,p) r_n! \, C(p)^n}{\sqrt{n!}} \left(\sum \frac{1}{(|m_1|+1)^2 \ldots (|m_{r(n)}|+1)^2}\right)^{1/2} \|\omega_n\|_{k,\infty}.$$

The sum which appears in (2.20) is smaller than

$$2^{r_n} \left(\sum_{i>0} \frac{1}{i^2}\right)^{r_n} \leq 2^{nd} K^{nd}$$

since $r_n \leq nd$. Thus, using (2.3) to estimate $r_n! \, \alpha(r_n, p)$, we get

(2.21) $$\qquad \|\sigma(\omega_n)\|_{L^{p,\alpha}} \leq \frac{K(p)^n}{\sqrt{n!}} \|\omega_n\|_{k,\infty}$$

and the first part of the theorem follows directly from (2.21). Part (2) of the theorem is proved in exactly the same way but using part (2) of Lemma (1.11) in place of part (1). □

We can get the same kind of theorem if the $\omega_{i,n}$ are matrices of forms on $M$. The statement is formally identical to Theorem (2.18) so we shall not repeat it in detail. This is particularly important for the study of the equivariant Chern character, [7] and [17]. We briefly recall the construction of the equivariant Chern character as a differential form on the smooth loop space $L^\infty M$ using Chen forms, for more details see [17].

Let $E$ be a Hermitian vector bundle over $M$ equipped with a Hermitian connection $\nabla_E$. Given an embedding of $E$ in the trivial bundle $M \times \mathbb{C}^N$ let $p(x)$ be the orthogonal projection from $\mathbb{C}^N$ to $E_x$ and let

$$\nabla_p = p \circ d \circ p + p^\perp \circ d \circ p^\perp$$

We choose the embedding of $E$ in $M \times \mathbb{C}^N$ so that the connection $\nabla_p$ preserves the splitting of $M \times \mathbb{C}^N$ as $E \oplus E^\perp$ and agrees with the given connection on $E$; this is possible by [32]. A simple calculation shows that the connection form of $\nabla_p$ is

$$A = p(dp) + p^\perp(dp^\perp),$$

that is $\nabla_p = d + A$, and the curvature of $\nabla_p$ is

$$F = dp \wedge dp.$$

Define a differential form $\mathcal{A}$ on $M \times S^1$ by

$$\mathcal{A} = A + F \wedge dt.$$

Let $\omega + \xi \wedge dt$ be a form on $M \times S^1$ where $\omega$ and $\xi$ are forms on $M$. Using such forms we can define a larger class of Chen forms as follows. Let $\alpha_i = \omega_i + \xi_i \wedge dt$, $1 \le i \le n$, be forms on $M \times S^1$ and let $\alpha = \alpha_1 \otimes \cdots \otimes \alpha_n$; then define

$$\tilde{\sigma}(\alpha) = \int_{\Delta^n} (\omega_1(dw,-)_{t_1} + \xi_1(-)_{w_{t_1}} dt_1) \wedge \cdots \wedge (\omega_n(dw,-)_{t_n} + \xi_n(-)_{w_{t_n}} dt_n).$$

This is a form on $L^\infty M$. Note that even if $\alpha$ is homogeneous this form on $L^\infty(M)$ is not homogeneous. Thus we can construct the matrix valued form, of infinite degree,

$$\sum_k p\,\tilde{\sigma}(\mathcal{A}^{\otimes k}).$$

It is proved in [17] that the form

$$\mathrm{Ch}(E, \nabla_E) = \sum \mathrm{trace}\left(p\,\tilde{\sigma}(\mathcal{A}^{\otimes k})\right)$$

on $L^\infty(M)$ is Bismut's equivariant Chern character of $(E, \nabla_E)$. It is not difficult to adapt the previous theory to show that $\mathrm{Ch}(E, \nabla_E)$ extends to a measurable form on $LM$ and to use the matrix analogue of (2.18) to show that $\mathrm{Ch}(E, \nabla_E)$ lies in $\mathcal{C}_{\infty,p}$. The important property of the $L^{\infty,p}$-norm which is needed here is that the Picard series of the solution of the equation 3.16 in [7] is absolutely convergent. See [9] and [35] for similar problems when the components are scalars rather than forms.

If, in the above notation, we define

$$\mathcal{F} = A(dw_t) + F_{w_t}(dt),$$

which is a sum of a 0-form and a 2-form on $L^\infty M$ with values in $M_N(\mathbb{C})$, then $\mathrm{Ch}(E, \nabla_E)$ is the time ordered exponential of $\mathcal{F}$. We conclude this section by making a few remarks on the theory of time ordered exponentials of forms.

The time ordered exponential of the form $\omega$ on $LM$ is the solution at time 1 of the stochastic differential equation, in the Stratonovitch sense,

$$(2.22) \qquad dH_t = H_t \wedge \omega(dw, -)_t, \qquad H_0 = 1.$$

We explain, briefly, why this equation makes sense. In terms of a basis $\beta_I$ for measurable forms we write

$$\int_0^t \omega(dw_s, -) = \sum_I \sigma_t(\omega, I)\beta_I, \qquad H_t = \sum H_t(K)\beta_K;$$

then

$$dH_t = \sum dH_t(K) \wedge \beta_K$$

$$H_t \wedge \omega(dw_t, -) = \sum_K \left( \sum_{(I,J)} H_t(I)d\sigma_t(\omega, J) \right) \beta_K$$

where the sum over $(I, J)$ is taken over those multi-indices $(I, J)$ with

$$I \cup J = K, \qquad I \cap J = \varnothing.$$

The components $H_t(K)$, for $t \leq 1$, are solutions of the system of linear equations

$$(2.23) \qquad \begin{aligned} dH_t(K) &= \sum_{(I,J)} H_t(I)d\sigma_t(\omega, J) \\ H_0(\varnothing) &= 1 \\ H_0(J) &= 0, \qquad \text{if } J \neq \varnothing. \end{aligned}$$

This system can be solved inductively on $|I|$.

Let $X_t$ and $Y_t$ be two continuous scalar processes. Then the solution of the equation, in the Stratonovitch sense,

$$dU_t = U_t dX_t + dY_t$$

is given by

$$\varphi_t \int_0^1 \varphi_s^{-1} dY_s$$

where $\varphi_s$ is the solution of the homogeneous equation starting from 1. This solution $\varphi_s$ can be computed by the Picard method. This allows us to solve (2.23) inductively in $|I|$.

A priori it is not clear that the Picard series, which is only defined for a fixed $t$, defines continuous processes $H_t(I)$ which are solutions of the equation (2.23). But in fact it follows from the Kolmogorov criterion that this equation does define a continuous process. In this case we have a true semi-martingale for the term $X_t$; we do not get an anticipative condition since $dX_t$ is given by the degree 1 component $\omega_1$ of $\omega$ and in the term $\omega_1(dw_t)$ no final condition in parallel transport appears. Thus we get a unique family of continuous processes $t \mapsto H_t(\omega, I)$ indexed by $I$.

It is also possible to show that the Picard series of (2.23) converges for all $t \leq 1$, instead of just $t = 1$, in all the $L^{p,\alpha}$ and so we get a form

$$\exp_w(\omega)(t)$$

of infinite degree for $t \leq 1$. Moreover the function

$$t \mapsto \exp_w(\omega)(t)$$

is continuous in all the $L^{p,\alpha}$, by using the basic estimate Lemma (1.10). By using Lemma (1.10) and the Kolmogorov criterion, it is also possible to show that if, for any continuous process $H_t$, we define $H_t(\omega, I)$ by

$$\exp_w(\omega)(t) = \sum H_t(\omega, I)\beta_I$$

then the $H_t(\omega, I)$ are solutions of (2.23).

## §3 $L^{B,p,\alpha}$-Chen forms

If $\mathcal{H}$ is a Hilbert space then there is a natural operator norm on the space $\Lambda^r(\mathcal{H}^*)$ of alternating multilinear functions on $\mathcal{H}$. Given $\beta \in \Lambda^r(\mathcal{H}^*)$ then $\|\beta\|_B$ is defined to be the smallest real number such that

$$\|\beta(x_1, \ldots, x_r)\| \leq \|\beta\|_B \|x_1\| \ldots \|x_r\|.$$

This defines a Banach space norm $\| - \|_B$ on $\Lambda^r(\mathcal{H}^*)$. Note that the norms $\| - \|_H$ and $\| - \|_B$ are inequivalent. Each occurs naturally, in different contexts, and so we develop the theory for both.

We can use the norm $\| - \|_B$ to define a weighted $L^p$-norm on the space of measurable forms on $PM$, $LM$, and $L_xM$. As in §2 we define the **random** $(B, p, \alpha)$-**norm** of the homogeneous measurable $r$-form $\sigma$ by

$$(3.1) \qquad \|\sigma\|_{B,p,\alpha,w} = \alpha(r,p)\|\sigma\|_{B,w}.$$

If $\sigma$ is not homogeneous then we define the random $(B, p, \alpha)$-norm of $\sigma$ to be the sum of the corresponding norms of the homogeneous components of $\sigma$. The $L^{B,p,\alpha}$-**norm** of $\sigma$ is defined by

$$
(3.2) \qquad \|\sigma\|_{L^{B,p,\alpha}} = \left( E\left[ \|\sigma\|_{B,p,\alpha}^p \right] \right)^{1/p}
$$

where the expectation is taken over $PM$, $LM$, or $L_xM$, according to the case, using the measures described at the beginning of §1. Once more there are infinite sums of homogeneous terms with finite $L^{B,p,\alpha}$-norm.

In this case the $B$ in the notation for these norms is there to emphasise that we are dealing with the $(p, \alpha)$-norms based on the Banach norm $\| - \|_B$. We will assume this and drop the $B$ from the notation. So if $\sigma$ is a measurable form, then, for the rest of this section, $\|\sigma\|$ will mean $\|\sigma\|_B$, $\|\sigma\|_{p,\alpha}$ will mean $\|\sigma\|_{B,p,\alpha}$, and $\|\sigma\|_{L^{p,\alpha}}$ will mean $\|\sigma\|_{L^{B,p,\alpha}}$.

If $p < q$, it follows from (2.1) that

$$
(3.3) \qquad \|\sigma\|_{p,\alpha} \leq \|\sigma\|_{q,\alpha}, \qquad \|\sigma\|_{L^{p,\alpha}} \leq \|\sigma\|_{L^{q,\alpha}}.
$$

The relation between the basic algebraic operations on forms and the $L^{p,\alpha}$-norm is given by the next two lemmas.

**Lemma 3.4.** *Let $\sigma$, $\tau$ be measurable forms on $PM$, $LM$, or $L_xM$, then:*

$$
\|\sigma \wedge \tau\|_{L^{p,\alpha}} \leq 2C_3(p) \|\sigma\|_{L^{2p,\alpha}} \|\tau\|_{L^{2p,\alpha}}.
$$

*Proof.* To begin with suppose that $\sigma$ is homogeneous of degree $r$ and $\tau$ is homogeneous of degree $s$ and then

$$
(\sigma \wedge \tau)(b_1, \ldots, b_{r+s}) = \sum_{(I,J) \in \Phi(r,s)} \text{sign}(I, J)\sigma(b_I)\tau(b_J).
$$

There are $(r+s)!/r!\, s!$ terms in this sum and each term has modulus smaller than

$$
\|\sigma\|\, \|\tau\| \prod_1^{r+s} \|b_i\|.
$$

It follows that

$$
\begin{aligned}
\|\sigma \wedge \tau\|_{L^{p,\alpha}} &\leq \frac{(r+s)!}{r!\, s!}\, \alpha(r+s, p) E\left[ (\|\sigma\|\, \|\tau\|)^p \right]^{1/p} \\
&\leq \frac{(r+s)!}{r!\, s!}\, \alpha(r+s, p) E\left[ \|\sigma\|^{2p} \right]^{1/2p} E\left[ \|\tau\|^{2p} \right]^{1/2p} \\
&\leq 2C_3(p)\alpha(r, 2p)\alpha(s, 2p) E\left[ \|\sigma\|^{2p} \right]^{1/2p} E\left[ \|\tau\|^{2p} \right]^{1/2p} \\
&= 2C_3(p)\|\sigma\|_{L^{2p,\alpha}} \|\tau\|_{L^{2p,\alpha}}.
\end{aligned}
$$

We have used (2.4) to obtain the last inequality and the expectations are taken over $PM$, $LM$, or $L_xM$, according to the case. A simple argument shows that this inequality is valid in the general, inhomogeneous, case. $\square$

**Lemma 3.5.** *Let $b$ be a measurable vector field and $\sigma$ a measurable form on $PM$, $LM$, or $L_xM$; then*

$$\|i_b(\sigma)\|_{L^{p,\alpha}} \leq \sqrt{C_1(p)}\|b\|_{L^{2p}}\|\sigma\|_{L^{2p,\alpha}}. -$$

*Proof.* We can assume that $\sigma$ is homogeneous of degree $r$ and then

$$|i_b(\sigma)(b_1,\ldots,b_{r-1})| = |\sigma(b,b_1,\ldots,b_{r-1})|$$

$$\leq \|\sigma\| \, \|b\| \prod_{i=1}^{r-1} \|b_i\|.$$

This shows that we have the inequality

$$\|i_b(\sigma)\| \leq \|\sigma\| \, \|b\|.$$

Now it follows that

$$\|i_b(\sigma)\|_{L^{p,\alpha}} = \alpha(r-1,p)\,(E\,[(\|i_b(\sigma)\|)^p])^{1/p}$$

$$\leq \sqrt{C_1(p)}\,\alpha(r,2p)\,(E\,[(\|\sigma\| \, \|b\|)^p])^{1/p}$$

$$\leq \sqrt{C_1(p)}\,\alpha(r,2p)\,\left(E\left[(\|\sigma\|)^{2p}\right]\right)^{1/2p}\left(E\left[(\|b\|)^{2p}\right]\right)^{1/2p}$$

$$= \sqrt{C_1(p)}\,\|b\|_{L^{2p}}\|\sigma\|_{L^{2p,\alpha}}.$$

In this computation we have used (2.2) and the expectations are computed over $PM$, $LM$, or $L_xM$ according to the case. □

We can get estimates for $\|\sigma\|$ when $\sigma$ is a Chen form by the following procedure. Choose an orthonormal basis $e_1(x),\ldots,e_d(x)$ for $T_xM$ such that each $e_i$ is a measurable function of $x$. Let $b$ be a measurable section of the field of Hilbert spaces $\mathcal{H}_w$. Then we define components $b^j$ by the following formula:

$$b(s,w) = \sum_{j=1}^d b^j(s,w)\tau_s(w)e_j(w_0).$$

For each $w$ the functions $b^j(s,w)$ are in the Hilbert space $\mathcal{H} = \mathcal{H}(\mathbb{R})$ defined in §1. It therefore follows that

$$(3.6) \qquad \int_0^1 |b^j(s,w)|^2\,ds \leq \|b\|^2, \qquad \int_0^1 \left(\frac{db^j(s,w)}{ds}\right)^2\,ds \leq \|b\|^2.$$

To simplify we will now work on the based loop space $L_xM$ and therefore we assume that $b(0,w) = b(1,w) = 0$. On the free loop space or the path space there are extra terms coming from the values of $b(s, w)$ when $s = 0, 1$. These extra terms are not difficult to estimate.

Now given measurable sections $b_1, \ldots, b_r$ of $T_w$ let us write the formula (1.6) in terms of the $b_i^j$. To do this in a reasonably compact form requires extra notation. Let $(I_1, \ldots, I_n)$ be a $(r_1, \ldots, r_n)$ shuffle of $\{1, \ldots, r\}$ where $r = r_1 + \cdots + r_n$. We write out $I_k$ as

$$I_k = \{i_{k,1}, \ldots, i_{k,r_k}\}.$$

Now let $J = \{j_1, \ldots, j_r\}$ be a set of indices for the components of the $b_i$ and define

$$j_{k,1} = j_{r_1 + \cdots + r_{k-1} + 1},$$

$$\vdots$$

$$j_{k,r_k} = j_{r_1 + \cdots + r_k},$$

$$J_k = \{j_{k,1}, \ldots, j_{k,r_k}\}$$

$$b_{I_k}^{J_k} = b_{i_{k,1}}^{j_{k,1}} \cdots b_{i_{k,r_k}}^{j_{k,r_k}}.$$

Using this notation we get the following formula:

$$(3.7) \qquad \sigma(\omega)(b_1, \ldots, b_r) = \sum_{(J; I_1, \ldots, I_n)} \varepsilon(I_1, \ldots, I_n) \lambda(J; I_1, \ldots I_n)$$

where

$$\lambda(J; I_1, \ldots I_n)_w =$$

$$\int_{\Delta^n} \omega_1(dw, \tau e_{J_1})_{s_1, w} b_{I_1}^{J_1}(s_1, w) \cdots \omega_1(dw, \tau e_{J_n})_{s_n, w} b_{I_n}^{J_n}(s_n, w).$$

The sum in (3.7) is taken over

$$|J| = r, \qquad (I_1, \ldots, I_n) \in \Phi(r_1, \ldots, r_n).$$

Note that $\lambda(J; I_1, \ldots, I_n)$ can be rewritten as
(3.8)

$$\int_{\Delta^n} \int_{t_1=0}^{s_1} \cdots \int_{t_n=0}^{s_n} \omega_1(dw, \tau e_{J_1})_{t_1} d\left(b_{I_1}^{J_1}\right)_{t_1} \cdots \omega_n(dw, \tau e_{J_n})_{s_n} d\left(b_{I_n}^{J_n}\right)_{t_n}.$$

Now let us compute (3.8) by interchanging the order of $t_i$ and $s_i$ in the integral: we get

$$(3.9) \qquad \int_{(t_1, \ldots, t_n) \in [0,1]^n} A(\omega; t; J) d\left(b_{I_1}^{J_1}\right)_{t_1} \cdots d\left(b_{I_n}^{J_n}\right)_{t_n}.$$

where $t = (t_1, \ldots, t_n)$ and $A(\omega; t; J)_w$ is given by

$$(3.10) \qquad \int_{s_1=t_1}^{1} \cdots \int_{s_n=t_n}^{1} \omega_1(dw, \tau e_{I_1})_{s_1,w} \ldots \omega_n(dw, \tau e_{I_n})_{s_n,w}.$$

Using (3.6) we get the following bound

$$\left( \int_{[0,1]^n} \left( \frac{db_{I_1}^{J_1}}{ds} \right)_{s_1}^{2} \cdots \left( \frac{db_{I_n}^{J_n}}{ds} \right)_{s_n}^{2} ds_1 \cdots ds_n \right)^{1/2} \leq \prod_{i=1}^{r} \|b_i\|$$

and then the above formulas give the following bound on $\|\sigma(\omega)\|$:

$$(3.11) \quad |\sigma(\omega)(b_1, \ldots, b_r)_w| \leq$$

$$\prod_{i=1}^{r} \|b_i\| \sum_{(J;I_1,\ldots,I_n)} \left( \int_{[0,1]^n} |A(\omega; s; J)|^2 ds_1 \cdots ds_n \right)^{1/2}$$

From (1.10) the $L^p$-norm of each of the $A(\omega; s; J)$ is bounded uniformly in $s$ and it follows that the $L^{p,\alpha}$ norm of $\sigma(\omega)$ is finite.

As in the previous section we can now define $L^{p,\alpha}$-Chen forms based on the Banach norm $\| - \|_B$. The formal properties of these spaces are identical to those for the corresponding spaces based on the Hilbert norm $\| - \|_H$.

**Definition 3.12.** The spaces of $L^{p,\alpha}$-**Chen forms** $C_{p,\alpha}(PM)$, $C_{p,\alpha}(LM)$, and $C_{p,\alpha}(L_xM)$ on $PM$, $LM$, and $L_xM$ are defined to be the the closure of the spaces $C(PM)$, $C(LM)$, and $C(L_xM)$ in the $L^{p,\alpha}$-norm.

Once more it is important to note that we are completing the space $C$, not just each of its homogeneous components so we allow certain infinite sums of homogeneous elements. It follows from (3.3) that

$$C_{q,\alpha} \subset C_{p,\alpha}, \qquad \text{if } q > p$$

and we define

$$C_{\infty,\alpha} = \bigcap_{p} C_{\infty,\alpha}.$$

The proofs of the next two theorems are easy modifications of the proofs of (2.15) and (2.16).

**Theorem 3.13.** If $\sigma, \tau \in C_{\infty,\alpha}$ then the exterior product $\sigma \wedge \tau$ is also in $C_{\infty,\alpha}$. $\square$

**Theorem 3.14.** *Suppose that $\sigma$ is in $\mathcal{C}_{\infty,\alpha}$ and that $b$ is a vector field which has finite $L^p$-norm for all $p$; then*

$$i_b(\sigma) \in \mathcal{C}_{\infty,\alpha}. \quad \square$$

We also get the analogue of Theorem (2.18) using the $L^{p,\alpha}$-norm based on $\|-\|_B$ rather than $\|-\|_H$. We formulate the result for the based loop space $L_x M$ since the proof we given below uses the inequality (3.11) and we have only established this in the case where of the based loop space. However, as we noted after (3.6) the extra terms introduced by working on the free loop space $LM$ or the path space $PM$ are not difficult to estimate and, as in Theorem (2.18), the analogous result is valid in these cases.

**Theorem 3.15.** *Let $\omega_{i,n}$, where $1 \le i \le n$, be differential forms on $M$ and define $\omega_n = \omega_{1,n} \otimes \cdots \otimes \omega_{n,n}$.*

(1) *Suppose that the power series*

$$\sum_{n=1}^{\infty} \left( \prod_{i=1}^{n} \|\omega_{i,n}\|_{k,\infty} \frac{z^n}{\sqrt{n!}} \right)$$

*has infinite radius of convergence for all $k \le 2d + 3$ where $d$ is the dimension of $M$; then*

$$\sum_{n=1}^{\infty} \sigma(\omega_n) \in \mathcal{C}_{\infty,\alpha}(L_x M).$$

(2) *Suppose that for all $u$ the diffusion $\tilde{w}_s(u)$ is hypoelliptic and that the power series*

$$\sum_{n=1}^{\infty} \left( \prod_{i=1}^{n} \|\omega_{i,n}\|_{1,\infty} \frac{z^n}{\sqrt{n!}} \right)$$

*has infinite radius of convergence; then*

$$\sum_{n=1}^{\infty} \sigma(\omega_n) \in \mathcal{C}_{\infty,\alpha}(L_x M).$$

*Proof.* It is enough to consider the case $p > 2$. Let $r_{i,n} = \deg \omega_{i,n} - 1$ and $r_n = r_{1,n} + \cdots + r_{n,n}$. From (3.11) we know that

$$|\sigma(\omega_n)(b_1, \ldots, b_{r_n})| \le$$

$$\prod_{i=1}^{r_n} \|b_i\| \sum_{(J;I_1,\ldots,I_n)} \left( \int_{[0,1]^n} |A(\omega_n); s; J)|^2 ds_1 \cdots ds_n \right)^{1/2}$$

and therefore

$$\|\sigma(\omega_n)\|_{L^{p,\alpha}} \leq$$

$$\sum_{(J;I_1,\ldots,I_n)} \alpha(r_n,p)E^{x,x}\left[\left(\int_{[0,1]^n}|A(\omega_n;s;J)|^2 ds_1\cdots ds_n\right)^{p/2}\right]^{1/p}.$$

But there are at most $r_n!$ shuffles so, from (2.3), it follows that

$$\|\sigma(\omega_n)\|_{L^{p,\alpha}} \leq$$

$$C_1(p)^{r_n}\sup_J E^{x,x}\left[\left(\int_{[0,1]^n}|A(\omega_n;s;J)|^2 ds_1\cdots ds_n\right)^{p/2}\right]^{1/p}$$

Now using part (1) Lemma (1.10) and Holder's inequality we get

$$E^{x,x}\left[\left(\int_{[0,1]^n}|A(\omega_n;s;J)|^2 ds_1\cdots ds_n\right)^{p/2}\right]^{2/p}$$

$$\leq \frac{K(p)^n\|\omega_{1,n}\|_{k,\infty}^2\cdots\|\omega_{n,n}\|_{k,\infty}^2}{n!}$$

where $k > 2d+3$. Putting these inequalities together we get

$$\|\sigma(\omega_n)\|_{L^{p,\alpha}} \leq \frac{K(p)^n\|\omega_{1,n}\|_{k,\infty}\cdots\|\omega_{n,n}\|_{k,\infty}}{\sqrt{n!}}$$

which proves that the series $\sum \sigma(\omega_n)$ is absolutely convergent in all the $L^{p,\alpha}$. This proves part (1) of the theorem and part (2) follows by using part (2) of Lemma (1.10). $\square$

We get exactly analogous theorems for $PM$ and $LM$ and also if the $\omega_{i,n}$ are matrices of forms and it follows that the equivariant Chern character belongs to all the $\mathcal{C}_{\infty,\alpha}$ in just the same way as in §2.

## §4 THE SCHWARTZ LEMMA

We now give the proof of Lemma (1.10), the version of a lemma of Schwartz [35] appropriate to our situation. To begin with we recall the necessary terminology and the statement we need to prove. Fix $x \in M$ and let $f_1,\ldots,f_r : [0,1] \to T_x M$ be functions which are continuous and have bounded variation. As pointed out in §2 we need some hypotheses on the

$f_i$ since we are using Stratonovitch integrals. These hypotheses are adequate for our applications but it is presumably possible to get away with weaker conditions. We choose a constant $C(f)$ such that $\|f_i\|_\infty \le C(f)$ for $1 \le i \le r$. Let

$$\omega : [0,1] \to \bar{\Omega}^{\otimes n}, \qquad s \mapsto \omega_{1,s} \otimes \cdots \otimes \omega_{n,s}$$

be a map which is piecewise constant and set

$$r_i = \deg \omega_i - 1$$
$$r = r_1 + \cdots + r_n$$
$$I_i = \{r_1 + \cdots + r_i + 1, \ldots, r_1 + \cdots + r_i + r_{i+1}\}$$
$$f = (f_1, \ldots, f_r);$$

as in §1, let

$$(4.1) \qquad A(f;\omega) = \int_{\Delta^n} \omega_1 \, (dw, \tau(f_{I_1}))_{s_1} \cdots \omega_n \, (dw, \tau(f_{I_n}))_{s_n} \,.$$

Recall that if $\omega_s$ is a form on $M$ which depends on $s$

$$\|\omega\|_{k,\infty} = \sup_s \left( \|\omega_s\|_\infty + \|\nabla \omega_s\|_\infty + \cdots + \|\nabla^k \omega_s\|_\infty \right).$$

Let $N$ be the frame bundle of $M$ and let $\pi : N \to M$ be the projection. Let $\tilde{w}_s(u)$, $s \ge 0$, $u \in N$, be the horizontal lift of the diffusion $w_s(x)$, $s \ge 0$, where $x = \pi(u)$. We can also consider the case where the diffusion $\tilde{w}_s(u)$ takes place on a sub-bundle of $N$.

**The Schwartz Lemma.**   (1) *For all $x \in M$ we have the estimate*

$$(4.2) \qquad E^{x,x} \left[ |A(f;\omega)|^p \right]^{1/p} \le \frac{C(f)^r C(p)^n}{\sqrt{n!}} \|\omega_1\|_{2d+3,\infty} \cdots \|\omega_n\|_{2d+3,\infty}.$$

*where $d$ is the dimension of $M$.*
   (2) *Suppose that for all $u$ the diffusion $\tilde{w}_s(u)$ is hypoelliptic. Then, for all $x \in M$, we have the estimate*

$$(4.3) \qquad E^{x,x} \left[ |A(f;\omega)|^p \right]^{1/p} \le \frac{C(f)^r C(p)^n}{\sqrt{n!}} \|\omega_1\|_{1,\infty} \cdots \|\omega_n\|_{1,\infty}.$$

   (3) *If each of the $\omega_i$ is a 1-form then (4.3) holds even if the diffusion $\tilde{w}_s(u)$ is not hypoelliptic.*

The forms $\omega_i$ are assumed to be piecewise constant but it should be possible to generalise to the case where the $\omega_i$ are smooth in $s$, or to follow the more general setting of the paper of Hu and Meyer [19]. As pointed out in §2 we can replace the constant $C(f)^r$ occuring in (4.2) and (4.3) by $\|f_1\|_\infty \cdots \|f_r\|_\infty$ but it is slightly easier, at several points in writing the proof, to choose the constant $C(f)$.

First we show how to reduce the general case to the special case where $p = 1$. It is enough to consider the case where $p = 2k$. Then, as in the formula for the product of iterated integrals, compare [17, §3] or [21], we can express $A(f;\omega)^p$ in terms of sums of iterated integrals of length $2nk$. By this procedure we get a sum of $(2nk)!/(n!)^{2k}$ terms of the type

$$\int_{\Delta^{2nk}} \psi_1(dw)_{s_1} \ldots \psi_{2nk}(dw)_{s_{2nk}}$$

where each term $\psi_j(dw)$ is given by

$$\psi_j(dw) = \omega_k(dw, \tau(f_{I_k}))$$

for some $k$ with $1 \le k \le n$, and exactly $2k$ of the $\psi_j(dw)$ are equal to $\omega_k(dw, \tau(f_{I_k}))$.

Recall Stirling's formula,

$$n! \simeq n^n e^{-n}\sqrt{2\pi n}, \qquad \text{as } n \to \infty;$$

it follows that

$$\frac{(2nk)!}{(n!)^{2k}} \simeq \frac{(2nk)^{2nk}\sqrt{4\pi nk}}{n^{2nk}(\sqrt{2\pi n})^{2k}} \le C(k)^n \qquad \text{as } n \to \infty$$

and this gives us an estimate on the number of terms. Combining this estimate of the number of terms with the estimate for each term

$$E^{x,x}\left[\int_{\Delta^{2nk}} \psi_1(dw)_{s_1} \ldots \psi_{2nk}(dw)_{s_{2nk}}\right]$$

given by the Schwartz Lemma in the case $p = 1$, the case of general $p$ follows from the case $p = 1$ and $n$ even.

Define $\Delta^l(s)$ to be the $l$-simplex

$$\Delta^l(s) = \{0 \le s_1 \le \cdots \le s_l \le s : s_i \in \mathbb{R}\};$$

we continue to use the notation $\Delta^n$ for the simplex $\Delta^n(1)$. We use the notation

$$\psi_j(dw) = \omega_j(dw, \tau(f_j)).$$

For any integer $l \leq n$ define $J(l, s)$ by

$$(4.4) \qquad J(l,s) = \int_{\Delta^l(s)} \psi_1(dw)_{s_1} \dots \psi_l(dw)_{s_l}.$$

Our aim is to get an estimate for $E^{x,x}[J(l,1)]$. Our method is to do this recursively and this requires us to get estimates for the expectation of $J(l,s)$ for $s \leq 1$. One reason for considering expectations of $J(l,s)$ is that we will get bounds in terms of "explicit formulas involved with heat kernels" in the manner of [20].

First we prove (4.3). Because of the hypoellipticity hypothesis we can estimate the expectation of $J(l,s)$ by working in the frame bundle $N$. Give $N$ the metric constructed from the Riemannian metric on $M$ and an invariant metric on the fibres, see [4, page 403] or [14]. Let $\tilde{X}_i$ be the canonical horizontal vector fields on $N$. Now consider the stochastic differential equation of the process $\tilde{w}_t(u)$ conditioned to end at $v$ after time $s$:

$$(4.5) \qquad d\tilde{w}_t(u,v) = \tilde{A}dB + \tilde{C}dt, \qquad \tilde{w}_0(u,v) = u, \qquad t < s.$$

In this equation

$$(4.6) \qquad \begin{aligned} \tilde{A}dB &= \sum_{i=1}^{d} \tilde{X}_i(\tilde{w}_t(u,v))dB_t^i \\ \tilde{C} &= \sum_{i=1}^{d} \tilde{X}_i(\tilde{w}_t(u,v)) \left\langle \tilde{X}_i(\tilde{w}_t(u,v)), \operatorname{grad} \log \tilde{p}_{s-t}(\tilde{w}_t(u,v),v) \right\rangle \end{aligned}$$

where $\tilde{p}_t$ is the heat kernel associated to $\tilde{w}_t$. The transition probability of $\tilde{w}_t(u,v)$ is

$$(4.7) \qquad \tilde{p}_{t,s}(u,r,v) = \frac{\tilde{p}_t(u,r)\tilde{p}_{s-t}(r,v)}{\tilde{p}_s(u,v)}.$$

We write $E_s^{u,v}[J(l,s)]$ for the expectation of $J(l,s)$ using the law of the process $\tilde{w}_t(u)$ conditioned to end at $v$ after time $s$.

Let $\pi : N \to M$ be the projection. Then we have the following recursive formula:

$$(4.8) \qquad J(l,s) = \int_0^s J(l-1,t)\psi_{l,t}(D\pi(\tilde{A}dB + \tilde{C}dt)).$$

This is a Stratonovitch stochastic integral and the first step is to write it in terms of Itô integrals. Let $\delta B_t^i$ be the Itô differential of the flat Brownian motion $B_t^i$. Then we decompose $J(l,s)$ into four terms

$$(4.9) \qquad J(l,s) = J_1(l,s) + J_2(l,s) + J_3(l,s) + J_4(l,s).$$

The first two terms are the natural ones:

(4.10)
$$J_1(l,s) = \int_0^s J(l-1,t)\psi_{l,t}(D\pi(\tilde{A}\delta B))$$
$$J_2(l,s) = \int_0^s J(l-1,t)\psi_{l,t}(D\pi(\tilde{C}dt)).$$

The term $J_3(l,s)$ arises from pairing $J(l-1,t)$ with $\psi_{l,t}(D\pi(\tilde{A}\delta B))$. It is given by a formula of the kind

(4.11)
$$J_3(l,s) = \int_0^s J(l-2,t)P_1(\psi_{l-1,t},\psi_{l,t})dt$$

where $P_1(\psi_{l-1,t},\psi_{l,t})$ is $\tilde{w}_t(u,v)$ measurable and

(4.12)
$$|P_1(\psi_{l-1},\psi_l)| \le C^2 C(f)^{d(l-1)+d(l)}\|\psi_{l-1}\|_{1,\infty}\|\psi_l\|_{1,\infty}.$$

Here, since $\psi_j = \omega_j\left(-,\tau\left(f_{I_j}\right)\right)$, $d(j)$ is the degree of the form $\omega_j$ and $\|\psi_j\|_{1,\infty} = \|\omega_j\|_{1,\infty}$. The last term $J_4(l,s)$ comes from the conversion of the Stratonovitch differential element $\psi_l(D\pi(\tilde{A}dB))$ into an Itô differential element. We get

(4.13)
$$J_4(l,s) = \int_0^s J(l-1,t)P_2(\psi_l)dt$$

where once more $P_2(\psi_l)$ is $\tilde{w}_t(u,v)$ measurable. In this case we have

(4.14)
$$|P_2(\psi_l)| \le C(f)^{d(l)}\|\psi_l\|_{1,\infty}.$$

Now we show that, for almost all $v$,

(4.15)
$$E_s^{u,v}[J_1(l,s)] = 0.$$

If the local martingale $t \mapsto J_1(l,t)$, $t \le s$ were a true martingale then, of course, (4.15) would follow. We will prove (4.15) for almost all $v$ by the following argument using Lemma (5.45) from §5 . It is enough to show that $E_s^{u,v}[|J(l,t|^p]]$ is bounded for all $p$ and for $t \le s$. Suppose, inductively, that $E_s^{u,v}[|J(l-1,t|^p]]$ is bounded for all $p$ and for $t \le s$. We get a formula analogous to (4.11)

$$J(l,t) = J_1(l,t) + J_2(l,t) + J_3(l,t) + J_4(l,t)$$

where the $J_i(l,t)$ are given by integrals identical to those in (4.11) except that $s$ in the upper limit is replaced by $t$ and $t$ inside the integral is replaced by $t'$. The only problem is to show that if, for $r < l$ and for all $q$, $E_s^{u,v}[|J_2(r,t)|^q]$ is bounded for $t \le s$ then so is $E_s^{u,v}[J_2(l,t)]$.

**Define**

(4.16) $\qquad \tilde{\Gamma}_i(t,s) = \left\langle \tilde{X}_i(\tilde{w}_t(u,v)), \operatorname{grad} \log \tilde{p}_{s-t}(\tilde{w}_t(u,v),v) \right\rangle$

then, by induction we have:

$$E_s^{u,v}\left[\left(\int_0^t \left|\tilde{\Gamma}_i(t',s)\right| |J(l-1,t')| \, dt'\right)^p\right]^{1/p}$$

$$\leq \int_0^t E_s^{u,v}\left[\left|\tilde{\Gamma}_i(t',s)\right|^p |J(l-1,t')|^p \, dt'\right]^{1/p}$$

(4.17)

$$\leq \int_0^t E_s^{u,v}\left[\left|\tilde{\Gamma}_i(t',s)\right|^{2p}\right]^{1/2p} E_s^{u,v}\left[|J(l-1,t')|^{2p} \, dt'\right]^{1/2p} dt'$$

$$\leq \int_0^s \frac{C(s,p)}{\sqrt{s-t'}} E_s^{u,v}\left[|J(l-1,t')|^{2p}\right]^{1/2p} dt'$$

$$< \infty$$

where we have used the fact that $\tilde{p}_t > 0$ and Lemma (5.45) to derive the third inequality. This completes the proof of (4.15). In fact (4.15) is true for all $v$ under the hypothesis of Remark (5.47).

From (4.10) – (4.14), we get the following estimates for all $v$:
(4.18)
$$|E_s^{u,v}[J_2(l,s)]| \leq$$

$$K_2 \int_0^s \int_N |E_t^{u,r}[J(l-1,t)]| \left(\sum_{i=1}^d \left|\left\langle \tilde{X}_i(r), \Phi(u,r,v)\right\rangle\right|\right) dr dt$$

$$|E_s^{u,v}[J_3(l,s)]| \leq$$

$$K_3\|\psi_{l-1}\|_{1,\infty}\|\psi_l\|_{1,\infty} \int_0^s \int_N |E_t^{u,r}[J(l-2),t)]| \, \tilde{p}_{t,s}(u,r,v) dr dt$$

$$|E_s^{u,v}[J_4(l,s)]| \leq$$

$$K_4 \int_0^s \int_N |E_t^{u,r}[J(l-1),t)]| \, \tilde{p}_{t,s}(u,r,v) dr dt.$$

Here the term $\Phi(u,r,v)$ occuring in the estimate for $J_2$ is given by

$$\Phi(u,r,v) = \frac{\tilde{p}_t(u,r)}{\tilde{p}_s(u,v)} \operatorname{grad} \tilde{p}_{s-t}(r,v)$$

and the constants $K_2$, $K_3$, and $K_4$ are given by

$$K_2 = C\|\psi_l\|_{1,\infty} C(f)^{d(l)}$$
$$K_3 = C^2 C(f)^{d(l)+d(l-1)}$$
$$K_4 = C\,C(f)^{d(l)}\|\psi_l\|_{1,\infty}.$$

Now let $\alpha = (\alpha_1, \ldots, \alpha_r)$ be a sequence with each $\alpha_j$ equal to 0 or 1 and at least $l - r$ terms equal to 0. We set

(4.19)
$$r = |\alpha|$$
$$q_{0,t}(u,v) = \tilde{p}_t(u,v)$$
$$q_{1,t}(u,v) = \sum \left| \left\langle \tilde{X}_i(u), \operatorname{grad} \tilde{p}_t(u,v) \right\rangle \right|.$$

Using (4.15), (4.18), and induction on $l$ it follows that, for almost all $v$,

(4.20) $\quad |E_s^{u,v}[J(l,s)]| \leq$
$$C^l C(f)^{d(1)+\cdots+d(l)} \|\psi_1\|_{1,\infty} \cdots \|\psi_l\|_{1,\infty} \sum_{l/2 \leq r \leq l} \sum_{|\alpha|=r} \tilde{I}(\alpha).$$

Here $\tilde{I}(\alpha)$ is given by the following integral:

$$\int_{\Delta^r(s)} d\underline{s} \int_{N \times \cdots \times N} \frac{\tilde{p}_{s_1}(u,v_1) q_{\alpha_1, s_2-s_1}(v_1,v_2) \ldots q_{\alpha_r, s-s_r}(v_r,v)}{\tilde{p}_s(u,v)} \, d\underline{v}$$

where

$$d\underline{s} = ds_1 \cdots ds_r, \qquad d\underline{v} = dv_1 \cdots dv_r.$$

This last formula is only true for almost all $v$ but this is not a serious problem.

Our goal is to estimate $E^{x,x}[J(n,1)]$ and

(4.21) $\qquad E^{x,x}[J(n,1)] = \int_{N_x} \frac{\tilde{p}_1(u,v)}{p_1(x,x)} E_1^{u,v}[J(n,1)] dv$

where $N_x = \pi^{-1}(x)$ is the fibre of $N$ over $x$. Thus (4.18) gives

$$E^{x,x}[J(n,1)] \leq C^n C(f)^n \|\omega_1\|_{1,\infty} \cdots \|\omega_n\|_{1,\infty} \sum_{n/2 \leq r \leq n} \sum_{\alpha} I(\alpha)$$

where $I(\alpha)$ is given by the following integral

$$\int_{\Delta^r(s)} d\underline{s} \int_{N_x} dv \int_{N \times \cdots \times N} \frac{\tilde{p}_{s_1}(u,v_1) q_{\alpha_1, s_2-s_1}(v_1,v_2) \ldots q_{\alpha_r, 1-s_r}(v_r,v)}{p_1(x,x)} \, d\underline{v}$$

This estimate is valid for almost all $x$ and by continuity it follows that it is true for all $x$.

Now we must estimate the integral $I(\alpha)$.

**Lemma 4.22.** *There is a constant $C$ such that*

$$(4.23) \quad \int_{N\times\cdots\times N} \tilde{p}_{s_1}(u,v_1)q_{\alpha_1,s_2-s_1}(v_1,v_2)\cdots q_{\alpha_r,1-s_r}(v_r,v)\,d\underline{v}$$

$$\leq C^r(s_2-s_1)^{-\alpha_1/2}(s_3-s_2)^{-\alpha_2/2}\cdots(1-s_r)^{-\alpha_r/2}$$

*Proof.* Since $N$ is compact we can use the short time expansion of [23] for the horizontal Laplacian of $N$. In terms of the associated sub-Riemannian distance $\tilde{d}(u,v)$ and the volume of balls, we get, for $t \leq 1$,

$$(4.24)$$
$$q_{0,t}(u,v) \leq \frac{C_1}{\tilde{V}(u,\sqrt{t})}\exp\left(\frac{-C_2\tilde{d}(u,v)^2}{t}\right)$$

$$q_{1,t}(u,v) \leq \frac{C_1'}{\sqrt{t}\tilde{V}(u,\sqrt{t})}\exp\left(\frac{-C_2'\tilde{d}(u,v)^2}{t}\right)$$

where $\tilde{V}(x,s) = \text{vol}\,\tilde{B}(x,s)$. These estimates are true for all $(u,v)$ and for some suitable constants $C_1, C_1', C_2, C_2' > 0$.

Using (4.24) we see that the integral in (4.23) is smaller than

$$(4.25) \quad C_3^r(s_2-s_1)^{-\alpha_1/2}(s_3-s_2)^{-\alpha_2/2}\cdots(1-s_r)^{-\alpha_r/2}\int_{N\times\cdots\times N} d\underline{v}$$

$$\frac{1}{\tilde{V}(u,\sqrt{s_1})}\exp\left(\frac{-C_4\tilde{d}(u,v_1)^2}{s_1}\right)\frac{1}{\tilde{V}(v_1,\sqrt{s_2-s_1})}\exp\left(\frac{-C_4\tilde{d}(v_1,v_2)^2}{s_2-s_1}\right)$$

$$\cdots\frac{1}{\tilde{V}(v_r,\sqrt{1-s_r})}\exp\left(\frac{-C_4\tilde{d}(v_r,v)^2}{1-s_r}\right)$$

for some suitable constants $C_3$ and $C_4$.

Let us apply the converse inequality of (4.24):

$$(4.26) \quad \frac{C_1''}{\tilde{V}(u,\sqrt{t})}\exp\left(\frac{-C_2''\tilde{d}(u,v)^2}{t}\right) \leq \tilde{p}_t(u,v)$$

and the doubling property of [23]: if $\rho \leq 1$, there is a constant $K$ such that

$$(4.27) \quad \tilde{V}(u,2\rho) \leq K\tilde{V}(u,\rho)$$

for all $u$. We conclude that there is a constant $\rho_0$, independent of $u$, such that the expression in (4.25) is smaller than

$$(4.28) \quad C^r(s_2-s_1)^{-\alpha_1/2}(s_3-s_2)^{-\alpha_2/2}\cdots(1-s_r)^{-\alpha_r/2}$$

$$\int_{N\times\cdots\times N} \tilde{p}_{\rho_0 s_1}(u,v_1)\tilde{p}_{\rho_0(s_2-s_1)}(v_1,v_2)\cdots\tilde{p}_{\rho_0(1-s_2)}(v_r,v)\,d\underline{v}.$$

The proof of (4.23) is completed using the Chapman-Kolmogorov equation. □

Now comparing the inequality (4.23) with the estimate of $E^{x,x}[J(l,1)]$ given by (4.21) we see that we are left to estimate the integral

$$(4.29) \qquad I_r(\beta) = \int_{\Delta^r} \frac{ds_1}{s_1^{\beta_1}} \frac{ds_2}{(s_2 - s_1)^{\beta_2}} \cdots \frac{ds_r}{(1 - s_r)^{\beta_r}}$$

where $\beta_i = \alpha_i/2$.

**Lemma 4.30.** *If $\beta < 1$ then*

$$(4.31) \qquad \int_0^1 \frac{s^m ds}{(1 - s)^\beta} \leq m^{\beta-1} \Gamma(-\beta + 1)$$

*Proof.* We have

$$\int_0^1 \frac{s^m ds}{(1 - s)^\beta} = \int_0^1 \frac{(1 - s)^m ds}{s^\beta} = \int_0^1 \frac{e^{m \log(1-s)} ds}{s^\beta}$$

and since $\log(1 - s) \leq -s$ when $0 < s < 1$ it follows that

$$\int_0^1 \frac{s^m ds}{(1 - s)^\beta} \leq \int_0^1 \frac{e^{-ms} ds}{s^\beta} \leq \int_0^\infty \frac{e^{-ms} ds}{s^\beta} = m^{\beta-1} \Gamma(1 - \beta) \quad \square$$

**Lemma 4.32.** *If $r \geq 2$ then*

$$(4.33) \qquad I_r(\beta) \leq I_1(\beta_1) \left( \prod_{k=2}^r \Gamma(1 - \beta_k) \right) \prod_{k=1}^{n-1} \left( k - \sum_{j=0}^k \beta_k \right)^{\beta_{k+1}-1}$$

*Proof.* Let us introduce the expression

$$I_r(\beta, s) = \int_{\Delta^r(s)} \frac{ds_1}{s_1^{\beta_1}} \frac{ds_2}{(s_2 - s_1)^{\beta_1}} \cdots \frac{ds_r}{(s - s_r)^{\beta_r}};$$

so

$$I_{r+1}(\beta) = \int_0^1 I_r(\beta, s) \frac{ds}{(1 - s)^{\beta_r+1}}.$$

Using the change of variables $s_i \mapsto s t_i$ we see that

$$I_r(\beta, s) = s^\nu I_r(\beta), \qquad \nu = r - \sum_{i=0}^r \beta_i.$$

Now, using Lemma (4.30), it follows that

$$I_{r+1}(\beta) = I_r(\beta) \int_0^1 s^\nu \frac{ds}{(1-s)^{\beta_{r+1}}} \leq I_r(\beta)\Gamma(1-\beta_{r+1})\nu^{\beta_{n+1}-1}.$$

This proves (4.33) by induction. □

Now we compute the actual integrals we need to deduce (4.3) from (4.21). Consider $I(\alpha/2)$ where each $\alpha_i$ is zero or 1, $\alpha_0 = 0$, at least $n-r$ of the $\alpha_i$ are zero and at most $2r-n$ of the $\alpha_i$ are equal to 1. In this case Lemma (4.32) implies that

$$I(\alpha/2) \leq \left( C(p)^n \prod_{k=1}^r k^{\alpha_k - 1} \right) \left( \frac{C(p)^n}{r!} \prod_{\alpha_k \neq 0} k^{1/2} \right)$$

$$\leq \frac{C(p)^n}{r!} \left( \frac{n!}{(n-r)!} \right)^{1/2}$$

$$\leq \frac{C(p)^n}{\sqrt{r!(n-r)!}}.$$

Now by using the binomial formula we get

$$\frac{1}{\sqrt{r!(n-r)!}} \leq \frac{C^n}{\sqrt{n!}}$$

and so

$$I(\alpha) \leq \frac{C(p)^n}{\sqrt{n!}}.$$

By combining this estimate for the integrals $I(\alpha)$ with (4.21), we complete the proof of (4.3). The proof in the case where all the forms $\omega_i$ are one-forms is similar but much simpler since parallel transport does not appear. This completes the proof of parts (2) and (3) of the Schwartz Lemma.

We now turn to the proof of part (1) of the Schwartz Lemma, that is (4.2). We estimate the quantity

$$E^{x,x}[J(n,1)],$$

where $J(n,1)$ is as in (4.4), by a different procedure based on ideas in [29]. As a result we obtain (4.2) in the case $p=1$. As we have already argued, this is enough to prove (4.2) in the general case.

As in [29] we consider the auxilliary measure on $M$

$$(4.34) \qquad \mu_n(f) = E^x[J(n,1)f(w_1)]$$

where the expectation is taken over the space of continuous paths $w_t$ which start at $x$. This measure has a density $q_n(y)$ given by

$$(4.35) \qquad q_n(y) = E^{x,y}[J(n,1)]\, p_1(x,y).$$

Since $p_1(x,y) > 0$, to get estimates for $E^{x,x}[J(n,1)]$, it is enough to get nice estimates of $q_n(y)$. In order to do this we will apply the Malliavin calculus. Let us introduce $Y_i$, a system of vector fields on $M$ such that the family $Y_i(x)$ span $T_x M$ at each point $x \in M$. If, for all multi-indices $\alpha$

$$(4.36) \qquad \mu_n[Y^\alpha f] \leq C_n(|\alpha|)\|f\|_\infty$$

we will get the following estimate:

$$\sup_y |q_n(y)| \leq K \sup_{|\alpha| \leq d+1} C_n(|\alpha|)$$

In order to show (4.36) we will apply the Malliavin calculus.

Since $J(n,1)$ is smooth in the Malliavin sense, and since the diffusion $w_t(x)$ is non-degenerate, we have an integration by parts formula

$$(4.37) \qquad \mu_n[Y^\alpha f] = E[C_\alpha(w)f(w_1)]$$

The expression $C_\alpha(w)$ is a polynomial in the derivatives of $J(n,1)$, of length $\leq 2|\alpha|$, the derivatives of $w_1(x)$, and in the inverse of the Malliavin matrix of $w_1(x)$. Here all the derivatives are in the Malliavin sense. Moreover $C_\alpha(w)$ is homogeneous of degree 1 in the derivative, in the Malliavin sense, of $J(n,1)$.

We now take the Schwartz construction of the Brownian motion

$$(4.38) \qquad dw_t(x) = \sum_{i=1}^m X_i(w_t(x))dB_t^i + X_0(w_t(x))dt, \qquad w_0(x) = x$$

for some suitable smooth vector fields on $M$. As in the previous proof we convert the Stratonovitch integrals which appear in $J(n,1)$ into Itô integrals. This gives us an expression for $J(n,1)$ as a sum of at most $4^n$ terms $\tilde{J}(r,1)$ of length $r$ where $n/2 \leq r \leq n$. Here a term of length $r$ means there are exactly $n-r$ contractions of terms of the form $\langle \delta B_t^i, \delta B_t^j \rangle$ and, therefore, there are exactly $n-r$ terms in $dt$ in the expression for $\tilde{J}(r,1)$; this is exactly as in the previous proof.

Recall the following formula from [21, page 107]: If $F$ and $G$ are Brownian functionals of the form

$$F = \int_0^1 u_s \delta B_s, \qquad G = \int_0^1 h_s ds$$

then

$$LF = \int_0^1 \left( Lu_s - \frac{1}{2}u_s \right) \delta B_s, \qquad LG = \int_0^1 Lh_s ds$$

where $L$ is the Ornstein-Uhlenbeck operator. These formulas lead to the following conclusion: By Meyer's inequality it is enough to apply $L^\alpha$ with $|\alpha| \le d+1$ to each term $\tilde{J}(r,1)$ to evaluate the contribution of the terms in the integration by parts formula which come from the derivatives. When we apply $L^\alpha$ to $\tilde{J}(r,1)$ we will get a sum of at most $C(n)^{|\alpha|}$ integrals of length $r$ with at least $n-r$ terms in $ds$. Each integrand depends on the derivatives of the forms $\omega_i$ up to order at most $2|\alpha|+1 \le 2d+3$. Here the $+1$ arises from the conversion of the Stratonovitch differential into an Itô differential.

The Ornstein-Uhlenbeck operator is an operator of order 2 in the Malliavin derivative. By using the chain rule it follows that the Malliavin derivative appears with order at most $2|\alpha| \le 2d+2$ for $w_t(x)$ and also for the parallel transport $\tau_t$. These quantities are bounded in $L^p$ but not uniformly so it is not possible to immediately apply the Schwartz method where the components of the integrals are bounded. But, since

$$\nabla(F_1 F_2) = \nabla(F_1)F_2 + F_1 \nabla(F_2),$$

in each term there are at most $2d+2$ terms which are not bounded. This fact will allow us to adapt the Schwartz method to this situation.

We will denote by $\tilde{J}(l,t)$ an integral which contains $\alpha(l) \le 2d+2$ unbounded terms. Let $l_0$ be the number of appearances of the $ds$ terms and $l_0'$ the number of appearances of $\delta B$ terms so that $l = l_0 + l_0'$. Let $\mathcal{J}(l)$ be the set of indices which occur in the forms appearing in the integral $\tilde{J}(l,t)$.

Suppose, by induction, that there exists a constant $C(\alpha(l),p)$ and two real numbers $n(l,l_0) > 0$, $k(l,l_0) > 0$ such that

$$(4.40) \qquad E^{x,x}\left[\tilde{J}(l,t)^p\right]^{1/p} \le C(\alpha(l),p) \prod_{j \in \mathcal{J}(l)} \|\omega_j\|_{2d+3} \frac{t^{n(l,l_0)}}{k(l,l_0)}.$$

To estimate the expectation $E^{x,x}\left[\tilde{J}(l+1,t)^p\right]^{1/p}$ we use the procedure given in [35, Lemma 1.1]. If

$$\tilde{J}(l+1,t) = \int_0^t \tilde{J}(l,s)U\delta B_s^i,$$

where $U$ is bounded, we have $\alpha(l+1) = \alpha(l)$. By using the Burkholder

inequality for some $k(l)$
(4.41)

$$E^{x,x}\left[|\tilde{J}(l+1,t)|^p\right]^{1/p} \leq C(p)\|\omega_{k(l)}\|_{2d+3}E^{x,x}\left[\left(\int_0^t|\tilde{J}(l,s)|^2ds\right)^{p/2}\right]^{1/p}$$

$$\leq C(p)\prod_{j\in\mathcal{J}(l+1)}\|\omega_j\|_{2d+3}\frac{C(\alpha(l),p)t^{\frac{2n(l,l_0)+1}{2}}}{k(l,l_0)\sqrt{2n(l,l_0)+1}}$$

If $U$ is not bounded, we have $\alpha(l+1) = \alpha(l)+1$ and by the Cauchy-Schwartz inequality
(4.42)

$$E^{x,x}\left[\left|\tilde{J}(l+1,t)\right|^p\right]^{1/p} \leq C(p)\|\omega_{k(l)}\|_{2d+3}\left(\int_0^t E^{x,x}\left[\tilde{J}^{2p}(l,s)\right]^{1/p}ds\right)^{1/2}$$

$$\leq C(2p)\prod_{j\in\mathcal{J}(l+1)}\|\omega_j\|_{2d+3}\frac{C(\alpha(l),2p)t^{\frac{2n(l,l_0)+1}{2}}}{k(l,l_0)\sqrt{2n(l,l_0)+1}}$$

If

$$\tilde{J}(l+1,t) = \int_0^t \tilde{J}(l,t)u\,ds$$

where $u$ is bounded then we have $\alpha(l+1) = \alpha(l)$. The $ds$ can appear from a contraction or from $X_0$ in (4.39). Let us consider the first case which is more complicated. For some integer $k(l)$

$$E^{x,x}\left[\left|\tilde{J}(l+1,t)\right|^p\right]^{1/p}$$

(4.43)

$$\leq C(p)\|\omega_{k(l)}\|_{2d+3}\|\omega_{k(l)-1}\|_{2d+3}\int_0^t E^{x,x}\left[J^p(l,s)\right]^{1/p}ds$$

$$\leq C(p)\prod_{j\in\mathcal{J}(l+1)}\|\omega_j\|_{2d+3}\frac{C(\alpha(l),p)t^{n(l,l_0)+1}}{k(l,l_0)(n(l,l_0)+1)}.$$

If $u$ is only bounded in $L^p$ we apply the Cauchy-Schwartz inequality to get

$$(4.44)\quad E^{x,x}\left[|\tilde{J}(l+1,t)|^p\right]^{1/p}$$

$$\leq C(2p)\prod_{j\in\mathcal{J}(l+1)}\|\omega_j\|_{2d+3}\frac{C(\alpha(l),2p)t^{n(l,l_0)+1}}{k(l,l_0)(n(l,l_0)+1)}.$$

From these recursive relations we can deduce that

$$C(\alpha(l),p) \leq \left|\prod_{i=0}^{|\alpha(l)|}C(2^kp)\right|^l$$

and since $\alpha(l)$ is bounded by $2d + 2$ we have

(4.45)                          $C(\alpha(l), p) \leq \tilde{C}(p)^l.$

From (4.41)–(4.45) we deduce that

$$n(l, l_0) = \frac{l}{2} + \frac{|l_0|}{2}$$

By using a similar argument to that used in the final step of the proof of (4.2) we can show that

(4.46)
$$k(l, l_0) \geq \left(\frac{1}{4}\right)^l \frac{\prod_{j=1}^l j}{\prod_{j=|l_0|+1}^l \sqrt{j}}$$
$$= \left(\frac{1}{4}\right)^l \sqrt{l!(l_0)!}$$

Let us return to the problem of estimating

$$E^{x,x}\left[|\tilde{J}(r, 1)|^p\right]^{1/p}.$$

In this case $\mathcal{J}(r)$ is the full set of indices $\{1, \ldots, n\}$. We have

(4.47)     $E\left[|\tilde{J}(r, 1)|^p\right]^{1/p} \leq \dfrac{C(p)^n}{\sqrt{n!(n-r)!}} \|\omega_1\|_{2d+3,\infty} \cdots \|\omega_l\|_{2d+3,\infty}$

since there are at least $n - r$ terms in $ds$ in these integrals. On the other hand the binomial formula implies that

(4.48)                  $\dfrac{1}{\sqrt{n!(n-r)!}} \leq \dfrac{2^n}{\sqrt{n!}} \leq \dfrac{C^n}{\sqrt{n!}}.$

This completes the proof of (4.2).

## §5 THE BRIDGE OF A HYPOELLIPTIC DIFFUSION

The goal of this section is to study the bridge of a hypoelliptic diffusion over a compact manifold $N$; the example we wish to apply the theory to is the frame bundle $N$ of the compact Riemannian manifold $M$. There are two cases to consider, the homogeneous case and the inhomogeneous case and we start with the homogeneous case.

Let $X_1, \ldots, X_m$ be $m$ vector fields on $N$. We make the following assumption on the $X_i$.

$(H_1)$    At each $x \in N$, the Lie algebra spanned by the $X_i$ is equal to $T_x N$.

The semi-group associated to the operator $\frac{1}{2} \sum X_i^2$ has a density $p_t(x,y)$ which is the density of the following stochastic differential equation in the Stratonovitch sense

$$(5.1) \qquad dw_t = \sum_{i=1}^{m} X_i(w_t) dB_t^i, \qquad w_0 = x$$

The equation of the associated Brownian bridge $w_t(x,y)$ starting at $x$ and ending at $y$ after time 1 is

$$(5.2) \qquad dw_t(x,y) = \sum_{i=1}^{m} X_i(w_t(x,y))(dB_t^i + \Gamma_t^i(x,y)dt)$$

where

$$(5.3) \quad \Gamma_t^i(x,y) = \sum_{i=1}^{m} X_i(w_t(x,y)) \langle X_i(w_t(x,y)), \operatorname{grad} \log p_{1-t}(w_t(x,y)) \rangle$$

**Theorem 5.4.** *If the vector fields $X_i$ satisfy $H_1$ then, for all $x,y$ and $t \in [0,1]$, the Brownian bridge $w_t(x,y)$ is a semi-martingale.*

*Proof.* It is enough to show that

$$\int_0^1 E\left[ |\Gamma_t^i(x,y)| \right] dt < \infty$$

The law of $w_t(x,y)$ has density

$$p_t(x,z,y) = \frac{p_t(x,z) p_{1-t}(z,y)}{p_1(x,y)}$$

so, using (5.3), it is enough to show that

$$\int_0^1 \int_N p_t(x,z) | \langle X_i(z), \operatorname{grad} p_{1-t}(z,y) \rangle | \, dz$$

is finite. Using estimates of Jerison-Sanchez [23] and Kusuoka-Stroock [27] we know that there are constants $C_1, C_2, C_3, C_4$ such that

$$(5.5) \quad \frac{C_1}{V(x,\sqrt{t})} \exp\left( -\frac{d(x,y)^2}{C_2 t} \right) \le p_t(x,y) \le \frac{C_3}{V(x,\sqrt{t})} \exp\left( -\frac{d(x,y)^2}{C_4 t} \right)$$

where $t \in [0,1]$, $(x,y) \in N \times N$, $d(x,y)^2$ is the hypoelliptic distance associated to the operator, and $V(x, \sqrt{t})$ is the volume of the hypoelliptic ball with centre $x$ and radius $\sqrt{t}$. We also know that there are constants $C_5$ and $C_6$ such that

$$(5.6) \qquad |\langle X_i(x), \operatorname{grad} p_t(x,y) \rangle| \leq \frac{C_5}{\sqrt{t}V(x,\sqrt{t})} \exp\left(-\frac{d(x,y)^2}{C_6 t}\right).$$

We will use the doubling inequality: there is a constant $C_7$ such that

$$(5.7) \qquad\qquad V(x, 2t) \leq C_7 V(x,t)$$

From (5.5), (5.6), and (5.7) we deduce that there exists an integer $p$ such that

$(5.8)$
$$p_t(x,z)|\langle X_i(z), \operatorname{grad} p_{1-t}(z,y)\rangle|$$

$$\leq \frac{C_3 C_5}{V(x,\sqrt{t})V(z,\sqrt{1-t})\sqrt{1-t}} \exp\left(-\frac{d(x,z)^2}{C_4 t}\right) \exp\left(-\frac{d(z,y)^2}{C_6(1-t)}\right)$$

$$\leq \frac{K}{V(x, 2^p\sqrt{t})V(z, 2^p\sqrt{1-t})\sqrt{1-t}} \exp\left(-\frac{d(x,z)^2}{2^p C_2 t}\right) \exp\left(-\frac{d(z,y)^2}{2^p C_6(1-t)}\right)$$

for some constant $K$. Together (5.5) and (5.8) show that

$$\int_M p_t(x,z)|\langle X_i(z), \operatorname{grad} p_{1-t}(z,y)\rangle|\,dz$$

$$(5.9) \qquad\qquad \leq \frac{K'}{\sqrt{1-t}} \int_N p_{2^p t}(x,z) p_{2^p(1-t)}(z,y)\,dz$$

$$\leq \frac{K' p_{2^p}(x,y)}{\sqrt{1-t}}$$

From this it follows that

$$\int_0^1 \int_N p_t(x,y)|\langle X_i(z), \operatorname{grad} p_{1-t}(z,y)\rangle|\,dz$$

is finite using the fact that

$$\int_0^1 \frac{dt}{\sqrt{1-t}} < \infty. \quad \square$$

Let us now consider the inhomogeneous case of the diffusion associated to

$$\frac{1}{2}X_0 + \frac{1}{2}\sum_{i=1}^m X_i^2.$$

In this case, in general, we do not have the Jerison-Sanchez estimates for $p_t(x, y)$ so we use another approach which follows computations of Kusuoka-Stroock [28, part 2]. Let us now make the following assumption on the $X_i$:

$(H_2)$ At each $x \in N$ the space spanned by the Lie brackets of length $\geq 2$ constructed using $X_i$, $i = 0, \ldots, m$ and $X_j$, $j = 1, \ldots, m$ is equal to $T_x N$.

The solution of the Stratonovitch equation

$$(5.10) \qquad dw_t(x) = X_0(w_t(x))dt + \sum_{i=1}^{m} X_i(w_t(x))dB_t^i, \qquad w_0(x) = x$$

has, at time 1, a smooth density $p_1(x, y)$. Let us denote by $P_1(dy)$ the law of probability $p_1(x, y)dy$. Let us consider the equation of the Brownian bridge arriving at $y$ after time 1

$$(5.11) \qquad dw_t(x, y) = X_0(w_t(x, y))dt + \sum_{i=1}^{m} X_i(w_t(x, y))(dB_t^i + \Gamma_t^i(x, y)dt),$$

$$w_0(x, y) = x$$

where $\Gamma_t^i(x, y)$ is as in (5.3). We have the following theorem.

**Theorem 5.12.** For almost all $y$ for $P_1(dy)$ the process $w_t(x, y)$ is a semi-martingale.

This theorem says that $w_t(x)$ is still a semi-martingale for the filtration augmented by $w_1(x)$.

*Proof.* Let us first give a stochastic interpretation of

$$F(x, y) = \langle X_i(x), \operatorname{grad} \log p_t(x, y) \rangle .$$

Consider the measure $\mu$ on $N$ defined by

$$\mu(f) = \int_N F(x, y)f(y)p_t(x, y)dy$$

$$(5.13) \qquad = \left\langle X_i(x), \int_N \operatorname{grad} p_t(x, y)f(y)dy \right\rangle$$

$$= \langle X_i(x), E[f(w_t(x))] \rangle$$

$$= E\left[ \left\langle df(w_t(x)), \frac{\partial w_t}{\partial x}(x)X_i(x) \right\rangle \right]$$

By using an integration by parts formula we get

$$(5.14) \qquad E\left[ \left\langle df(w_t(x)), \frac{\partial w_t}{\partial x}(x)X_i(x) \right\rangle \right] = E[\xi_t(x)f(w_t(x))]$$

and this shows us that

(5.15)
$$|\langle X_i(x), \text{grad} \log p_t(x,y)\rangle| = |E\left[\xi_t(x) : w_t(x) = y\right]|$$
$$\le E\left[|\xi_t(x)| : w_t(x) = y\right].$$

Now let us estimate the quantity

$$E^{w_1(x)}\left[E^{w.(x,y)}\left[|\Gamma_t^i(x,y)|\right]\right].$$

We see that
(5.16)
$$E^{w_1(x)}\left[E^{w.(x,y)}\left[|\Gamma_t^i(x,y)|\right]\right]$$
$$\le \int_{N\times N} \frac{p_t(x,z)p_{1-t}(z,y)}{p_1(x,y)} E\left[|\xi_{1-t}(z)| : w_{1-t}(z) = y\right] p_1(x,y)\,dy\,dz$$
$$= \int_{N\times N} p_t(x,z)p_{1-t}(z,y)E\left[|\xi_{1-t}(z)| : w_{1-t}(z) = y\right]\,dy\,dz$$
$$= \int_N p_t(x,z)dz \int_N E\left[|\xi_{1-t}(z)| : w_{1-t}(z) = y\right]p_{1-t}(z,y)\,dy$$
$$= \int_N p_t(x,z)E\left[|\xi_{1-t}(z)|\right]dz$$

where the last equality follows from the definition of the conditional expectation. So, to prove the theorem, it is enough to show that

$$E\left[|\xi_{1-t}(z)|\right] \le \frac{C}{\sqrt{1-t}}$$

for some constant $C$ independent of $z$.

We follow the method of [28, Chapter 2]. Let $\alpha = (\alpha_1, \ldots, \alpha_n)$ be a sequence of $n$ integers between 0 and $n$ and define $\|\alpha\|$ to be the number of $\alpha_i$ which are non-zero plus twice the number of $\alpha_i$ which are zero. Now we put

$$X_{(\alpha,i)} = [X_i, X_\alpha],$$
$$X_{(i)} = X_i, \qquad \text{for } i \ne 0,$$

and then define

(5.17)
$$G_l(t,x) = \sum_{i=1}^m \sum_{\|\alpha\|\le l-1} t^{\|\alpha\|+1} X_{(\alpha,i)}(x) \otimes X_{(\alpha,i)}(x)$$

By hypothesis there is an integer $l_0$ such that $G_{l_0}(t, x)^{-1}$ exists and is smooth on $N$. From the compactness of $N$ there exist

$$a_{\alpha,\beta} \in C^\infty(N, \mathbb{R})$$

such that

$$(5.18) \qquad X_{\langle\alpha\rangle} = \sum_{\beta \neq 0, \|\beta\| \leq l_0} a_{\alpha,\beta}(x) X_{\langle\beta\rangle}(x)$$

for any $\alpha$ with $\|\alpha\| \geq l_0$. This follows since, for any vector field $X(x)$,

$$(5.19) \qquad X(x) = \sum_{\alpha \neq 0, \|\alpha\| \leq l_0} \langle X(x), G_{l_0}(1, x)^{-1} X_{\langle\alpha\rangle}(x) \rangle \, X_{\langle\alpha\rangle(x)}.$$

Let $\varphi_t(w, x)$ be the stochastic flow associated to (5.10); then it follows from [27] that

$$(5.20) \qquad \left(\frac{\partial \varphi_t}{\partial x}\right)^{-1} X_{\langle\alpha\rangle}(x) = \sum_{\beta \neq 0, \|\beta\| \leq l_0} c_{\alpha,\beta}(t, x) X_{\langle\beta\rangle}(x)$$

where

$$(5.21) \qquad E\left[\,|c_{\alpha,\beta}(t, x)|^p\,\right]^{1/p} \leq C(p)(\sqrt{t})^{\|\beta\| - \|\alpha\|}.$$

Now let

$$M_t(x) = \langle Dw_t(x), Dw_t(x) \rangle$$

be the Malliavin matrix associated to $w_t(x)$. Classically

$$M_t(x) = \int_0^t \sum_{i=0}^m \left\langle \left(\frac{\partial \varphi_s}{\partial x}\right)^{-1} X_i(w_s(x)), - \right\rangle^2 ds$$

and, moreover,

$$(5.22) \qquad M_t(x) = \sum_{\alpha \neq 0, \|\alpha\| \leq l_0} \sum_{\beta \neq 0, \|\beta\| \leq l_0} e_{\alpha,\beta}(t, x) X_{\langle\alpha\rangle}(x) \otimes X_{\langle\beta\rangle}(x)$$

where

$$(5.23) \qquad \sup_x E\left[|e_{\alpha,\beta}(t, x)|^p\right]^{1/p} \leq C(p)(\sqrt{t})^{\|\beta\| - \|\alpha\|}.$$

We get the following lemma, analogous to Theorem (2.8) of [28].

**Lemma 5.24.** Let

$$\lambda(t, x) = \inf \left\{ \langle \xi, M_t(x)\xi \rangle : \xi \in T_x N, \langle \xi, G_{l_0}(t, x)\xi \rangle = 1 \right\};$$

then there exist constants $C$, $c > 0$, and $\nu > 0$ such that

(5.25)          $$P \left\{ \lambda(t, x) < \frac{1}{K} \right\} \leq C \exp\left(-cK^\nu\right))$$

*Proof.* Since

$$\langle \xi, M_{t/K}(x)\xi \rangle = \sum_{i=1}^d \int_0^{t/K} \left\langle \xi, \left( \frac{\partial \varphi_s}{\partial x} \right)^{-1} X_i(w_s(x)) \right\rangle^2 ds$$

it follows that $\langle \xi, M_{t/K}(x)\xi \rangle$ is bounded below by
(5.26)

$$\sum_{i=1}^d \int_0^{t/K} \left( \sum_{\|\alpha\| \leq l_0 - 1} \theta^\alpha(t) \langle \xi, X_{\langle \alpha, i \rangle}(x) \rangle \right)^2 ds - \sum_{i=1}^d \int_0^{t/K} \langle \xi, R_i(s, x) \rangle^2 ds$$

where the $\theta_\alpha$ are given by [27, Appendix]. The $\theta_\alpha$ satisfy

(5.27)          $$\left( \frac{\partial \varphi_s}{\partial x} \right)^{-1} X_i(w_s(x)) = \sum_{\|\alpha\| \leq l_0 - 1} \theta^\alpha(s) X_{\langle \alpha, i \rangle}(x) + R_i(s, x)$$

with

(5.28)          $$R_i(s, x) = \sum_{\|\beta\| \leq l_0} r_i^{(\beta)}(s, x) X_{\langle \beta \rangle}(x).$$

Now

$$\int_0^{t/K} \left( \sum_{\|\alpha\| \leq l_0 - 1} \theta^\alpha(s) \langle \xi, X_{\langle \alpha, i \rangle}(x) \rangle \right)^2 ds$$

$$= \int_0^{1/K} \left( \sum_{\|\alpha\| \leq l_0 - 1} \theta^\alpha(s) \left\langle \xi, t^{\|\alpha\|/2} X_{\langle \alpha, i \rangle}(x) \right\rangle \right)^2 ds$$

and so from Theorem A.6 in the appendix of [27] we deduce the following estimate:
(5.29)

$$P \left\{ \inf_{\xi \in S} \left\{ \sum_{i=1}^d \int_0^{t/K} \left( \sum_{\|\alpha\| \leq l_0 - 1} \theta^\alpha(s) \langle \xi, X_{\langle \alpha, i \rangle}(x) \rangle \right)^2 ds \right\} < 2K^{-l_0 - 1/3} \right\}$$

$$\leq C'' \exp(-c'' K^{\nu'})$$

where
$$S = \{\xi : \langle \xi, G_{l_0}(t,x)\xi \rangle = 1\}.$$

On the other hand

(5.30)
$$|\langle \xi, R_i(s,x)\rangle| \leq \sum_{\|\beta\| \leq l_0} \left| r_i^\beta(s,x)\langle \xi, X_{\langle \beta \rangle}(x)\rangle \right|$$

$$\leq \frac{t^{-l_0+1}}{K} \sum_{\|\beta\| \leq l_0} \left| \sup_{s \leq t/K} r_i^\beta(s,x) \right|.$$

So we need only to find an upper bound for

$$P\left\{ \sum_{i=1}^d \sum_{\|\beta\| \leq l_0} \sup_{0 \leq s \leq t/K} |r_i^\beta(s,x)| > \left(\frac{t}{K}\right)^{l_0 - 1/3} \right\}.$$

By the computations of the appendix of [27] this is smaller than

$$C'' \exp(-c'' K^{\nu''}) \qquad \square$$

From the previous lemma we deduce the following lemma which is analogous to Corollary (2.10) of [28].

**Lemma 5.31.** *There is a constant $C(p)$ such that for all $x$ and $t \leq 1$*

(5.32)
$$E\left[ \left| \langle X_{\langle \alpha \rangle}(x), M_t(x)^{-1} X_{\langle \beta \rangle}(x)\rangle \right|^p \right]^{1/p} \leq C(p) t^{-\frac{\|\alpha\| + \|\beta\|}{2}}.$$

*for $\alpha \neq 0$, $\beta \neq 0$, $\alpha \leq l_0$, and $\beta \leq l_0$.*

*Proof.* We have

(5.33) $\langle X_{\langle \alpha \rangle}(x), M_t(x)^{-1} X_{\langle \beta \rangle}(x)\rangle^p$

$$\leq \langle X_{\langle \alpha \rangle}(x), M_t(x)^{-1} X_{\langle \alpha \rangle}(x)\rangle^{p/2} \langle X_{\langle \beta \rangle}(x), M_t(x)^{-1} X_{\langle \beta \rangle}(x)\rangle^{p/2}$$

so it is enough to show (5.32) when $\alpha = \beta$. Let us write

$$G_{l_0}(t,x) = UU^t,$$

where $U^t$ is the transpose of $U$, and

$$M_t(x) = U \begin{pmatrix} \lambda_1 \dots 0 \\ \vdots \ddots \vdots \\ 0 \dots \lambda_d \end{pmatrix} U^t.$$

Then lemma (5.24) shows that

$$
\begin{pmatrix}
\lambda_1 \dots 0 \\
\vdots \ddots \vdots \\
0 \dots \lambda_d
\end{pmatrix}^{-1}
$$

belongs to all the $L^p$ uniformly in $x$ and $t$. So it is enough to show that

(5.34) $$E\left[\left|U^{-1}X_{(\alpha)}(x)\right|^p\right]^{1/p} \leq \frac{C}{t^{\|\alpha\|/2}}$$

However we have

(5.35) $$
\begin{aligned}
UU^t &= \sum_{\alpha \neq 0, \|\alpha\| \leq l_0} t^{\|\alpha\|} X_{(\alpha)}(x) \otimes X_{(\alpha)}(x) \\
&= \sum_{\alpha \neq 0, \|\alpha\| \leq l_0} t^{\|\alpha\|} X_{(\alpha)}(x) X_{(\alpha)}^t(x).
\end{aligned}
$$

But

(5.36) $$
\begin{aligned}
U^{-1}UU^t(U^t)^{-1} = 1 &= \sum_{\alpha \neq 0, \|\alpha\| \leq l_0} t^{\|\alpha\|} U^{-1}X_{(\alpha)}(x) X_{(\alpha)}^t(x)(U^t)^{-1} \\
&= \sum_{\alpha \neq 0, \|\alpha\| \leq l_0} t^{\|\alpha\|} U^{-1}X_{(\alpha)}(x) \otimes U^{-1}X_{(\alpha)}(x)
\end{aligned}
$$

and this shows that

$$\left|U^{-1}X_{(\alpha)}(x)\right| \leq Ct^{-\|\alpha\|/2}.$$

This proves (5.34) and so completes the proof of Lemma (5.31). $\square$

As a consequence of Lemma (5.31) we get the following result. We use the notation $\| - \|_{p,1}$ for the Sobolev $(p,1)$-norm.

**Lemma 5.37.** For all $x$, $t \leq 1$, $\alpha \neq 0$, $\beta \neq 0$ where $\|\alpha\| \leq l_0$ and $\|\beta\| \leq l_0$

(5.38) $$\|\langle X_{(\alpha)}(x), M_t(x)^{-1}X_{(\beta)}(x)\rangle\|_{p,1} \leq C(p)t^{-\frac{\|\alpha\|+\|\beta\|}{2}}$$

*Proof.* Using (5.23) and computing derivatives we get

(5.39) $$D\langle X_{(\alpha)}(x), M_t(x)^{-1}X_{(\beta)}(x)\rangle =$$

$$\sum De_{\gamma_1,\gamma_2}(t,x)\langle X_{(\alpha)}(x), M_t(x)^{-1}X_{(\gamma_1)}(x)\rangle \langle X_{(\gamma_2)}(x), M_t(x)^{-1}X_{(\beta)}(x)\rangle$$

The sum in this formula is taken over the set of $\gamma_1$ and $\gamma_2$ such that

$$\gamma_i \neq 0, \quad \|\gamma_i\| \leq l_0.$$

We can improve (5.23) to the following estimate

$$(5.40) \qquad \sup_{x} \|e_{\alpha,\beta}(t,x)\|_{p,1} \leq C(p,1)(\sqrt{t})^{\|\beta\|+\|\alpha\|}$$

The lemma now follows from (5.40) and (5.22). $\square$

We can now compute explicitly the term $\xi_t(x)$ which appears in (5.14). Let us recall (5.13) and write, for $j \neq 0$

$$(5.41) \qquad X_j(x) = \sum_{j=1}^{m} \int_0^t \left(\frac{\partial\varphi_s}{\partial x}\right)^{-1} X_i(w_s(x))h_{i,j}(s)ds$$

with

$$(5.42) \qquad h_{i,j}(s) = \left\langle \left(\frac{\partial\varphi_s}{\partial x}\right)^{-1} X_i(w_s(x)), M_t(x)^{-1}X_j(x) \right\rangle$$

We have

$$(5.43) \qquad \begin{aligned} X_j E\left[f(w_t(x))\right] &= E\left[\langle Df(w_t(x)), h\rangle\right] \\ &= E\left[f(w_t(x))\delta h\right] \end{aligned}$$

where $\delta h$ is the Skorohod integral of $h$ [31]. Let us recall the following fact from [31]. If $h$ is a process in the Cameron-Martin space, $F$ a scalar Wiener functional

$$(5.43) \qquad \delta(hF) = F\delta h - \langle h, DF\rangle$$

By using the fact that the Skorohod integral of a non-anticipative process is nothing more than the Itô integral and by using lemma (5.?) and (5.20) we get

$$(5.44) \qquad E\left[|\delta h|^p\right]^{1/p} \leq \frac{C}{\sqrt{t}}$$

**Lemma 5.45.** *The finite variational part of the semi-martingale $w_t(x,y)$, denoted by*

$$\int_0^t V_s(x,y)\,ds$$

*satisfies*

$$E^{x,(x,y)}\left[|V_s(x,y)|^p\right]^{1/p} \leq \frac{C}{\sqrt{1-s}}.$$

Note that this is exactly the Lemma required to deduce the inequality

$$\int_0^t E_s^{u,v}\left[\left|\tilde{\Gamma}_i(t',s)\right|^{2p}\right]^{1/2p} E_s^{u,v}\left[|J(l-1,t')|^{2p}\,dt'\right]^{1/2p} dt' \leq$$
$$\int_0^s \frac{C(s,p)}{\sqrt{s-t'}}\, E_s^{u,v}\left[|J(l-1,t')|^{2p}\right]^{1/2p} dt'$$

used in the proof of (4.15). This gives a slight improvement on Theorem (5.12). Our technique will give the analogue of [28, (2.18)] in our situation.

*Proof of Lemma* (5.45). We can in fact improve (5.16)
(5.46)
$$E^{w_1(x)}\left[E^{w.(x,y)}\left[\Gamma_{1-t}^i(x,y)^p\left(\sqrt{1-t}\right)^p\right]\right]$$
$$= \int_{N\times N} p_t(x,z)E\left[(\sqrt{1-t})\xi_{1-t}(z)\,|\,x_{1-t}(z)=y\right]^p p_{1-t}(z,y)dy$$
$$\leq \int_N p_t(x,z)dz \int_N E\left[(\sqrt{1-t})^p\,\xi_{1-t}^p(z)\,|\,x_{1-t}(z)=y\right] p_{1-t}(z,y)dy$$
$$= \int_N p_t(x,z)dz E\left[(\sqrt{1-t})^p\,\xi_{1-t}^p(z)\right] \leq C$$

So we have shown that for almost all $y$

$$w_t(x,y) = \text{Martingale} + \int_0^1 V_s(x,y)ds$$

where

$$E^{x.(x,y)}\left[|V_s(x,y)|^p\right]^{1/p} \leq \frac{C}{\sqrt{1-s}}$$

for all $p$.

*Remark 5.47.* Suppose that the Lie brackets of length 2 span $T_xN$ for all $x$ and that the horizontal space over $N$ is of constant dimension $d'$. We are of course in the situation of Theorem (5.12). In that case, it follows from the Nagel-Stein-Wainger estimates [33] that the volume of the hypoelliptic ball is bounded by

$$C(\sqrt{t})^{2d-d'} \leq V(x,\sqrt{t}) \leq C'(\sqrt{t})^{2d-\delta} = C'(\sqrt{t})^\nu$$

uniformly for $x \in M$ and $t \leq 1$. From a recent result of [36], we get for all $\varepsilon > 0$ and for $x,y \in M$ and $t \leq 1$

$$\frac{C'(\varepsilon)}{V(x,\sqrt{t})}\exp\left(-\frac{d(x,y)^2}{2t(1-\varepsilon)}\right) \leq p_t(x,y) \leq \frac{C(\varepsilon)}{V(x,\sqrt{t})}\exp\left(-\frac{d(x,y)^2}{2t(1+\varepsilon)}\right).$$

From [28, Theorem (2.18)] we deduce, for any $\delta < 1$, the existence of $C(\delta)$ such that

$$|\langle X_i(x), p_t(x,y)\rangle| = \left| \int \langle X_i(x), p_{t\delta}(x,z)\rangle \, p_{t(1-\delta)}(z,y) dz \right|$$

$$\leq \frac{C(\delta)}{\sqrt{t}} \left( \int_M p_{t\delta}(x,z) p_{t(1-\delta)}(z,y)^2 dz \right)^{1/2} \quad (5.48)$$

By substituting the above bounds for $V(x, \sqrt{t})$ and $p_t(x,y)$ in the inequality (5.48) we get

$$|\langle X_i(x), p_t(x,y)\rangle|$$

$$\leq \frac{C(\delta, \varepsilon)}{(\sqrt{t})^{1+\nu/2}} \left( \int_M \exp\left( -\frac{d(x,z)^2}{2\delta t(1+\varepsilon)} \right) \exp\left( -\frac{d(z,y)^2}{(1-\delta)t(1+\varepsilon)} \right) dz \right)^{1/2}$$

$$\leq \frac{C(\delta, \varepsilon)}{(\sqrt{t})^{1+\nu/2}} p^{\frac{1}{2}}_{t(\frac{1+\varepsilon}{1-\varepsilon})(\frac{1}{2}+\frac{\delta}{2})}(x,y)$$

$$\leq \frac{C(\delta, \varepsilon)}{(\sqrt{t})^{1+\nu}} \exp\left( -\frac{d(x,y)^2}{2t\left(\frac{1+\varepsilon}{1-\varepsilon}\right)(1+\delta)} \right)$$

From this we deduce that for any $\varepsilon$

$$|\langle X_i(x), p_t(x,y)\rangle| \leq \frac{C(\delta, \varepsilon)}{(\sqrt{t})^{1+N}} \exp\left( -\frac{d(x,y)^2}{2t(1+\varepsilon)} \right)$$

and that

$$E^{w.(x,y)} \left[ |V_{1-t}(x,y)|^p \left( \sqrt{1-t} \right)^p \right] < C(p).$$

## REFERENCES

1. M.F. Atiyah, *Circular symmetry and stationary phase approximation*, Astérisque **131** (1985), 311-323.

2. R. Azencott, *Densité des diffusions en temps petit*, Seminaire de probabilité XVII, Lecture Notes in Mathematics, vol. 1059, Springer-Verlag, Berlin-Heidelberg-New York, 1984, pp. 402-499.

3. R. Azencott, A. Bellaïch, C. Bellaïch, P. Bougerol, M. Maurel, P. Baldi, L. Elie, and J. Granara, *Geodesiques et diffusions en temps petit*, Astérisque **84-85** (1981).

4. G. Ben-Arous, *These de troisième cycle*, Université Paris VI, 1981.

5. G. Ben-Arous and R. Leandre, *Decroissance exponentielle du noyau de la chaleur sur le diagonale*, Probability theory and related fields (to appear).

6. J.-M. Bismut, *Mecanique aléatoire*, Lecture Notes in Mathematics, vol. 866, Springer-Verlag, Berlin-Heidelberg-New York, 1981.

7. J.-M. Bismut, *Index theorem and equivariant cohomology on the loop space*, Comm. Math. Phys. **98** (1985), 213-237.

8. J.-M. Bismut, *Large deviations and the Malliavin calculus*, Progress in mathematics, vol. 45, Birkhäuser, Basel, 1989.

9. E. Carlen and P. Kree, *$L^p$-estimates on iterated stochastic integrals* (to appear).

10. M. Chaleyat-Maurel and T. Jeulin, *Grossisement gaussien de la filtration Brownienne*, Grossisement de filtrations: exemplement et applications (T. Jeulin and M. Yor, eds.), Lecture Notes in Mathematics, vol. 1118, Springer-Verlag, Berlin-Heidelberg-New York, 1984, pp. 59-109.

11. K.T. Chen, *Iterated path integrals of differential forms and loop space homology*, Annals of Math. **97** (1983), 217-246.

12. A. Connes, *Non-commutative differential geometry*, Publ. Math. IHES **62** (1985), 257-360.

13. A. Connes, *Entire cyclic cohomology of Banach algebras and characters of θ-summable Fredholm modules*, K-theory **1** (1988), 519-548.

14. D. Elworthy, *Stochastic differential equations on manifolds*, London Mathematical Society Lecture Notes, vol. 70, Cambridge University Press, Cambridge, 1982.

15. M. Emery and R. Leandre, *Sur une formule de Bismut*, Seminaire de Probabilité XXIV, Lecture Notes in Mathematics, vol. 1426, Springer-Verlag, Berlin-Heidelberg-New York, 1990, pp. 448-452.

16. E. Getzler, *Cyclic homolgy and the path integral of the Dirac operator* (to appear).

17. E. Getzler, J.D.S. Jones, and S.B. Petrack, *Differential forms on loop spaces and the cyclic bar complex*, Topology (to appear).

18. E. Getzler and A. Szenes, *On the Chern character of a theta summable Fredholm module*, Journal of Functional Analysis **89** (1989), 343–358.

19. Y.Z. Hu and P.A. Meyer, *Chaos de Wiener et integrale de Feynmann*, Seminaire de Probabilités XXII (J. Azema, P.A. Meyer, and M. Yor, eds.), Lecture Notes in Mathematics, vol. 1321, Springer-Verlag, Berlin-Heidelberg-New York, 1988, pp. 51–72.

20. Y.Z. Hu and P.A. Meyer, *Sur les integrales multiples de Stratonovitch*, Seminaire de Probabilités XXII (J. Azema, P.A. Meyer, and M. Yor, eds.), Lecture Notes in Mathematics, vol. 1321, Springer-Verlag, Berlin-Heidelberg-New York, 1988, pp. 72–82.

21. Y.Z. Hu, *Calculs formels sur les equations de Stratonovitch*, Seminaire de Probabilités XXIV (J. Azema, P.A. Meyer, and M. Yor, eds.), Lecture Notes in Mathematics, vol. 1426, Springer-Verlag, Berlin-HeidelbergNew York, 1990, pp. 453–461.

22. N. Ikeda and S. Watanabe, *Stochastic differential equations and diffusion processes*, North Holland, 1981.

23. D. Jerison and A. Sanchez, *Estimates for the heat kernel for the sum of squares of vector fields*, Indiana Univ. Math. Jour. **35** (1986), 835–859.

24. J.D.S. Jones, *Cyclic homology in geometry and topology*, Bull. Lond. Math. Soc (to appear).

25. J.D.S. Jones and S.B. Petrack, *The fixed point theorem in equivariant cohomology*, Transactions of the A.M.S. **322** (1990), 35–49.

26. D. Kastler, *Introduction to entire cyclic cohomology*, Stochastics, algebra, and analysis in classical and quantum mechanics (S. Albeverio, Ph. Blanchard, and D. Testard, eds.), Kluwer, Academic Press, Dordrecht-Boston-London, 1990, pp. 73–153.

27. S. Kusuoka and D. Stroock, *Applications of the Malliavin calculus, Part II*, J. Fac. Science University Tokyo **32** (1985), 1–76.

28. S. Kusuoka and D. Stroock, *Applications of the Malliavin calculus, Part III*, J. Fac. Science University Tokyo **34** (1987), 391–442.

29. R. Leandre and P.A. Meyer, *Sur le developpement en chaos de Wiener d'une diffusion*, Seminaire de Probabilités XXIII (J. Azema, P.A. Meyer, and M. Yor, eds.), Lecture Notes in Mathematics, vol. 1372, Springer-Verlag, Berlin-Heidelberg-New York, 1989, pp. 161–165.

30. P. Malliavin, *Stochastic calculus of variation and hypoelliptic operators*, Proceedings of International Conference in Stochastic Differential Equations (K. Itô, eds.), Wiley, 1978, pp. 195–263.

31. E. Nualart and E. Pardoux, *Stochastic calculus with anticipating integrands*, Probability theory and related fields **78** (1988), 535–581.

32. M.S. Narasimhan and S. Ramanan, *The existence of universal connections*, American J. Math. **83** (1961), 563–572.

33. A. Nagel, E. Stein, and M. Wainger, *Balls and metrics defined by vector fields*, Acta Math. **155** (1985), 103–197.

34. L. Schwartz, *Construction directe d'une diffusion sur une varieté*, Seminaire de Probabilités XIX (J. Azema, P.A. Meyer, and M. Yor, eds.), Lecture Notes in Mathematics, vol. 1123, Springer-Verlag, Berlin-Heidelberg-New York, 1984, pp. 91–113.

35. L. Schwartz, *La convergence de la serie de Picard pour les equations differentielle stochastiques*, Seminaire de Probabilités XXIII (J. Azema, P.A. Meyer, and M. Yor, eds.), Lecture Notes in Mathematics, vol. 1372, Springer-Verlag, Berlin-Heidelberg-New York, 1984, pp. 343-355.

36. N. Varopoulos, *Small time Gaussian estimates of heat diffusion kernels II. The theory of large deviations*, Journal of Functional Analysis **93** (1990), 1–34.

MATHEMATICS INSTITUTE,
UNIVERSITY OF WARWICK,
COVENTRY CV4 7AL,
ENGLAND


DÉPARTEMENT DE MATHÉMATIQUES,
ULP,
RUE DESCARTES,
67084 STRASBOURG,
FRANCE

# Convex Geometry and Nonconfluent Γ-Martingales I: Tightness and Strict Convexity.

Wilfrid S. Kendall.

Department of Statistics, University of Warwick.

## 1. Introduction.

In Kendall (1990) it is explained how three nonlinear Dirichlet problems are closely connected to a problem about the existence of a certain convex surrogate distance function. Here we consider an aspect of these relationships in a setting more general than the Riemannian case of Kendall (1990). The problems are as follows. Suppose M is a smooth manifold equipped with a connection Γ and separately with a reference Riemannian structure (the connection need *not* be compatible with the metric!). Consider $\mathcal{B}$ a closed region in M. (In the sequel $\mathcal{B}$ is generally compact, but we prefer to state the following properties as applicable to a general region.)

(A): *Does $\mathcal{B}$ have Γ-convex geometry?* That is to say, does there exist a (product-connection) convex function $\mathcal{Q} : \mathcal{B} \times \mathcal{B} \to [0,1]$ vanishing only on the diagonal $\Delta = \{(x,x) : x \in \mathcal{B}\}$? Here the "Γ-" in "Γ-convex" refers to the use of the connection Γ to build the product-connection, instead of the Levi–Civita connection supplied by the reference Riemannian metric. (In the rest of the paper the prefix "Γ-" is omitted; by "convex" we mean "Γ-convex" unless indicated otherwise.)

(B): *Dirichlet problem for Γ-martingales lying in $\mathcal{B}$.* This problem requires one to find Γ-martingales $X$ (under a given filtration) attaining a given terminal value $X(\infty)$. In the following the heading (B) refers specifically to whether the Dirichlet problem is *well–posed* and has *unique solutions*. By "well–posed" we mean a non-uniform interpretation: if $X^1$, $X^2$, ..., $X^\infty$ are Γ-martingales converging in $\mathcal{B}$ as time tends to infinity, subject to the same filtration, and such that $X^n(\infty) \to X^\infty(\infty)$ in probability then

$$\lim_{n \to \infty} \sup_{t \in [0,\infty)} \text{dist}(X^n(t), X^\infty(t)) \quad = \quad 0$$

using convergence in probability. We do not use well–posedness for **(B)** in the following, but it will play a part in later work. Note that the question of uniqueness is also described in the literature as the problem of *nonconfluence* of $\Gamma$-martingales for $\mathcal{B}$, explaining the title of this paper.

**(C)**: *Dirichlet problem for $\Gamma$-harmonic maps with image in $\mathcal{B}$*. This problem requires one to find $\Gamma$-harmonic map extensions $F : \mathcal{O} \to \mathcal{B}$ of the boundary-value data $\partial F : \partial \mathcal{O} \to \mathcal{B}$, where $\mathcal{O}$ is an open subset of a Riemannian manifold and $\partial \mathcal{O}$ is its (potential–theoretic) boundary. By "well–posed" here we mean: if $F^1$, $F^2$, ..., $F^\infty$ are $\Gamma$-harmonic maps from $\mathcal{O}$ to $\mathcal{B}$ such that $\partial F^n \to \partial F^\infty$ almost surely in $\mathcal{O}$-harmonic measure then

$$\lim_{n \to \infty} \sup \{\text{dist}(F^n(m), F^\infty(m)) \quad : \quad m \in \mathcal{K}\} \quad = \quad 0$$

for each compact set $\mathcal{K} \subset \mathcal{O}$. Again in the following the heading **(C)** refers specifically to whether the Dirichlet problem is *well–posed* and has *unique solutions*.

**(D)**: *Dirichlet problem for $\Gamma$-geodesics lying in $\mathcal{B}$*. This problem requires one to find $\Gamma$-geodesics $\gamma$ connecting given pairs of points in $\mathcal{B}$. Again in the following the heading **(D)** refers specifically to whether the Dirichlet problem is *well–posed* and has *unique solutions*.

Problem **(D)** can be viewed as a special case of problems **(B)** and **(C)**. The whole setting generalizes Kendall (1990) because a general connection $\Gamma$ is used rather than the Levi-Civita connection supplied by the reference Riemannian metric. The metric is used simply as a convenience; definitions and constructions are all essentially independent of the choice of metric.

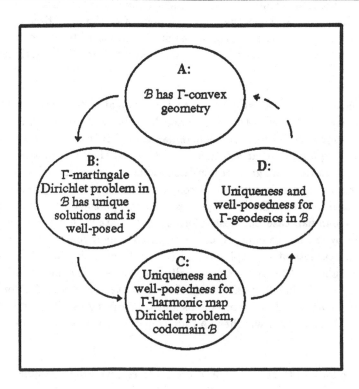

**Figure:** Three implications established in Kendall (1990) for a compact convex region $\mathcal{B}$ (in the Riemannian case; the case of general Γ is a straightforward extension), together with the Emery conjecture (dotted arrow).

The figure illustrates some elementary and fundamental logical relationships between these Dirichlet problems in the case of compact and convex $\mathcal{B}$. Suppose first that $\mathcal{B}$ is compact. Kendall (1990) treats the Riemannian case (Γ is the Levi–Civita connection for the metric) and establishes implications corresponding to the three solid arrows in the figure. In particular convex geometry (**A**) implies that if solutions exist for the problems (**B**) and (**C**) then they are unique and depend continuously on the data. Indeed, in the Riemannian case (Γ the Levi-Civita connection) Kendall (1990) shows that a consequence of (**A**) is that solutions *exist* for (**B**) and (**C**). The major part of the work of the 1990 paper concerns existence arguments rather than uniqueness; the three uniqueness implications are straightforward. They extend directly to the non–Riemannian case (indeed the existence implications also extend, since a check shows that the argument of section 8 of Kendall, 1990, allows one to use a reference Riemannian metric unrelated to the connection). In particular *convex geometry* of $B$ implies *nonconfluence* of Γ-martingales in $B$.

Emery, in a separate investigation into Γ-martingales (for example Emery, 1985; Emery and Meyer, 1989; Emery and Mokobodzki, preprint), has made the bold conjecture that in the Riemannian case if $B$ is *convex* and compact (convexity required so that solutions to (D) exist) then uniqueness for (D) implies (A). In the more general non-Riemannian case considered here he has a counterexample which suggests his conjecture should be strengthened to require in addition that (D) be well–posed (Emery, personal communication). In the Riemannian case our best information to date is that the conjecture is verified by explicit calculation of the required convex function for the case of compact subsets of hemispheres and (a slight generalization of this case) for closed subsets of regular geodesic balls lying in general complete Riemannian manifolds (Kendall, 1990, 1991). The next step is to consider compact regions in two-dimensional paraboloids and to verify by explicit (and probably tedious) calculation whether (A) holds in those instances for which (D) holds.

Were some form of the conjecture to prove correct, we would be able to close the circle of implications of the figure. This demonstrates the importance of Emery's conjecture — the extent to which it is true has profound implications for the nonlinear elliptic variational theory of harmonic maps! The figure suggests a plan for an indirect attack on Emery's conjecture: instead of attempting direct calculations one might try to reverse each in turn of the solid arrows of implication in the figure. In this approach the Emery conjecture is broken up into three possibly more manageable problems, listed as conjectures in section 4 below and each possessing an intrinsic interest.

The purpose of this note is to advertise this picture of Emery's conjecture and to make a small step towards fulfilling this indirect programme of attack. We shall show below (theorem 3.2) that under a natural condition on the compact region $B$ (that $B$ supports a strictly convex function) the uniqueness part of (B) implies (A). Thus if $B$ is strictly–convex–supporting and compact then *nonconfluence* of Γ-martingales in $B$ implies *convex geometry* of $B$. The property (B) can be viewed as defining a problem in stochastic control involving Γ-martingales. Our method is to show that the convex geometry required by (A) is provided by the value of the stochastic control problem.

Note that, in a discussion of set–valued barycentres, Emery and Mokobodzki (preprint) obtain results closely related to those given below. In particular it is possible to prove theorem 3.2 using their methods. Note also that Picard (1991) gives an alternative approach to existence aspects of the Dirichlet problem for Γ-martingales.

The following note was added just before this paper went to press: I believe I can now construct a counterexample to Emery's conjecture (uniqueness for **(D)** implies **(A)** for convex and compact $B$). Details will be published elsewhere.

The comments of an anonymous referee were helpful in improving the clarity of the exposition.

## 2. Weak Convergence and Γ-martingales.

In this preliminary section we establish an elementary criterion for the weak convergence of Γ-martingales. Our approach is based on a tightness criterion for "Itô processes" given in Stroock and Varadhan (1979). An alternative approach to Γ-martingale tightness is implicit in the treatment of set-valued barycentres in Emery and Mokobodzki (preprint); this uses the Skorokhod representation of weak convergence and starts with Rebolledo's theorem on weak convergence of continuous Euclidean martingales (Rebolledo, 1976).

Let $\mathcal{D}$ be a closed region in a smooth manifold **N**, such that **N** is furnished with a connection Γ and a reference Riemannian metric as given in the introduction. (Actually we have in mind $\mathcal{D} = B \times B \subset \mathbf{N} = \mathbf{M} \times \mathbf{M}$, with the product connection and metric.) Let $\omega$ denote the canonical continuous sample path stochastic process in $\mathcal{D}$, so that

$$\omega \quad \in \quad C([0, \infty); \mathcal{D}).$$

We endow $C([0, \infty); \mathcal{D})$ with the metric

$$\overline{d}(\omega, \omega') \quad = \quad \sum_{r=1}^{\infty} \left( 2^{-r} \sup_{t \in [0.r]} \frac{\mathrm{dist}(\omega(t), \omega'(t))}{1 + \mathrm{dist}(\omega(t), \omega'(t))} \right).$$

This makes $C([0, \infty); \mathcal{D})$ into a Polish space. We shall consider the weak convergence of Borel probability measures on this Polish space $C([0, \infty); \mathcal{D})$;

if $\mathbf{P}^n$ weakly converges to $\mathbf{P}$ in this sense then we write $\mathbf{P}^n \Rightarrow \mathbf{P}$. Note that the *topology* for $C([0, \infty); \mathcal{D})$ (and hence the weak convergence topology for probability measures on $C([0, \infty); \mathcal{D})$) does not depend on the reference Riemannian metric used for $\text{dist}(\omega(t), \omega'(t))$ – indeed the same topology and weak convergence of probability measures result if dist is replaced by the Euclidean metric inherited from a smooth embedding of $\mathcal{D}$ into Euclidean space (a fact exploited in the proof of Theorem 2.5 below).

For the notion of a Γ-martingale see Emery and Meyer (1989) or the introductory material in Kendall (1987, 1988, 1990). A Γ-martingale $Z$ on $\mathcal{D}$ is a manifold-valued semimartingale whose Γ-development is a Euclidean local martingale. Recall that the Γ-development $M$ of $Z$ is defined via the Γ-parallel transport $\Xi$ using the Stratonovich form of the equations of Cartan development and parallel transport:

$$\begin{aligned} d_S Z &= \Xi \, d_S M \\ d_S \Xi &= H_\Xi \, d_S Z \end{aligned} \qquad (1)$$

where $H_\Xi$ is the horizontal subspace sitting above $\Xi$ of the tangent frame bundle $T(\text{GL}(\mathcal{D}))$, given by and characterizing the connection Γ.

It is convenient to have the following notation for exit times from geodesic balls.

DEFINITION 2.1. *Exit times. For each $u > 0$ define the following stopping times for the canonical process $\omega$ of $C([0, \infty); \mathcal{D})$:*

$$\begin{aligned} T^{K;s} &= \inf\{u > s : \omega(u) \notin K\} \\ T^K &= \inf\{u > 0 : \omega(u) \notin K\} \end{aligned}$$

*under the convention, infima of empty sets are interpreted as $\infty$.*

We summarize the relationship between Γ-martingales and convexity in a theorem.

THEOREM 2.2. *The probability law $\mathbf{P}$ renders the canonical process $\omega$ as a Γ-martingale if and only if for all $t > s > 0$, all closed geodesic balls $B \subset \mathbf{N}$, all convex $\phi : B \to [0, 1]$*

$$\mathbf{E}^{\mathbf{P}}[\phi(\omega(T^{B;s} \wedge t))|F_s] \geq \phi(\omega(s)) \qquad \mathbf{P}\text{-almost surely.} \qquad (2)$$

*Moreover to test for the Γ-martingale property it suffices to establish the inequality only for smooth convex $\phi$, only for a dense subset of all possible closed geodesic balls B (under the topology induced by the parametrization $B \mapsto$ (centre, radius)), and only for a dense subset of possible $t > s > 0$.*

We give no proof, as this theorem follows immediately from results of Darling (1982) and Emery and Zheng (1984). Darling discusses the case of smooth convex $\phi$. The essence of the matter is to apply Itô's formula via (1) to $\phi(\omega)$, and to use the rich supply of locally strictly convex functions furnished by exponential coordinates additively perturbed by multiples of $x \mapsto \text{dist}(x, z)^2$ for $x$ near to $z$. Emery and Zheng discuss the case of general (nonsmooth) convex functions. Finally, it is enough to test only for a dense subset of geodesic balls because the Γ-martingale property is a local property, and a straightforward convergence argument shows that a countable dense subset of $t > s > 0$ suffices.

Our next task is to show that the space of all Γ-martingale laws is closed in the topology of weak convergence of probability measures.

DEFINITION 2.3. *The space of Γ-martingale laws. Let $\mathcal{D} \subset \mathbf{N}$ be as above. Define $\Gamma\mathcal{M}(\mathcal{D})$ to be the set of probability laws on the path–space $C([0, \infty); \mathcal{D})$ making the canonical process $t \mapsto \omega(t)$ into a Γ-martingale. Furthermore define $\Gamma\mathcal{M}_z(\mathcal{D})$ by*

$$\Gamma\mathcal{M}_z(\mathcal{D}) = \{\mathbf{P} \in \Gamma\mathcal{M}(\mathcal{D}) : \mathbf{P}[\omega(0) = z] = 1\}.$$

THEOREM 2.4. *Γ-martingales and weak convergence. The space of probability measures $\Gamma\mathcal{M}(\mathcal{D})$ is weakly closed in the space of probability measures on $C([0, \infty); \mathcal{D})$.*

*Proof:* Suppose $\mathbf{P}^n \Rightarrow \mathbf{P}$ with $\mathbf{P}^n \in \Gamma\mathcal{M}(\mathcal{D})$ for all $n$. We have to show the canonical process $\omega$ is a Γ-martingale under $\mathbf{P}$.

From $\mathbf{P}^n \in \Gamma\mathcal{M}(\mathcal{D})$ it follows

$$\mathbf{E}^{\mathbf{P}^n}[\phi(\omega(T^{B;s} \wedge t))|F_s] \geq \phi(\omega(s)) \qquad \mathbf{P}^n\text{-almost surely.} \tag{3}$$

for all $t > s > 0$, all geodesic balls $B = \text{ball}(z, R)$ in $\mathbf{N}$ and all convex $\phi : B \to [0, 1]$. The inequality persists in the limit precisely when $f^{B;s,t} : \omega \mapsto$

$\phi(\omega(T^{B;s} \wedge t))$ is continuous at $\omega$ for $\mathbf{P}$-almost all $\omega$. Now this latter function is in general only lower–semicontinuous. However for each fixed $z$, for all but a countable number of $R$ (the choice of which depends on $z$, $t$, and $s$), the function $f^{\text{ball}(z,R);s,t}$ is continuous at $\omega$ for $\mathbf{P}$-almost all $\omega$ (see Stroock and Varadhan, 1979, Lemma 11.1.2).

Thus for a countable dense subset of $t > s > 0$, for a countable dense set of geodesic balls $B \subset \mathbf{N}$ and all convex $\phi : B \to [0,1]$ it follows

$$\mathbf{E}^{\mathbf{P}}\left[\phi(\omega(T^{B;s} \wedge t))|F_s\right] \geq \phi(\omega(s)) \qquad \mathbf{P}\text{-almost surely.} \qquad (4)$$

This establishes the theorem. □

THEOREM 2.5. *Γ-martingales and tightness. Suppose $\mathcal{D}$ is compact and $\{Z^n : n \geq 1\}$ is a sequence of Γ-martingales in $\mathcal{D}$ all begun at $z \in \mathcal{D}$, with $Z^n$ having Γ-parallel transport $\Xi^n$ and Γ-development $M^n$. Suppose further that for some constant $\lambda > 0$ the differentials of the "intrinsic time" measures $\tau^n$ of the $Z^n$ satisfy the bound*

$$\begin{aligned} d\tau^n &= \operatorname{tr}\left((\Xi^n \otimes \Xi^n) \cdot d[M^n, M^n]\right) \\ &= \sum d\left[\int \langle (\Xi^n)^T u^i, dM^n \rangle, \int \langle (\Xi^n)^T u^i, dM^n \rangle\right] \leq \lambda \, dt \end{aligned} \qquad (5)$$

*where $(\Xi^n)^T$ is the adjoint of $\Xi^n$ and the $u^i$ form an fixed orthonormal basis field over all $\mathcal{D}$. Then the sequence $\{\mathbf{P}^n : n \geq 1\}$ of probability measures, formed by the laws of the $Z^n$, is tight.*

**Remark:** It suffices that the initial laws $\{Z^n(0) : n \geq 1\}$ form a tight sequence of probability measures.

**Remark:** Because of the compactness of $\mathcal{D}$ the choice of reference metric (which intervenes in the choice of the $\{u^i\}$ and thence the definition of the trace) does not alter the condition (5).

*Proof:* We employ the tightness criterion described in Stroock and Varadhan (1979, Theorem 1.4.6). Adapted to our situation this is as follows: embed $\mathcal{D}$ smoothly in Euclidean space. For each nonnegative smooth real function $f$

there should be a constant $A_f$ such that the process $t \mapsto f(Z^n(t)) + A_f t$ is a submartingale for each $n$. The Stroock and Varadhan criterion applies when $A_f$ can be chosen to be the same for all translates of $f$. By compactness of $\mathcal{D}$ this is automatic if $A_f$ depends continuously on $f$ and its derivatives and the connection coefficients of Γ.

Apply Itô's formula via (1) to $f(Z^n)$. The Itô differential of $f(Z^n)$ turns out to be

$$d_I(f(Z^n)) \quad = \quad \langle \operatorname{grad} f, \Xi^n d_I M^n \rangle + \tfrac{1}{2} \langle \operatorname{Hess} f(\Xi^n, \Xi^n), d[M^n, M^n] \rangle \quad (6).$$

Here Hess $f$ is the second covariant derivative of $f$, defined of course by using the connection Γ. The second term is bounded by $A_f = \tfrac{1}{2} \| \operatorname{Hess} f \| \lambda \, dt$, applying the intrinsic time bound given in the theorem. Note the definition of $\| \operatorname{Hess} f \|$ depends on the choice of reference metric. Now $\| \operatorname{Hess} f \|$ is finite since $f$ is smooth and $\mathcal{D}$ is compact. Furthermore Hess $f$ depends continuously only on $f$ and its derivatives and the connection coefficients of Γ. Hence the tightness criterion applies, yielding the theorem. □

## 3. Stochastic Control and Γ-martingales.

Consider, for a given compact region $\mathcal{B}$ as above, the *value* of the stochastic control problem

$$\mathcal{V}(x, y) \quad = \quad \sup \{ \mathbf{P}[X(\infty) = Y(\infty)] \; : \; (X, Y) \in \Gamma \mathcal{M}_{(x,y)}(\mathcal{B} \times \mathcal{B}) \} \quad (7).$$

(Note one or both of the limits may not exist, in which case we take them to be not equal.) A standard argument shows that $\mathcal{V}(X, Y)$ defines a supermartingale for any Γ-martingale $(X, Y) \in \Gamma \mathcal{M}_{(x,y)}(\mathcal{B} \times \mathcal{B})$ (note $(X, Y)$ should more properly be referred to as a "product-connection-martingale", but the slight abuse of terminology causes no ambiguity). Now $X$ and $Y$ can be produced by composing Γ-geodesics with real–valued continuous martingales. Hence $\mathcal{Q}(x, y) = 1 - \mathcal{V}(x, y)$ is a convex function, since this consideration shows it yields (conventional Euclidean) convex functions when composed with product-connection geodesics in $\mathcal{B} \times \mathcal{B}$ (which are simply products of Γ-geodesics from $\mathcal{B}$). So $\mathcal{Q}$ is a (product-connection) convex function $\mathcal{Q}$ on $\mathcal{B} \times \mathcal{B}$.

It is clear that $Q$ vanishes on the diagonal $\Delta$ of $B \times B$. It therefore supplies convex geometry for $B$ precisely when it vanishes nowhere else. Suppose now that uniqueness holds for problem (B), so that Γ-martingales in $B$ are uniquely defined by their terminal values. We shall show the supremum in the definition of $\mathcal{V}$ above is always attained if $\mathcal{V}(x, y) = 1$, and hence in such a case $x = y$. It follows $Q$ vanishes precisely on $\Delta$, and so $B$ has convex geometry.

Here is a preliminary lemma showing the effect on Γ-martingales of a strictly–convex–supporting property for $B$.

LEMMA 3.1. *Suppose the compact region $B$ supports a strictly convex and smooth function $f : B \to [0, 1]$. Let $\kappa^2 > 0$ provide a lower bound for the Hessian of $f$ via*

$$\text{Hess } f \quad \geq \quad \kappa^2 g \tag{8}$$

*where $g$ is the metric tensor for the reference Riemannian structure. Suppose $Z$ is a Γ-martingale in $B$ with Γ-development $M$ and Γ-parallel transport $\Xi$. Then the expectation of the total intrinsic time of $Z$ is bounded by*

$$\mathbf{E}\left[\int_0^\infty \text{tr}\left((\Xi \otimes \Xi) \cdot d[M, M]\right)\right] \quad \leq \quad \frac{2 \sup f}{\kappa^2} \tag{9}.$$

*Proof:* From equation (6)

$$\mathbf{E}\left[f(Z(\infty))\right] - f(Z(0)) \quad = \quad \tfrac{1}{2}\mathbf{E}\left[\int_0^\infty \langle \text{Hess } f(\Xi, \Xi), d[M, M]\rangle\right]$$

The lemma follows by applying inequality (8). □

Remark: Because of the Darling–Zheng Γ-martingale convergence theorem (Darling, 1983; Zheng, 1983), it follows that the strictly–convex–supporting property for a compact region $B$ implies that all Γ-martingales in $B$ converge at time $\infty$.

THEOREM 3.2. *Suppose the compact region $B$ supports a strictly convex function. Suppose further that solutions to the Γ-martingale problem for $B$ are unique when they exist. Then $\mathcal{V}$ equals 1 only on the diagonal $\Delta$ and consequently $B$ has convex geometry.*

*Proof:* Suppose $\mathcal{V}(x,y) = 1$. Let $\{(X^n, Y^n) : n \geq 1\}$ form a sequence of Γ-martingales in $\mathcal{B} \times \mathcal{B}$ such that $X^n(0) = x$, $Y^n(0) = y$ and

$$\mathbf{P}\left[X^n(\infty) = Y^n(\infty)\right] \quad \rightarrow \quad 1.$$

Let $\Xi^n$, $\Psi^n$ be the Γ-parallel transports, and $M^n$, $N^n$ the Γ-developments, of $X^n$, $Y^n$ respectively.

The problem is unaffected by time–changes and so we may suppose the differential of the intrinsic time measure of $(X^n, Y^n)$ is given by $1_{[0,T^n]}dt$, where $T^n$ is a Markov time for the filtration for $(X^n, Y^n)$ and $1_{[0,T^n]}$ is 1 when $0 \leq t \leq T^n$ and 0 otherwise. In particular both $X^n$ and $Y^n$ stop moving after time $T^n$. Applying the Γ-martingale tightness criterion of theorem 2.5, and selecting a subsequence if necessary, we may suppose $(X^n, Y^n)$ converges weakly to a Γ-martingale $(X, Y)$.

The differential of the intrinsic time measure of $(X^n, Y^n)$ is bounded above:

$$
\begin{aligned}
1_{[0,T^n]}dt \;&=\; \operatorname{tr}\left(((\Xi^n, \Psi^n) \otimes (\Xi^n, \Psi^n)) \cdot d[(M^n, N^n), (M^n, N^n)]\right) \\
&\leq\; 2\,\operatorname{tr}((\Xi^n \otimes \Xi^n) \cdot d[M^n, M^n]) \;+\; 2\,\operatorname{tr}((\Psi^n \otimes \Psi^n) \cdot d[N^n, N^n]).
\end{aligned}
$$

Consequently by lemma 3.1, and using the notation of that lemma,

$$\mathbf{E}\left[T^n\right] \quad \leq \quad \frac{8 \sup f}{\kappa^2}$$

and so for each $h > 0$

$$\mathbf{P}\left[T^n > h\right] \quad \leq \quad \frac{8 \sup f}{\kappa^2} h^{-1}.$$

Hence $T^n < \infty$ with probability one, and so the limits $X^n(\infty)$ and $Y^n(\infty)$ genuinely exist. A similar argument holds for the intrinsic time of the weak limit Γ-martingale $(X, Y)$, so that $(X, Y)$ also converges at infinity.

By the boundedness of $\mathcal{B}$ and a dominated convergence argument

$$\mathbf{E}\left[\operatorname{dist}(X^n(\infty), Y^n(\infty))\right] \quad \rightarrow \quad 0.$$

Since $X^n$ and $Y^n$ stop at time $T^n$ we deduce from the upper bound on $\mathbf{P}\left[T^n > h\right]$ that

$$\mathbf{E}\left[\operatorname{dist}(X^n(h), Y^n(h))\right] \quad \leq \quad \mathbf{E}\left[\operatorname{dist}(X^n(\infty), Y^n(\infty))\right] + \frac{8 \sup f}{\kappa^2}(\operatorname{diam}\mathcal{B}) h^{-1}$$

and so it follows

$$\limsup_{n \to \infty} \mathbf{E}\left[\operatorname{dist}(X^n(h), Y^n(h))\right] \quad \leq \quad \frac{8 \sup f}{\kappa^2}(\operatorname{diam}\mathcal{B}) h^{-1}$$

and so by dominated convergence again

$$\mathbf{E}\left[\mathrm{dist}(X(h), Y(h)\right] \quad \leq \quad \frac{8 \sup f}{\kappa^2} (\mathrm{diam}\,\mathcal{B})\, h^{-1}.$$

Taking the limit as $h \to \infty$ we see

$$\mathbf{E}\left[\mathrm{dist}(X(\infty), Y(\infty))\right] \quad = \quad 0.$$

By uniqueness of the $\Gamma$-martingale Dirichlet problem for $\mathcal{B}$ it now follows $X_t = Y_t$ for all $t$ and hence in particular $x = y$ as required. The theorem follows. □

## 4. Discussion.

If the strictly–convex–supporting condition is removed in theorem 3.2 then one may still construct a $\mathcal{V}$-supremum–attaining sequence of $\Gamma$-martingales $\{(X^n, Y^n) : n \geq 1\}$ with $(X^n, Y^n) \Rightarrow (X, Y)$ and $\mathbf{P}\left[X^n(\infty) = Y^n(\infty)\right] \to 1$, but control is lost over the convergence of $(X^n(\infty), Y^n(\infty))$. It is natural to conjecture that the use of well–posedness might recover some control:

CONJECTURE 4.1. $(\mathbf{B}) \Rightarrow (\mathbf{A})$. *Let $\mathcal{B}$ be a compact region for which the $\Gamma$-martingale Dirichlet problem has unique and well–posed solutions (when they exist). Then $\mathcal{B}$ has convex geometry.*

**Remark:** This conjecture has now been verified. A proof will appear in Kendall (preprint).

Note that $(\mathbf{B}) \Rightarrow (\mathbf{A})$ is immediate if well–posedness holds uniformly. Of course strengthening $(\mathbf{B})$ to require uniform well–posedness would make the $(\mathbf{C}) \Rightarrow (\mathbf{B})$ below still harder!

We sketch some further considerations which may be informative.

Suppose that $\mathcal{B}$ is a *convex* compact region. Set

$$\mathcal{Z} \quad = \quad \{(x, y) : \mathcal{V}(x, y) = 1\}.$$

Then $\Delta \subset \mathcal{Z}$ and $\mathcal{Z}$ is a convex closed set in $\mathcal{B} \times \mathcal{B}$, because if $(x, y)$ lies on a geodesic between two members of $\mathcal{Z}$ then a simple pasting argument yields a $\mathcal{V}$-supremum–attaining sequence of $\Gamma$-martingales for $(x, y)$ built from the two $\mathcal{V}$-supremum–attaining sequences for the two members of $\mathcal{Z}$ (it is at this point that we need convexity for $\mathcal{B}$).

The geometry of $\mathcal{Z}$ is intriguing. Any point $z \in \mathcal{Z} \setminus \Delta$ has running through it a $\Gamma$-martingale $Z$ staying entirely in $\mathcal{Z}$, obtained as a weak limit of a supremum–attaining sequence constructed as in the proof of theorem 3.2. The special form of the intrinsic time measures of the limiting sequence, and the requirement for each of the members of the sequence to move from $z$ to $\Delta$, allow one to argue that $Z$ is non–trivial. But this means that at every point $z$ of $(\partial \mathcal{Z}) \setminus \Delta$ there must be at least one $\Gamma$-geodesic with contact of second–order. For otherwise one can construct a convex smooth function $f$ defined in a neighbourhood of $z$ such that $f(z) = 0$ and $f < 0$ where it is defined on $\mathcal{Z}$, and a contradiction follows from consideration of the submartingale properties of $f(Z)$. So if there are non–trivial $\mathcal{Z}$ then their boundaries have an interesting second–order geometry.

For two-dimensional $\mathcal{B} \times \mathcal{B}$ it follows that $(\partial \mathcal{Z}) \setminus \Delta$ is the union of images of $\Gamma$-geodesics connecting points in $\Delta$. It follows from uniqueness for (**B**) (and hence for (**D**)) that $\partial \mathcal{Z} \subset \Delta$ and hence $\mathcal{Z} \subset \Delta$. Unfortunately $\dim(\mathcal{B} \times \mathcal{B}) = 2$ means $\mathcal{B}$ is one–dimensional and hence geometrically trivial.

However the whole line of argument of this paper also applies to an analogous set of problems in which the special case of two dimensions is not so trivial. These are problems centering around whether there are no non-trivial $\Gamma$-martingales converging to deterministic limits: we indicate them in the following list.

(**A′**): *Do all singleton point-sets of $\mathcal{B}$ occur as zero sets of nonnegative convex functions?*

(**B′**): $\Gamma$-*martingale Liouville property.* No non-trivial $\Gamma$-martingales converge to deterministic limits, moreover a $\Gamma$-martingale with an almost deterministic limit is almost trivial (that is to say, is almost constant in time).

(**C′**): $\Gamma$-*harmonic map Liouville property.* No non-trivial $\Gamma$-harmonic maps $F : \mathcal{O} \to \mathcal{B}$ have constant boundary-value data $\partial F : \partial \mathcal{O} \to \mathcal{B}$, moreover a $\Gamma$-harmonic map with almost constant boundary data is almost trivial (that is to say, is almost a constant map).

(**D′**): Γ-*geodesic Liouville property*. There are no nontrivial Γ-geodesics with coincident endpoints. Moreover a Γ-geodesic with almost coincident endpoints is almost trivial (that is to say, is almost a constant geodesic).

As before it is straightforward to show (**A′**) ⇒ (**B′**) ⇒ (**C′**) ⇒ (**D′**) (for example (**B′**) ⇒ (**C′**) is described in Kendall, 1988 Theorem 6). We can argue much as in section 3 to show (**B′**) ⇒ (**A′**) under the additional condition that $B$ is strictly–convex–supporting. From the considerations given above it can be shown that (**B′**) ⇒ (**A′**) without any further conditions when $B$ is convex and two–dimensional. Emery conjectures (**D′**) ⇒ (**A′**) when $B$ is convex and compact.

Reverting to Dirichlet problems, for completeness we mention the two other conjectures raised by the figure.

CONJECTURE 4.2. (**C**) ⇒ (**B**). *Let $B$ be a compact region which is a co-domain for a Γ-harmonic map Dirichlet problem with unique and well–posed solutions (when they exist). Then the same is true for the Γ-martingale Dirichlet problem for $B$.*

CONJECTURE 4.3. (**D**) ⇒ (**C**). *Let $B$ be a compact convex region for which the Γ-geodesic Dirichlet problem has unique and well–posed solutions. Then the same is true for the Γ-harmonic map Dirichlet problem for codomain $B$ when its solutions exist.*

Progress on the (**C**) ⇒ (**B**) conjecture awaits a study of the fundamental problem of approximations of Γ-martingales by images of Riemannian Brownian motions under Γ-harmonic maps. The (**D**) ⇒ (**C**) conjecture appears almost as difficult as the Emery conjecture itself, and it may be better to try to show (**D**) ⇒ (**B**). For example, under (**D**) an inductive argument shows (**B**) for the restricted class of *dyadic Γ-martingales* built from iterated barycentres by Emery and Mokobodzki (preprint; see proof of Théorème 2).

The three conjectures of this section would all be resolved affirmatively if the Emery conjecture were shown to be true. A major difficulty in making progress towards resolving the Emery conjecture is that direct checks on specific cases typically require non–obvious and tedious calculation. These three conjectures define a useful *indirect* programme for an attack on Emery's fundamental question, of whether uniqueness and well–posedness for (**D**) implies (**A**). Moreover they have intrinsic interest, so that the indirect approach offers rewards even if the Emery conjecture proves false. Theorem 3.2 of this paper makes the first and simplest step of this indirect programme, by giving a partial resolution of the (**B**) ⇒ (**A**) conjecture. (But see the note added at the end of the introduction, section 1).

**References.**

R. W. R. Darling (1982), "Martingales in Manifolds – Definition, Examples and Behaviour under Maps." *Séminaire de Probabilitiés XVII*, Lecture Notes in Mathematics 921, Berlin, Springer, 217–236.

R.W.R.Darling (1983), "Convergence of Martingales on a Riemannian Manifold" *Publ. R.I.M.S Kyoto Univ.* **19** 753–763.

M.Emery (1985), "Convergence des Martingales dans les Variétés." Colloque en l'Honneur de Laurent Schwartz (Volume 2), *Ásterisque* **132**, 47–63.

M.Emery and P.A.Meyer (1989), *Stochastic calculus in manifolds.* Berlin, Springer.

M.Emery and G.Mokobodzki (preprint), "Sur le barycentre d'une probabilité dans une variété." Submitted to *Séminaire de Probabilitiés XXV*.

M.Emery and W.A.Zheng (1984), "Fonctions Convexes et Semimartingales dans une Variété." *Séminaire de Probabilitiés XVIII*, Lecture Notes in Mathematics 1059, Berlin, Springer, 501–538.

W.S.Kendall (1987), "Stochastic Differential Geometry: An Introduction." *Acta Applicandae Math.* **9**, 29–60.

W.S.Kendall (1988), "Martingales on Manifolds and Harmonic Maps." *Contemporary Mathematics* **73**, 121–157 (but see also the correction sheet available from the author).

W.S.Kendall (1990), "Probability, convexity, and harmonic maps with small image I: uniqueness and fine existence." *Proceedings of the London Math. Soc.* (3) **61**, 371–406.

W.S.Kendall (1991), "Convexity and the hemisphere." To appear, *Journal of the London Math. Soc.*

W.S.Kendall (preprint), "Convex Geometry and Nonconfluent Γ-Martingales II: well–posedness and Γ-martingale convergence."

J.Picard (1991), "Martingales on Manifolds with Prescribed Limits." To appear *Journal Funct. Analysis.*

R.Rebolledo (1976), "Convergence en loi des martingales continues." *C. R. Acad. Sc. Paris* **282**, 483–485.

D.Stroock and S.R.S.Varadhan (1979), *Multidimensional Diffusion Processes.* Berlin, Springer.

W.A.Zheng (1983), "Sur la Théorème de Convergence des Martingales dans une Variété Riemanniene" *Z. Warscheinlichkeitstheorie verw. Gebiete* **63**, 511–515.

Department of Statistics
University of Warwick
Coventry CV4 7AL, UK.

# Some Caricatures of Multiple Contact Diffusion-Limited Aggregation and the $\eta$-model

HARRY KESTEN*

Department of Mathematics, Cornell University, Ithaca, N.Y. 14853

Abstract. We consider some variants of DLA which shift the distribution of the place where a new particle is added in a very strong way to the points of maximal harmonic measure. As a consequence these variants can grow like "generalized plus signs", with the aggregate containing only points on the coordinate axes at all times.

## 1. Introduction and statement of results.

We construct connected lattice sets $A_n$, $n = 1, 2, \ldots$, by two procedures, both of which are variants of common procedures in DLA (Diffusion Limited Aggregation). The original DLA model was introduced by Witten and Sander [22]; see also [13] and [21, Sect. 6] for a general introduction to DLA. We only consider lattice models, so that $A_n$ is a connected subset of $\mathbb{Z}^d$. We always take $A_1 = \{0\}$ and $A_n$ will contain exactly $n$ sites. The site added to $A_n$ to make $A_{n+1}$ is denoted by $y_n$, so that $A_{n+1} = A_n \cup \{y_n\}$. $y_n$ is chosen from $\partial A_n$, the boundary of $A_n$, which is the collection of sites adjacent to $A_n$, but not in $A_n$. To describe the distribution of $y_n$ we introduce some notation. Let $S_k$, $k \geq 0$, be a simple symmetric nearest neighbor random walk on $\mathbb{Z}^d$. $P_x$ is the probability measure governing random walk paths which start at $x$; $E_x$ denotes expectation with respect to $P_x$. For any finite set of vertices $B$ define

$$T(B) = \inf\{n \geq 0 : S_n \in B\},$$

$$H_B(x, y) = P_x\{S_{T(B)} = y\} \qquad \text{if} \quad y \in B,$$

*Research supported by the NSF through a grant to Cornell University.

Typeset by $\mathcal{A}_{\mathcal{M}}\mathcal{S}$-TEX

$$\mu_B(y) = \lim_{x \to \infty} P_x\{S_{T(B)} = y \mid T(B) < \infty\}$$

$$= \lim_{x \to \infty} \frac{H_B(x, y)}{\sum\limits_{z \in B} H_B(x, z)} \qquad \text{if} \quad y \in B.$$

Both $H_B(x, y)$ and $\mu_B(y)$ are taken as zero when $y \notin B$. The existence of the limit in the definition of $\mu_B$ was proved by Spitzer ([19, Theorem 14.1 and Prop. 2.6.2]). For $d = 2$ one does not need to condition on $T(B) < \infty$ since this event has probability 1. For $d \geq 3$ $\mu_B(y)$ equals (cf. [19, Prop. 26.2], [6])

$$\frac{Es_B(y)}{\sum\limits_{z \in B} Es_B(z)},$$

where

$$Es_B(y) = P_y\{S_n \notin B \quad \text{for all} \quad n \geq 1\}.$$

($Es_B$ is the so-called escape probability from B.)

We now describe the two procedures.

**Procedure 1.** This procedure is related to the method of noise reduction or multiple contact DLA, which is used widely in simulations ([5], [12], [13], [14], [17], [20], [21, Sect. 9.2.2]). When $A_n$ has been determined, we associate with each $z \in \partial A_n$ a *score*, $s(z)$. We start all these scores at zero and, in a sequence of steps increase one stochastically selected score by one. In each step the probability of increasing $s(z)$ by one (and leaving the other scores alone) is $\mu_{\partial A_n}(z)$. Thus we can view the sites in $\partial A_n$ as urns. In this description each step corresponds to "releasing a random walk at infinity and letting it run until it falls in one of the urns." $s(z)$ counts how many random walkers have been collected in the urn for $z$. We continue such steps until one of the scores for the first time reaches a predetermined value $m(n)$. In our version we have $m(n) \sim C_0 \log n$ with a $C_0$ to be determined later (cf. (2.72)). If $s(z_k)$ is the first score to reach the value $m(n)$, then we take $y_n = z_k$ and $A_{n+1} = A_n \cup \{z_k\}$. Next we reset all scores to zero and also attach scores zero to the new boundary sites (i.e., the points of $\partial A_{n+1} \setminus \partial A_n$). We then follow the same procedure to select the next point $y_{n+1}$, etc..

In the above process, $m(n)$ random walks have to land on one site before it is selected for addition to the aggregate. The original Witten-Sander model of DLA is the special case $m(n) \equiv 1$. For the simulations one has so far only considered $m(n) = m$, a constant. However, Meakin and Tolman ([12], [14]) have done simulations with $m = 100$, $n \leq 10^5$ (even some simulations with $m = 10^4$, $n = 5175$ in [14]). Unless the constant $C_0$ in $m(n)$ has to be taken much larger than we expect, then $100 \geq C_0 \log(10^5)$, so that one cannot distinguish simulations done so far with large constant $m$ from simulations with $m(n) \geq C_0 \log n$.

There is, however, a very significant difference between our version and the one usually simulated. In the latter versions the scores $s(z)$, $z \in \partial A_n$, are *not* reset to zero when $y_n$ has been chosen. Only the scores of the new boundary points (points in $\partial A_{n+1} \setminus \partial A_n$) are started at zero for the selection of $y_{n+1}$. The other scores start with their values at the time $y_n$ was chosen, when $y_{n+1}$ is to be selected. For small $n$ both versions (with or without resetting the scores to zero) behave similarly with all particles being added at the tips only (see [1]) and Theorem 1 below.

**Procedure 2.** This is a version of the so-called $\eta$-model for dielectric breakdown (cf. [16], [4], [11]). In this procedure we take

$$(1.1) \qquad P\{y_n = y \mid A_n\} = Z_n^{-1}(\eta(n))[\mu_{\partial A_n}(y)]^{\eta(n)}, \quad \text{for} \quad y \in \partial A_n$$

$$\text{and} \quad 0 \quad \text{otherwise} ,$$

where

$$(1.2) \qquad Z_n(\eta(n)) = \sum_{y \in \partial A_n} [\mu_{\partial A_n}(y)]^{\eta(n)}.$$

Again the usual version is to take $\eta(n)$ constant. We will let $\eta(n)$ grow with $n$ so that $\eta(n) \sim C_0 \log n$ for a constant $C_0$ to be determined later (see (2.72)). Taking $\eta(n)$ in this fashion is akin to taking the limit $\eta \to \infty$ before $n$ goes to $\infty$. This kind of limit was already mentioned in [16].

The effect of requiring $m(n)$ contacts in Procedure 1 or of raising $\mu_{\partial A_n}$ to a power $\eta(n)$ in Procedure 2 is to make it more likely that the new particle $y_n$ is added near a "tip" of $A_n$, where $\mu_{\partial A_n}$ is relatively high. In fact, in both procedures we will choose the proportionality constant $C_0$ in the growth rate of $m(n)$ and $\eta(n)$, respectively, so large that with overwhelming probability $y_n$ has to be one of the tips of $A_n$ itself. To make the notion of "tip" precise, we restrict ourselves to a special class of $A$'s which consist of intervals of some of the coordinate axes. We say that $A$ is a *generalized plus sign* if

$$(1.3) \qquad A = \bigcup_{i=1}^{d} \{ke_i : -k(i,-) \leq k \leq k(i,+)\}$$

for some $k(i, \varepsilon) \geq 0$, with $k(i, \varepsilon) > 0$ for at least two choices of $(i, \varepsilon) \in \{1, \ldots, d\} \times \{-, +\}$. (The terminology is of course inspired by the case $d = 2$ and $k(1, -) = k(1, +) = k(2, -) = k(2, +)$, which represents an ordinary plus sign.) We denote the set $A$ of (1.3) by $A(k(i, \varepsilon))$. Note that we allow $k(i, \varepsilon) = 0$ for various $(i, \varepsilon)$. Our principal result says that with the exception of $d = 2$ or 3 generalized plus signs with more or less equal arms occur with positive probability.

THEOREM 1. *Let $S$ be any subset of $\{1, \ldots, d\} \times \{-, +\}$ with $|S| \geq 2$ ($|S|$ denotes the cardinality of $S$). Let $E(S)$ be the following event:*

(1.4)   $E(S) = \{$For all $n$, $A_n$ is a generalized plus sign,
$$A(k(i, \varepsilon, n)) \text{ say, with } k(i, \varepsilon, n) = 0 \text{ for } (i, \varepsilon) \notin S$$
$$\text{and } \frac{1}{n} k(i, \varepsilon, n) \rightarrow \frac{1}{|S|} \text{ for } (i, \varepsilon) \in S\}.$$

*Then for sufficiently large $C_0$ (cf. (2.72) below)*

(1.5)                          $P\{E(S)\} > 0$

*for Procedure 1 if $d \geq 3$, and for Procedure 2 if $d \geq 4$. For both procedures (1.5) continues to hold in $d = 2$ or 3 if $S$ has the special form $\{(i, -), (i, +)\}$.*

In the case where $S = \{(i, -), (i, +)\}$, $E(S)$ says that $A_n$ is a line segment along the $i$-th coordinate axis, asymptotically symmetric with respect to the origin. We suspect that in dimension 2 these are the only $E(S)$ with positive probability. We have no proof of this for $|S| = 3$ or 4, but the next theorem shows that $A_n$ cannot have an asymptotic L-shape.

THEOREM 2. *For $d = 2$ and both procedures with any $C_0 > 0$*

(1.6)     $P\{E(S)\} = 0$ *when* $S = \{(1, \varepsilon_1), (2, \varepsilon_2)\}(\varepsilon_i \in \{-, +\})$.

**Remarks.**

(i) We do not know whether $P\{E(S)\} > 0$ for Procedure 2 in dimension 3.

(ii) Theorem 1 shows that with strictly positive probability $A_n$ never contains a point off all coordinate axes, and grows only (and at more or less equal rate) along the half-axes indexed by $S$. One expects that there is also a strictly positive probability of obtaining finite perturbations of this. More specifically, let $F$ be a finite connected lattice set containing the origin and define

(1.7)    $E(S, F) = \Big\{$ For all large $n, A_n = A(k(i, \varepsilon, n)) \cup F$

$\qquad\qquad$ for some $k(i, \varepsilon, n)$ with $k(i, \varepsilon, n) = 0$ if $(i, \varepsilon) \notin S$

$\qquad\qquad\qquad$ and $\dfrac{1}{n} k(i, \varepsilon, n) \longrightarrow \dfrac{1}{|S|}$ if $(i, \varepsilon) \in S.\Big\}$

We can then indeed prove that

(1.8)    $$P\{E(S, F)\} > 0$$

for any finite connected $F$ containing 0 in the following cases:

(1.9)    $d = 3$,    Procedure 1 and    $\{(i, -), (i, +)\} \subset S$    for some    $i$.

(1.10)    $d = 2$,    Procedure 1 or 2 and    $S = \{(i, -), (i, +)\}$.

It is even possible in dimension 2 to obtain a "halfline with a finite side branch". Specifically, for $k_0$ and $C_0$ sufficiently large and $d = 2$, we have

(1.11)    $P\Big\{$ For all large $n, A_n = \{je_i : 0 \le j \le n - k_0\}$

$\qquad\qquad\qquad \cup \{\ell e_{3-i} : 0 \le \ell \le k_0\}\Big\} > 0,$

for both Procedure 1 and 2 and $i = 1, 2$.

(iii) The proof of Theorem 2 will show that if for some $n$

(1.12)    $A_n = \{je_1 : 0 \le j \le k(1, +, n)\} \cup \{je_2 : 0 \le j \le k(2, +, n)\},$

then either at some later time $m$, $A_m$ will no longer be of the form (1.12), or if $A_m$ has the form (1.12) for all $m \ge n$, then

$$\frac{k(i, +, n)}{k(3 - i, +, n)} \to 0 \quad \text{for } i = 1 \text{ or } 2 ,$$

i.e., one arm will grow much faster than the other. If $C_0$ is sufficiently large, then one arm will even stop growing at some time. In this case $k(1, +, n)$ or $k(2, +, n)$ will be constant for some time $n$ on, and $A_n$ behaves as in (1.11).

We summarize in the table below which configurations are possible or impossible.

| d | Procedure 1 | Procedure 2 |
|---|---|---|
| $\geq 4$ | Any generalized plus sign possible | Any generalized plus sign possible |
| 3 | Any generalized plus sign. Even finite perturbation possible if plus sign contains a symmetric line segment (i.e. as in (1.9)). | Symmetric line segment with finite perturbation possible. |
| 2 | Symmetric line segment with finite perturbation, and halfline with finite side branch possible. L-shape *not* possible. | Symmetric line segment with finite perturbation, and halfline with finite side branch possible. L-shape *not* possible. |

*Table of possible and impossible asymptotic shapes.*

**Note added in proof:**   Greg Lawler has shown that for Procedure 2 in $d = 3$ certain $L$-shapes are possible.

## 2. Proof of Theorem 1.

We shall concentrate on Procedure 1 in dimension 3. The case $d \geq 4$ is considerably simpler, while much of the preparation for $d = 2$ will have to be left for another publication ([8], [9]). We define the *tips* of the generalized plus sign $A(k(i, \varepsilon))$ to be all the points of the form

$$(k(i, +) + 1)e_i \quad \text{if} \quad k(i, +) > 0,$$

and

$$-(k(i, -) + 1)e_i \quad \text{if} \quad k(i, -) > 0.$$

(See Figure 1.)   Note that we defined the

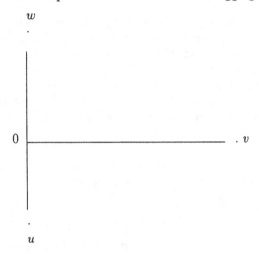

Figure 1. A generalized plus sign in dimension 2. In this example
$k(1, -) = 0$ and the tips are $u$, $v$ and $w$.

tips to be points of $\partial A$, rather than of $A$ itself. We also introduce the notation

$$\overline{B} = B \cup \partial B = \{v : v \in B \text{ or } v \text{ is adjacent to } B\},$$

for any lattice set $B$. Its hitting and return time are defined as

$$(2.1) \qquad T(B) = \inf\{n \geq 0 : S_n \in B\} \quad \text{and} \quad R(B) = \inf\{n \geq 1 : S_n \in B\},$$

respectively, and its escape probability is

$$(2.2) \qquad\qquad Es_B(v) = P_v\{S_n \notin B \quad \text{for all } n \geq 1\}.$$

We shall use $K_i$ to denote strictly positive, finite constants whose precise value is unimportant for our purposes. In fact, their values may vary from appearance to appearance.

$$a \wedge b = \min\{a, b\}, \quad a \vee b = \max\{a, b\}.$$

The idea of the proof of Theorem 1 is to prove two types of inequalities. The first one (see (2.4), (2.6), (2.61) and (2.67)) says that if $A_n$ is a generalized plus sign with arms of more or less equal length, then $\mu_{\partial A_n}(t)$ is approximately the same for all tips $t$ of $A_n$. The second inequality says that $\mu_{\partial A_n}(w)$ is less than a fixed multiple $(1-a) < 1$ of $\mu_{\partial A_n}(t)$, whenever $w$ is not a tip and $t$ is a tip (cf. (2.5), (2.62), (2.69) and (2.70)). Since our procedures shift mass to the points with high $\mu_{\partial A_n}$, the second inequality can be used to show that for large $C_0$ this effect is sufficiently strong to guarantee that $A_n$ only grows at the tips forever. The first inequality is then used to show that with positive probability the different (non zero) arms grow at the same asymptotic rate.

PROPOSITION 1. *Let $d \geq 3$. Then there exist constants $a > 0$, $K_1, K_2 < \infty$, depending on $d$ only, with the following properties (2.4), (2.5): For any generalized plus sign $A(k(i, \varepsilon))$ which satisfies*

(2.3)    *for some   $k > 0$   either   $k(i, \varepsilon) = 0$   or   $k \leq k(i, \varepsilon) \leq 2k$*
$$\text{for all   } (i, \varepsilon) \in \{1, \ldots, d\} \times \{-, +\},$$

*we have*

(2.4)
$$\left| \frac{\mu_{\partial A}(t_2)}{\mu_{\partial A}(t_1)} - 1 \right| \leq K_1 (\log k)^{-1}$$

*if $t_1, t_2$ are tips of $A$ and $k \geq K_2$, and also*

(2.5)
$$\frac{\mu_{\partial A}(w)}{\mu_{\partial A}(t)} \leq 1 - a$$

*if $t$ is a tip of $A$, but $w \in \partial A$ is not a tip of $A$ and $k \geq K_2$ ($w$ is otherwise arbitrary in $\partial A$).*

*If $d \geq 4$, then (2.4) can be sharpened to*

(2.6)
$$\left| \frac{\mu_{\partial A}(t_2)}{\mu_{\partial A}(t_1)} - 1 \right| \leq K_1 k^{3-d}.$$

We break up the proof into several lemmas. In Lemmas 2–6 we take $d = 3$.

LEMMA 2. *Let*
$$I_k = \{-je_1 : 0 \leq j \leq k\}.$$
*For each fixed $v \in \mathbb{Z}^3$ there exists a constant $C_1(v) > 0$ such that*

(2.7)
$$Es_{I_k}(v) \sim C_1(v)(\log k)^{-1/2}, \qquad k \to \infty.$$

*Moreover, for some constant $K_3$ (independent of $k, j$) we have*

$$(2.8) \quad Es_{I_k}(-je_1) \quad \leq \quad K_3(\log k)^{-1/2}\{\log((j \wedge (k-j)) + 2)\}^{-1/2},$$

$$k \geq 2, \quad 0 \leq j \leq k.$$

**Proof:** We closely follow the arguments for Lemmas 5 and 6 in [7]. Let $\sigma_0 = 0$ and

$$\sigma_{\ell+1} = \inf\{n > \sigma_\ell : S_n \text{ belongs to the first coordinate axis}\}.$$

Thus, $\{S_{\sigma_\ell}\}$ is the imbedded random walk on the first coordinate axes. We set

$$(2.9) \qquad\qquad T_\ell = \text{first coordinate of } S_{\sigma_\ell},$$

$$Y_{\ell+1} = T_{\ell+1} - T_\ell.$$

Note that the $Y_\ell$, $\ell \geq 2$, are i.i.d.. We write $Y$ for a random variable with the same distribution as these variables. If $S_0$ is on the first coordinate axis, then even $Y_1$ has the same distribution as $Y$.

We begin with calculating the characteristic function of $Y$, $\Psi(\theta)$. For simplicity we take $S_0 = 0$; then $Y$ has the distribution of the first coordinate of $S_{\sigma_1}$. Let $0 \leq \tau_1 < \tau_2 < \ldots$ be the successive values of $t$ for which $S_{t+1} - S_t \in \{\pm e_2, \pm e_3\}$. The $S_{\tau_{j+1}} - S_{\tau_j}$, $j \geq 1$, are i.i.d. The part of $S_{\tau_{j+1}} - S_{\tau_j}$ in the second and third coordinate direction equals $\pm e_2$ or $\pm e_3$, each possibility occurring with probability $1/4$. On the other hand, the first coordinate of

$$S_{\tau_{j+1}} - S_{\tau_j}$$

is the sum of a geometrically distributed number of steps of a symmetric simple random walk on $\mathbf{Z}$ (and independent of the second and third coordinate). From this it is easy to find

$$(2.10) \qquad \zeta(\theta) \quad := \quad E\{\exp i\theta(S_{\tau_{j+1},1} - S_{\tau_j,1})\}$$

$$= \quad \sum_{\ell=1}^{\infty} \frac{2}{3}\left(\frac{1}{3}\right)^{\ell-1} (\cos\theta)^{\ell-1}, \qquad j \geq 1.$$

Furthermore, $S$ can return to the first coordinate axis for the first time, either at the first step (when $S_1 = \pm e_1$), or at the time $\tau_\nu$, where

$$\nu = \min\{j : S_{\tau_j + 1} \in \text{first coordinate axis}\}$$

(when $S_1 \neq \pm e_1$ and $\tau_1 = 0$). In the first case $Y = \pm 1$, each event occurring with probability $1/6$. In the second case, given that $\nu = k$, the conditional characteristic function of $Y$ is $[\zeta(\theta)]^{k-1}$. In total we find

$$\Psi(\theta) := E e^{i\theta Y} = \frac{1}{3} \cos\theta + \frac{2}{3} \sum_{k=1}^{\infty} P_\theta \{\text{a two dimensional simple}$$

symmetric random walk returns first to 0 at time $2k\} [\zeta(\theta)]^{2k-1}$.

Let $V_0 = 0, V_1, \ldots$ be a simple symmetric two-dimensional random walk, starting at the origin, and set

$$f_{2k} = P\{\text{first return to 0 by } V \text{ is at time } 2k\},$$

$$u_{2k} = P\{U_{2k} = 0\}, \qquad U(s) = \sum_0^\infty u_{2k} s^{2k}.$$

Then standard renewal theory [2, Theorem XIII.3.1] and the above show that

$$\Psi(\theta) \quad = \quad \frac{1}{3} \cos\theta + \frac{2}{3} \sum_1^\infty f_{2k} [\zeta(\theta)]^{2k-1}$$

$$= \quad \frac{1}{3} \cos\theta + \frac{2}{3\zeta(\theta)} \{1 - [U(\zeta(\theta))]^{-1}\}.$$

Now it is easy to see from (2.10) that

$$1 - \zeta(\theta) \sim \frac{1}{4}\theta^2, \qquad \theta \to 0.$$

Also $u_{2n} \sim (\pi n)^{-1}$ ([2, Sect. XIV.7]) and the Abelian theorem ([3, Theorem XIII.5.5]) show that

$$U(s) \sim \frac{1}{\pi} \log \frac{1}{1-s}, \qquad s \uparrow 1.$$

Consequently

(2.11) $$1 - \Psi(\theta) \sim \frac{4\pi}{3} \left[\log \frac{1}{|\theta|}\right]^{-1}, \qquad \theta \to 0.$$

This takes the place of (2.17) in [7].

The same kind of argument shows that

$$(2.12) \quad P\{Y \geq n\} = P\{Y \leq -n\} = \frac{2}{3} \sum_{1}^{\infty} f_{2k} P\{W_{2k-1} \geq n\}, \quad n > 1,$$

where $W_{2k-1}$ is the sum of $(2k - 1)$ independent random variables, each with characteristic function $\zeta$. In particular, by the central limit theorem,

$$(2.13) \qquad P\{W_{2k-1} \geq n\} \rightarrow \frac{1}{2} \quad \text{as} \quad k/n^2 \rightarrow \infty.$$

Moreover,

$$\sum_{k=1}^{\infty} s^k \sum_{\ell=k}^{\infty} f_\ell = \frac{s}{1-s} \left\{ 1 - \sum_{1}^{\infty} f_\ell s^\ell \right\}$$

$$= \frac{s}{1-s} \frac{1}{U(s)} \sim \pi \left\{ (1-s) \log \frac{1}{1-s} \right\}^{-1}, \quad s \uparrow 1.$$

Hence, by the Tauberian theorem ([3, Theorem XIII.5.5]

$$(2.14) \qquad r_{2k} := \sum_{\ell=2k}^{\infty} f_\ell \sim \pi (\log k)^{-1}, \quad k \rightarrow \infty.$$

We can now find the asymptotic behavior (as $n \rightarrow \infty$) of (2.12). For any $\varepsilon > 0$ the sum over $k < n^2/\varepsilon$ is, by Chebyshev's inequality, at most

$$\frac{2}{3} \sum_{k \leq n^2/\varepsilon} f_{2k} \frac{k}{n^2} = \frac{2}{3} \sum_{k < n^2/\varepsilon} r_{2k} \frac{1}{n^2} - \frac{2}{3} r_{2n^2/\varepsilon} \frac{1}{\varepsilon} \quad \text{(summation by parts)}$$

$$= o\left( (\log n)^{-1} \right) \quad \text{(by (2.14))}.$$

For small $\varepsilon$, the sum over $k \geq n^2/\varepsilon$ is by (2.13) and (2.14) approximately

$$\frac{2}{3} \cdot r_{2n^2/\varepsilon} \cdot \frac{1}{2} \sim \frac{\pi}{6} (\log n)^{-1}.$$

Therefore,

$$(2.15) \qquad P\{Y \geq n\} = P\{Y \leq -n\} \sim \frac{\pi}{6} (\log n)^{-1}, \quad n \rightarrow \infty.$$

This replaces (2.27) in [7].

We now proceed as in [7]. $v(n)$ is defined by

$$\sum_0^\infty v(n)r^n = \exp\left\{ \sum_1^\infty \frac{1}{k} r^{\frac{(\sum_2^{k+1} Y_i)}{2}} \; ; \; \sum_2^{k+1} Y_i > 0 \right\} + \frac{1}{2}P\left\{ \sum_2^{k+1} Y_i = 0 \right\}$$

$$= \exp\left\{ -\frac{1}{4\pi} \int_{-\pi}^{+\pi} \frac{1-r^2}{1+r^2 - 2r\cos\theta} \log[1 - \Psi(\theta)] \, d\theta \right\}.$$

By (2.11), when $z = e^{i\theta}$ and $\theta \to 0$,

$$(1 - \psi(\theta))|1 + \log(1 - e^{i\theta})^{-1}| = (1 - \psi(\theta))|1 + \log(1 - z)^{-1}| \to \frac{4\pi}{3}.$$

Therefore, by Poisson's formula and Fatou's theorem (cf. [18, Theorems 11.30b with proof and 11.23]

$$\lim_{r\uparrow 1} \left| 1 + \log\frac{1}{1-r} \right|^{-1/2} \sum_0^\infty v(n)r^n$$

$$= \lim_{r\uparrow 1}\exp\left\{ -\frac{1}{4\pi} \int_{-\pi}^{+\pi} \frac{1-r^2}{1+r^2 - 2r\cos\theta} \right.$$

$$\left. \log\left[ (1 - \psi(\theta)) \, |1 + \log(1 - e^{i\theta})^{-1}| \right] \, d\theta \right\}$$

$$= \left( \frac{4\pi}{3} \right)^{1/2}.$$

Equivalently,

$$\sum_0^\infty v(n)r^n \sim \left[ \left( \frac{4\pi}{3} \right) \log\frac{1}{1-r} \right]^{1/2}, \quad r \uparrow 1,$$

and, again by the Tauberian theorem ([3, Theorem XIII.5.5])

$$(2.16) \qquad \sum_0^n v(k) \sim \left( \left( \frac{4\pi}{3} \right) \log n \right)^{1/2}, \quad n \to \infty.$$

Now assume $S_0 = je_1$, a point of the positive first coordinate axis, so that $T_0 = j$ (see (2.9) for $T_\ell$). Define

$$\rho = \min\{n : T_n \le 0\},$$

$$G(j, \ell) = E\{\text{number of } 0 \le n < \rho \text{ with } T_n = \ell | T_0 = j\}$$

$$(2.17) \qquad = \sum_{n=1}^{j \wedge \ell} v(j - n) v(\ell - n)$$

(cf. [19, Prop. 19.3]). Then

$$P\{S_{\sigma_\rho} \notin I_k | S_0 = je_1\} = P\{T_\rho < -k | T_0 = j\}$$

$$= \sum_{\ell=1}^{\infty} G(j, \ell) P\{Y < -(k + \ell)\}$$

$$\sim \sum_{n=0}^{j-1} v(n) \sum_{\ell=0}^{\infty} v(\ell) \frac{\pi}{6} [\log(k + \ell + j - n)]^{-1}$$

$$\sim 2 \left(\frac{\pi}{3}\right)^{3/2} \sum_{n=0}^{j-1} v(n)(\log k)^{-1/2}, \quad k \to \infty.$$

(use (2.16) and partial summation for the last step; compare [7, Lemma 6]).

Now we first prove (2.7) for $v = je_1$, $j > 0$ fixed. Clearly

$$Es_{I_k}(je_1) \le P\left\{S_{\sigma_\rho} \notin I_k | S_0 = je_1\right\}$$

$$(2.18) \qquad \sim 2 \left(\frac{\pi}{3}\right)^{3/2} \sum_{n=0}^{j-1} v(n)(\log k)^{-1/2}.$$

On the other hand, for any constant $A$

$$ES_{I_k}(je_1) \ge P\left\{S_{\sigma_\rho} \notin I_{Ak} \text{ and } S_n \notin I_k \text{ for } n \ge \sigma_\rho | S_0 = je_1\right\}$$

$$= \sum_{r=Ak+1}^{\infty} P\{T_\rho = -r | T_0 = j\}$$

$$\left(1 - P\{S_n \text{ visits } I_k \text{ at some time } | S_0 = -re_1\}\right).$$

But, if

$$(2.19) \qquad g(x, y) = E_x\{\text{number of visits to } y \text{ by } S.\}$$

is Green's function for simple random walk on $\mathbf{Z}^3$, then it is well known ([19, Props. 25.5 and 26.1], [10, Theorem 1.5.4]) that

$$(2.20) \qquad g(S_{n \wedge T\{y\}}, y) \text{ is a positive martingale}$$

and that

$$(2.21) \qquad g(x,y) \; = \; g(0, y - x) \; \sim \; \frac{3}{2\pi} |y - x|^{-1}.$$

Therefore,

$$P_x\{S. \text{ ever visits } y\} \; \sim \; C_2 |y - x|^{-1}, \quad y - x \to \infty$$

for $C_2 = 3\{2\pi g(0,0)\}^{-1}$.  Hence for $r \geq Ak$

$$P_{-re_1}\{S. \text{ ever visits } I_k\} \; \leq \; 2C_2 \sum_{j=0}^{k} \frac{1}{|r - j|} \; \leq \; \frac{3C_2}{A - 1}.$$

Thus

$$Es_{I_k}(je_1) \geq \left(1 - \frac{3C_2}{A - 1}\right) P\left\{S_{\sigma_\rho} \notin I_{Ak} | S_0 = je_1\right\}$$

$$\sim \; \left(1 - \frac{3C_2}{A - 1}\right) 2 \left(\frac{\pi}{3}\right)^{3/2} \sum_{n=0}^{j-1} v(n) \, (\log Ak)^{-1/2}.$$

It follows that

$$(2.22) \quad \lim_{k \to \infty} (\log k)^{1/2} ES_{I_k}(je_1) \; = \; 2 \left(\frac{\pi}{3}\right)^{3/2} \sum_{n=0}^{j-1} v(n)$$

$$\left(\sim 4 \left(\frac{\pi}{3}\right)^2 \{\log j\}^{1/2} \quad \text{for large} \quad j\right).$$

This proves (2.7) for $v = je_1$, $j > 0$.

For general $v$ one can prove (2.7) by a decomposition with respect to $S_R$ where

$$R = \min\{n \geq 1 : S_n \in \text{ first coordinate axis}\},$$

more or less as in Lemma 6 of [7]. We skip the details. We will, however, prove (2.8) and this proof illustrates most of the ideas.

Let $v = -je_1$, $0 \leq j \leq k/2$. Then most of the previous considerations remain valid. However, this time we replace (2.18) by

$$Es_{I_k}(x) \leq P_v\left\{S_{\sigma_1} \notin I_k, S_{\sigma_\rho} \notin I_k\right\}$$

$$\leq P_v\left\{S_{\sigma_1} \in \{-\ell e_1 : \ell > k\}\right\} + \sum_{\ell=1}^{\infty} P_v\left\{S_{\sigma_1} = \ell e_1\right\} P_{\ell e_1}\left\{S_{\sigma_\rho} \notin I_k\right\}$$

$$= P\{Y \leq -(k - j)\} + \sum_{\ell=1}^{\infty} P\{Y = j + \ell\} \sum_{r=1}^{\infty} G(\ell, r) P\{Y < -r - k\}.$$

The first term in the right hand side is $O((\log k)^{-1})$ for $j \leq k/2$, by (2.15). Also, by (2.17) and (2.15), for large $k$

$$\sum_{r=1}^{\infty} G(\ell,r)P\{Y < -r - k\} \leq \sum_{r=1}^{\infty} \sum_{n=1}^{\ell \wedge r} v(\ell - n)\, v(r - n)\frac{\pi}{6}(\log(r + k))^{-1}$$

$$= \frac{\pi}{6}\sum_{n=1}^{\ell} v(\ell - n)\sum_{s=0}^{\infty} v(s)\,(\log(s + n + k))^{-1}.$$

If we now use partial summation and (2.16) (as in the proof of Lemma 6 of [7]) we find that

$$\sum_{r=1}^{\infty} G(\ell,r)P\{Y < -r - k\}$$

$$\leq K_4 \sum_{n=1}^{\ell} v(\ell - n)\sum_{s=0}^{\infty} (1 + \log s)^{1/2}(s + n + k)^{-1}(\log(s + n + k))^{-2}$$

$$\leq K_5 \sum_{n=1}^{\ell} v(\ell - n)(\log(n + k))^{-1/2}$$

$$\leq 2K_5 \left\{\frac{\log(\ell + 1)}{\log k}\right\}^{1/2}.$$

Of course, we also have for all $\ell \geq 1$

$$\sum_{r=1}^{\infty} G(\ell,r)P\{Y < -r - k\} = P_{\ell e_1}\{S_{\sigma_\rho} \notin I_k\} \leq 1.$$

Finally, with another partial summation

$$Es_{I_k}(v) \leq K_6\,(\log k)^{-1} + \sum_{\ell=1}^{\infty} P\{Y = j + \ell\} \min\left\{2K_5\left(\frac{\log(\ell + 1)}{\log k}\right)^{1/2}, 1\right\}$$

$$\leq K_6\left\{\log(j + 2).\log k\right\}^{-1/2}.$$

This proves (2.8) when $0 \leq j \leq k/2$. The case $k/2 \leq j \leq k$ follows now by symmetry with respect to the midpoint of $I_k$. ∎

Unfortunately we have to work with escape probabilities from $\overline{I}_k$ rather than from $I_k$ itself. The next lemma shows that $Es_{\overline{I}_k}$ is of the same order as $Es_{I_k}$.

LEMMA 3. *There exists a constant $C_3 > 0$ such that*

$$(2.23) \qquad \lim_{k \to \infty} (\log k)^{1/2} Es_{\overline{I}_k}(e_1) = C_3.$$

**Proof:** We use a last exit decomposition (with respect to $\overline{I}_k$) :

$$(2.24) \qquad Es_{I_k}(je_1) = Es_{\overline{I}_k}(je_1) + P_{je_1}\{S \text{ visits } \overline{I}_k \text{ at some}$$
$$\text{time} \geq 1, \text{ but not } I_k\}.$$

The second term here may be decomposed as (see (2.1) for $R$)

$$\sum_{v \in \overline{I}_k \setminus I_k} P_{je_1}\{R(\overline{I}_k) < \infty, \ S_{R(\overline{I}_k)} = v\} Es_{I_k}(v).$$

We want to show that

$$(2.25) \qquad \lim_{k \to \infty} (\log k)^{1/2} \sum_{v \in \overline{I}_k \setminus I_k} P_{je_1}\{R(\overline{I}_k) < \infty, S_R(\overline{I}_k) = v\} Es_{I_k}(v)$$

exists and can be calculated by taking the limit inside the sum. The main step for this is to estimate, for large $L$

$$(2.26) \qquad (\log k)^{1/2} \sum_{\ell=L}^{k} P_{je_1}\{R(\overline{I}_k) < \infty,$$
$$S_{R(\overline{I}_k)} = -\ell e_1 + e_2\} Es_{I_k}(-\ell e_1 + e_2).$$

In view of (2.8) this sum is at most

$$(2.27)$$
$$K_3 \sum_{\ell=L}^{k} P_{je_1}\{R(\overline{I}_k) < \infty, S_{R(\overline{I}_k)} = -\ell e_1 + e_2\} \{\log((\ell \wedge (k - \ell)) + 2)\}^{-1/2}$$

$$\leq 6K_3 \sum_{\ell=L}^{k} P_{je_1}\{R(\overline{I}_k) < \infty, S_{R(\overline{I}_k)} = -\ell e_1 + e_2, S_{R(\overline{I}_k)+1} = S_{R(I_k)} = -\ell e_1\}$$
$$(\log 2)^{-1/2}$$

$$\leq K_4 P_{je_1}\{R(I_k) < \infty, S_{R(I_k)} = -\ell e_1 \text{ for some } L \leq \ell \leq k\}$$

$$\leq K_4 P_{je_1}\{S \text{ hits negative first coordinate axis first in } \{-\ell e_1 : \ell \geq L\}\}.$$

By taking $L$ large we can make the right hand side of (2.27) small. This gives an estimate for (2.26) which is uniform in $k$.

Now $\overline{I}_k \setminus I_k$ consists of the points $e_1$, $(-k-1)e_1$, $-\ell e_1 \pm e_2$, $-\ell e_1 \pm e_3$, $0 \le \ell \le k$. The sums over $v = -\ell e_1 - e_2$ or $v = -\ell e_1 \pm e_3$ are equal to the sum over $v = -\ell e_1 + e_2$. The term

$$(\log k)^{1/2} P_{je_1} \left\{ R(\overline{I}_k) < \infty,\ S_{R(\overline{I}_k)} = -(k+1)e_1 \right\} Es_{I_k}(-(k+1)e_1)$$

is negligable, since, by symmetry,

$$Es_{I_k}(-(k+1)e_1) = Es_{I_k}(e_1) \sim C_1(e_1)(\log k)^{-1/2}.$$

These estimates justify interchanging the limit and sum in (2.25). If we set

$$(2.28) \qquad \overline{R} = \inf \left\{ n \ge 1 : S_n \in \overline{\{-je_1 : j \ge 0\}} \right\},$$

then we obtain from (2.24)

$$(2.29) \quad \lim_{k \to \infty} (\log k)^{1/2} Es_{\overline{I}_k}(je_1)$$

$$= C_1(je_1) - 4 \sum_{\ell \ge 0} P_{je_1} \left\{ S_{\overline{R}} = -\ell e_1 + e_2 \right\} C_1(-\ell e_1 + e_2).$$

Thus the limit in (2.23) exists and we merely have to show that it is strictly positive. Of course it suffices to show (2.29) strictly positive for some large $j$, since

$$(2.30) \quad Es_{\overline{I}_k}(e_1) \ge P_{e_1} \left\{ S. \text{ visits } je_1 \text{ before returning to } \overline{I}_k \right\} Es_{\overline{I}_k}(je_1)$$

and the first factor in the right hand side here is bounded away from 0 as $k \to \infty$.

To show the positivity of (2.29) we go back to (2.24). Note that for $v = -\ell e_1 \pm e_2$ or $-\ell e_1 \pm e_3$

$$Es_{I_k}(-\ell e_1) \ge P_{-\ell e_1}\{S_1 = v\} Es_{I_k}(v) = \frac{1}{6} Es_{I_k}(v).$$

Thus, by (2.7) and (2.8), there exists a constant $K_7$ such that

$$(\log k)^{1/2} Es_{I_k}(v) \le K_7 \quad \text{for all} \quad v \in \overline{I}_k \setminus I_k.$$

Thus, by (2.24)

$$(2.31) \qquad (\log k)^{1/2} Es_{\overline{I}_k}(je_1) \ge (\log k)^{1/2} Es_{I_k}(je_1) - K_7.$$

Now (2.22) shows that the right hand side here is strictly positive for large $j$.

■

LEMMA 4. *Let* $A = A(k(i,\varepsilon))$ *be a generalized plus sign. Assume that* $d = 3$ *and that for some* $k > 0$ *(2.3) holds. Let* $t = \varepsilon(k(i,\varepsilon) + 1)e_i$ *be a tip of* $A$ *and* $w = 0$ *or* $w = -\varepsilon \ell e_i \pm e_j$ *for some fixed* $\ell \geq 0$, $j \neq i$, *so that* $t + w \in \partial A$. *Correspondingly let* $\tilde{w} = 0$ *or* $\tilde{w} = -\ell e_1 + e_2$. *Then for some constants* $K_i$ *(which may depend on* $w$ *but not on* $k$)

$$\left| Es_{\overline{A}}(t + w) - Es_{\overline{I}_k}(e_1 + \tilde{w}) \right| \leq K_1 (\log k)^{-3/2}$$

(2.32)
$$\leq K_2 (\log k)^{-1} Es_{\overline{A}}(t + w)$$

*for* $k$ *sufficiently large. In particular*

(2.33)
$$Es_{\overline{A}}(t) \sim C_3 (\log k)^{-1/2}, \quad k \to \infty.$$

**Proof:** Once we prove the first inequality in (2.32), (2.33) follows from (2.23) (take $w = 0$). It is also easy to obtain from (2.23) (use an estimate like (2.30)) that

$$Es_{\overline{I}_k}(e_1 + \tilde{w}) \geq K_3 (\log k)^{-1/2}$$

for some $K_3 = K_3(w) > 0$. Together with the first inequality of (2.32) this will then also give the second inequality of (2.32). Thus it suffices to prove the first inequality of (2.32). To this end we introduce the notation

$$J(i, +) = \{ je_i : 0 \leq j \leq k(i, +) \},$$
$$J(i, -) = \{ je_i : -k(i, -) \leq j \leq 0 \}.$$

Also, just as in the case $d = 2$ in [7], let $\mathcal{C}(s)$ be the discrete sphere of radius $s$

$$\mathcal{C}(s) = \{ v \in \mathbf{Z}^d : |v| > s, \text{ but } v \text{ is the endpoint of an edge}$$
$$\text{whose other endpoint, } v', \text{ satisfies } |v'| \leq s \}.$$

To begin we prove

(2.34)
$$P_z \{ R(\overline{I}_k) < \infty \} \leq \frac{K_4}{\log k} \sum_{v \in \overline{I}_k} g(z, v),$$

where, as in (2.19), $g(\cdot, \cdot)$ is Green's function for simple random walk. (2.34) rests on the observation

(2.35)
$$\sum_{v \in \overline{I}_k} g(z, v) = E_z \{ \text{number of visits by } S \text{ to } \overline{I}_k \}$$

$$= I \left[ z \in \overline{I}_k \right] + \sum_{u \in \overline{I}_k} P_x \left\{ R(\overline{I}_k) < \infty, \, S_{R(\overline{I}_k)} = u \right\}$$

$$\cdot E_u \{ \text{number of visits by } S \text{ to } \overline{I}_k \}.$$

Now it follows from (2.21) that for any $u \in \overline{I}_k$

$$E_u\{\text{number of visits by } S. \text{ to } \overline{I}_k\} = \sum_{w \in \overline{I}_k} g(u, w)$$

$$\geq K_5 \sum_{1 \leq j \leq k/2} \frac{1}{j} \geq K_4 \log k.$$

Thus (2.34) is immediate from (2.35). As a consequence of (2.34) we find with the help of (2.21) that

(2.36)        $$P_z\{R(\overline{I}_k) < \infty\} \leq K_6 \frac{k}{\log k} [d(z, I_k) + 1]^{-1}.$$

Next, with $T(\cdot)$ as in (2.1), it follows that

$$ES_{\overline{I}_k}(v) \geq P_v\Big\{T(\mathcal{C}(2k)) < R(\overline{I}_k) \text{ and } S. \text{ does not hit } \overline{I}_k \text{ after } T(\mathcal{C}(2k))\Big\}$$

$$\geq P_v\Big\{T(\mathcal{C}(2k)) < R(\overline{I}_k)\Big\} \inf_{z \in \mathcal{C}(2k)} P_z\{R(\overline{I}_k) = \infty\}$$

$$\geq P_v\Big\{T(\mathcal{C}(2k)) < R(\overline{I}_k)\Big\} \left(1 - \frac{K_6}{\log k}\right).$$

Thus, uniformly in $v$

(2.37)        $$P_v\{T(\mathcal{C}(2k)) < R(\overline{I}_k)\} \leq 2Es_{\overline{I}_k}(v).$$

In particular

$$P_{e_1+\tilde{w}}\Big\{T(\mathcal{C}(k/2)) < R(\overline{I}_{k/2})\Big\} \leq P_{e_1+\tilde{w}}\{T(\mathcal{C}(k/2)) < R(\overline{I}_{k/4})\}$$

$$\leq 2Es_{\overline{I}_{k/4}}(e_1 + \tilde{w})$$

(2.38)        $$\leq K_7(\log k)^{-1/2} \quad \text{(by Lemma 2)}.$$

Now assume for the sake of argument that $k(1, +) \geq k$ and that $t$ is the tip $(k(1, +) + 1)e_1$ of $A$ and $\tilde{w} = w$. Note that for large $k$, $t + w$ lies inside the sphere $t - e_1 + \mathcal{C}(k(1, +)/2)$, and $\overline{A} \setminus \overline{J}(1, +)$ lies outside this sphere. Therefore a random walk starting at $t + w$ cannot hit $\overline{A} \setminus \overline{J}$ before it hits $t - e_1 + \mathcal{C}(k(1, +)/2)$. By a decomposition with respect to

$T(t - e_1 + C(k(1,+)/2))$ we find

(2.39)
$$0 \le Es_{\overline{J}(1,+)}(t+w) - Es_{\overline{A}}(t+w) \quad (\text{since } \overline{J} \subset \overline{A})$$

$$= P_{t+w}\{S. \text{ does not hit } \overline{J}(1,+), \text{ but does hit } \overline{A}\}$$

$$= P_{t+w}\{T(t - e_1 + C(k(1,+)/2)) < R(\overline{J}(1,+)), \text{ but } S \text{ hits} \\ \overline{A} \setminus \overline{J}(1,+) \text{ after } R(t - e_1 + C(k(1,+)/2))\}$$

$$\le P_{t+w}\{T(t - e_1 + C(k(1,+)/2)) < R(\overline{J}(1,+))\} \\ \max_{z \in (t-e_1+C(k(1,+)/2)) \setminus \overline{J}(1,+)} \left\{1 - Es_{\overline{A}\setminus\overline{J}(1,+)}(z)\right\}.$$

A translation by $k(1,+)e_1$ shows that the first factor in the last member of (2.39) is at most

(2.40)     $$P_{e_1+w}\left\{T(C(k(1,+)/2)) < R(\overline{I}_{k/2})\right\} \le K_7(\log k)^{-1/2}.$$

As for the second factor,

(2.41)     $$1 - Es_{\overline{A}\setminus\overline{J}(1,+)}(z) \le P_z\{S. \text{ hits one of the arms } \overline{J}(1,-) \\ \text{or } \overline{J}(i,\pm), i \ge 2\}.$$

But for $z \in t - e_1 + C(k(1,+)/2))$

$$d(z, J(1,-)) \ge \frac{1}{2}k(1,+) - 1 \text{ and } d(z, J(i,\pm)) \ge \frac{1}{2}k(1,+) - 1$$

for $i \ge 2$. Thus, (2.36) and obvious symmetry considerations show that

(2.42)          $$1 - Es_{\overline{A}\setminus\overline{J}(1,+)}(z) \le 2(2d-1)K_6(\log k)^{-1},$$

uniformly for $z \in C(k(1,+)/2))$. It follows from (2.39)–(2.41) that

$$\left|Es_{\overline{J}(1,+)}(t+w) - Es_{\overline{A}}(t+w)\right| \le 2(2d-1)K_6K_7(\log k)^{-3/2}.$$

However, the same argument as used in (2.39)–(2.41) shows that for $r = k(1,+)$

$$0 \le Es_{\overline{I}_k+re_1}(t+w) - Es_{\overline{J}(1,+)}(t+w) \le 2K_6K_7(\log k)^{-3/2}.$$

Since
$$Es_{\overline{I}_k+re_1}(t+w) = Es_{\overline{I}_k}(e_1+w),$$

(2.32) follows when $t = (k(1,+)+1)e_1$ and $w = \tilde{w}$. For the other tips of $A$ and for other $w$, (2.32) follows by symmetry. ∎

Lemma 4 proves (2.4) for $d = 3$, because

(2 .43) $$\mu_{\partial A}(y) = \{C(\overline{A})\}^{-1}Es_{\overline{A}}(y).$$

where $C(B) =$ capacity of $B$ (see [19, Prop. 26.2], [7, equation (1.4)] ).

We leave it to the reader to check that for $d = 3$ and any fixed finite set $F$, even

$$\left|Es_{\overline{A\cup F}}(t+w) - Es_{\overline{I}_k}(e_1+\tilde{w})\right| \leq K_1(\log k)^{-3/2}$$

(2.44) $$\leq K_1(\log k)^{-1}Es_{\overline{A\cup F}}(t+w),$$

where $A, t$ and $w$ are as in (2.32), but $K_i$ may now depend on $F$ as well. Basically the only change to be made in the proof is to replace $\overline{A}$ by $\overline{A\cup F}$ in (2.39)–(2.42) , and $\overline{J}(1,-)$ by $\overline{J(1,-)\cup F}$ in the right hand side of (2.41). (2.44) allows us to generalize (2.4) to

(2.45) $$\left|\frac{\mu_{\partial(A\cup F)}(t_2)}{\mu_{\partial(A\cup F)}(t_1)} - 1\right| \leq K_1(\log k)^{-1}$$

with $A, t_1, t_2$ as in (2.4), $F$ a fixed finite set.

It is also easy to see how to modify the proof of Lemma 4 for $d \geq 4$ to obtain (2.6). We now simply use the lower bound

$$E_u\{\text{number of visits by } S. \text{ to } \overline{I}_k\} \geq 1$$

for $u \in \overline{I}_k$, and the asymptotic relation

$$g(x,y) \sim a_d|x-y|^{2-d}, \quad x-y \to \infty$$

(cf. [10, Theorem 1.5.4]) to obtain

(2.46) $$P_z\{R(\overline{I}_k) < \infty\} \leq K_6k[d(z,I_k)+1]^{2-d}.$$

This replaces (2.36). We then have for

$$N := \{-je_1 : j \geq 0\} = \text{negative first coordinate axis,}$$

and $v \notin N$

$$Es_{\overline{I}_k}(v) \rightarrow Es_{\overline{N}}(v) > 0 \quad \text{as} \quad k \rightarrow \infty$$

and also

$$P_v\Big\{T(C(k/2)) < R(\overline{I}_{k/4})\Big\} \rightarrow Es_{\overline{N}}(v) > 0$$

It is now easy to imitate the argument from (2.39) on (with (2.46) replacing (2.36)) to obtain (2.6).

The next lemma begins the proof of (2.5). Again we restrict ourselves to $d = 3$.

LEMMA 5. *Let $d = 3$. There exists a constant $0 < a < 1$ such that for any generalized plus sign $A(k(i,\varepsilon))$ which satisfies (2.3) with $k(1,+) \geq k$ and any $w$ of the form $w = \pm e_i - je_1$, $i = 2, 3$, $j \geq 0$ fixed, one has*

$$(2.47) \qquad \frac{\mu_{\partial A}(k(1,+)e_1 + w)}{\mu_{\partial A}((k(1,+)+1)e_1)} = \frac{Es_{\overline{A}}(k(1,+)e_1 + w)}{Es_{\overline{A}}((k(1,+)+1)e_1)}$$

$$\leq 1 - a \quad \text{for all sufficiently large } k.$$

**Proof:** The equality in (2.47) is again obvious form (2.43). In addition, by (2.32), it suffices to prove

$$(2.48) \qquad \frac{Es_{\overline{I}_k}(w)}{Es_{\overline{I}_k}(e_1)} \leq 1 - a \quad \text{for large } k.$$

Now write $J_k$ for $\{-je_1 : -1 \leq j \leq k\}$. Note that $J_k = I_k \cup \{e_1\}$ and

$$(2.49) \qquad \overline{J}_k \setminus \overline{I}_k = \{2e_1, e_1 + e_2, e_1 - e_2, e_1 + e_3, e_1 - e_3\}.$$

For any $w$ a decomposition with respect to the last exit from $\overline{J}_k$ g ives

$$(2.50) \quad Es_{\overline{I}_k}(w) = Es_{\overline{J}_k}(w)$$
$$+ \sum_{v \in \overline{J}_k \setminus \overline{I}_k} \sum_{n=1}^{\infty} P_w\{S_n = v, R(\overline{I}_k) > n\} \, Es_{\overline{J}_k}(v).$$

Also by a shift of $-e_1$

$$Es_{\overline{J}_k}(v) = Es_{\overline{I}_{k+1}}(v - e_1),$$

and just as in (2.24)

$$Es_{\overline{I}_k}(v - e_1) = Es_{\overline{I}_{k+1}}(v - e_1)$$
$$+ \sum_{z \in \overline{I}_{k+1} \setminus \overline{I}_k} P_{v-e_1}\Big\{R(\overline{I}_{k+1}) < \infty, \; S_{R(\overline{I}_{k+1})} = z\Big\} Es_{\overline{I}_k}(z).$$

Since

$$\overline{I}_{k+1} \setminus \overline{I}_k = \{-(k+2)e_1, -(k+1)e_1 \pm e_2, -(k+1)e_1 \pm e_3\},$$

and (by (2.7) and (2.8))

$$Es_{\overline{I}_k}(z) \leq K_1 (\log k)^{-1/2}, \quad z \in \overline{I}_{k+1} \setminus \overline{I}_k,$$

we have

(2.51)
$$Es_{\overline{J}_k}(v) = Es_{\overline{I}_{k+1}}(v - e_1) = Es_{\overline{I}_k}(v - e_1) + o(\log k)^{-1/2}$$

$$\sim Es_{\overline{I}_k}(v - e_1), \quad k \to \infty,$$

for any fixed $v$ for which there is some $j > 0$ with

(2.52)
$$P_{v-e_1}\{S. \text{ hits } je_1 \text{ before } \overline{R}\} > 0$$

(see (2.28) for $\overline{R}$). Indeed if (2.52) holds for some $j > 0$, then it holds for all $j > 1$ and

$$Es_{\overline{I}_k}(w - e_1) \geq P_{w-e_1}\{S \text{ hits } je_1 \text{ before } \overline{R}\} Es_{\overline{I}_k}(je_1),$$

and we saw in (2.31) that for large enough $j$

$$Es_{\overline{I}_k}(je_1) \geq C_j (\log k)^{-1/2}$$

for some $C_j > 0$. In particular (2.51) holds for all $v \in \overline{J}_k \setminus \overline{I}_k$ (see (2.49)).

Now we first apply (2.50) with $w = e_1$. Observe that by symmetry

$$P_{e_1}\{S_n = e_1 + u, \; R(\overline{I}_k) > n\} Es_{\overline{J}_k}(e_1 + u)$$

has the same value for $u = \pm e_2, \pm e_3$. We therefore obtain from (2.50)–(2.52) (and $Es_{\overline{J}_k}(e_1) = 0$) that

(2.53)
$$Es_{\overline{J}_k}(2e_1) \sim Es_{\overline{I}_k}(e_1)$$

$$= \sum_{n=1}^{\infty} P_{e_1}\{S_n = 2e_1, R(\overline{I}_k) > n\} \, Es_{\overline{J}_k}(2e_1)$$

$$+ 4\sum_{n=1}^{\infty} P_{e_1}\{S_n = e_1 + e_2, \, R(\overline{I}_k) > n\} \, Es_{\overline{J}_k}(e_1 + e_2)$$

$$\sim Es_{\overline{I}_k}(e_1) \sum_{n=1}^{\infty} P_{e_1}\{S_n = 2e_1, \, R(\overline{I}_k) > n\}$$

$$+ 4Es_{\overline{I}_k}(e_2) \sum_{n=1}^{\infty} P_{e_1}\{S_n = e_1 + e_2, \, R(\overline{I}_k) > n\}.$$

Thus (2.48) for $w = e_2$ (or $-e_2$ or $\pm e_3$) will follow once we show

(2.54) $$\sum_{n=1}^{\infty} P_{e_1}\{S_n = 2e_1, \, R(\overline{I}_k) > n\}$$

$$+ 4\sum_{n=1}^{\infty} P_{e_1}\{S_n = e_1 + e_2, \, R(\overline{I}_k) > n\} \geq 1 + b$$

for some $b > 0$, independent of $k$. Simple counting of paths shows

$$P_{e_1}\{S_1 = 2e_1, \, R(\overline{I}_k) > 1\} = P_{e_1}\{S_1 = e_1 + e_2, \, R(\overline{I}_k) > 1\} = 6^{-1}$$

$$P_{e_1}\{S_3 = 2e_1, \, R(\overline{I}_k) > 3\} = 9.6^{-3}$$

$$P_{e_1}\{S_3 = e_1 + e_2, \, R(\overline{I}_k) > 3\} = 7.6^{-3}.$$

Thus the left hand side of (2.54) exceeds

$$6^{-1} + 9.6^{-3} + 4.6^{-1} + 28.6^{-3} = 217.6^{-3} = 1 + 6^{-3};$$

(2.54) holds for $b = 6^{-3}$ and hence (2.48) holds for $w = \pm e_2, \pm e_3$.

It is now easy to obtain (2.48) for any $w = \pm e_i - je_1$, $i = 2, 3, j \geq 0$, by induction on $j$. Indeed, (2.50) and (2.51) for $w = e_2 - je_1$ show as above that

$$Es_{\overline{I}_k}(e_2 - je_1) \sim Es_{\overline{I}_k}(e_2 - (j+1)e_1)$$

$$+ \sum_{v \in J_k \backslash \overline{I}_k} \sum_{n=1}^{\infty} P_w\{S_n = v, \, R(\overline{I}_k) > n\}Es_{\overline{I}_k}(v - e_1).$$

In particular

$$\limsup_{k \to \infty} \frac{Es_{\overline{I}_k}(e_2 - (j+1)e_1)}{Es_{\overline{I}_k}(e_2 - je_1)} \leq 1.$$

Thus if (2.48) holds for $w = e_2 - je_1$ and some $a > 0$, then it holds for $w = e_2 - (j+1)e_1$ and $a$ decreased by an arbitrarily small amount. By induction we therefore have (2.48) for $w = e_2 - je_1$, $j \geq 0$ and then also for $w = \pm e_i - je_1$, $i = 1, 2$, $j \geq 0$. ∎

**Remark.** It is easy to refine this proof to show that the left hand side of (2.48) has a limit as $k \to \infty$, but this is not important for us.

LEMMA 6. *(2.5) holds.*

**Proof:** Let $A$ satisfy (2.3), and for the sake of argument assume again that $k \leq k(1, +) \leq 2k$. Then for $w = e_2 + \ell e_1$, $0 \leq \ell \leq k(1, +)$, we have from (2.8)

$$Es_{\overline{A}}(w) \leq Es_{J(1,+)}(w) \leq K_3(\log k)^{-1/2}\{\log(\ell \wedge (k(1,+) - \ell) + 2)\}^{-1/2}.$$

If we take $L$ such that

$$K_3\{\log(L+2)\}^{-1/2} \leq \frac{1}{4}C_3,$$

then we find (see (2.33))

$$(2.55) \qquad Es_{\overline{A}}(e_2 + \ell e_1) \leq \frac{1}{2} Es_{\overline{A}}(t)$$

for $L \leq \ell \leq k(1, +) - L$ and $t$ any tip of $A$. By obvious symmetry considerations the same argument works for any $w$ of the form

$$(2.56) \qquad w = \pm e_i + \ell e_j, \quad \text{if} \quad k \leq k(j, +) \leq 2k,$$

$$L \leq \ell \leq k_j(t) - L, \quad i \neq j \quad \text{or if}$$

$$k \leq k(j, -) \leq 2k \quad \text{and} \quad -k(j, -) + L \leq \ell \leq -L, \ i \neq j.$$

Thus for all such $w$ we have

$$\frac{\mu_{\partial\overline{A}}(w)}{\mu_{\partial\overline{A}}(t)} \leq \frac{1}{2}$$

if $t$ is a tip of $A$ and $k \geq K_4$ for some $K_4$ independent of $w$.

With $L$ fixed as above we can now take care of the remaining $w \in \partial A$ which are not tips by means of (2.47). Indeed

$$\frac{\mu_{\partial \overline{A}}(w)}{\mu_{\partial \overline{A}}(t)} \leq 1 - a$$

for any $w$ of the form

(2.57)        $w = \pm e_i + \ell e_j$  if  $k \leq k(j,+) \leq 2k$,

$k(j,+) - L \leq \ell \leq k(j,+)$,  $i \neq j$,  or if

$k \leq k(j,-) \leq 2k$  and  $-k(j,-) \leq \ell \leq -k(j,-) + L$,

provided $k \geq K_5$ for some $K_5$ (which may depend on $L$, but we have fixed $L$ now). This is immediate from (2.47) (plus (2.4) and symmetry considerations). Finally, if

(2.58)        $w = \pm e_i + \ell e_j$  with  $k \leq k(j,+) \leq 2k$,

$0 \leq \ell \leq L$  and  $i \neq j$,  or with

$k \leq k(j,-) \leq 2k$,  $-L \leq \ell \leq 0$  and  $i \neq j$,

then we use the following symmetry argument. For the sake of argument take once again $k \leq k(1,+) \leq 2k$, $w = e_2 + \ell e_1$, $0 \leq \ell \leq L$. Then

$$\begin{aligned}
Es_{\overline{A}}(w) &\leq Es_{\overline{J}(1,+)}(w) \\
&= Es_{\overline{J}(1,+)}(e_2 + \ell e_1) \\
&= Es_{\overline{J}(1,+)}(e_2 + (k(1,+) - \ell)e_1) \\
&\leq (1-a)Es_{\overline{J}(1,+)}((k(1,+) + 1)e_1)
\end{aligned}$$

for $k \geq K_5$.

The reader can check that (2.56)–(2.58) take care of all $w \in \partial A$ which are not tips of $A$. Note that even the point $-e_j$ for a $j$ with $k(j,-) = 0$ is taken care of, since this is of the form $-e_j + 0.e_i$ for some $i$ with $k(i,+) \geq k$ or $k(i,-) \geq k$, $i \neq j$ (recall that a generalized plus sign has at least two arms, i.e., $k(i,\varepsilon) > 0$ for at least the choices of $(i,\varepsilon) \in \{1,\ldots,d\} \times \{+,-\}$). We therefore have (2.5) for $K_2 = \max\{K_4, K_5\}$.        ■

Again it is easy for $d = 3$ to generalize (2.5) with $A$ replaced by $A \cup F$ for some fixed finite set $F$, provided that for some $i$, $k(i, +) \geq k$ and $k(i, -) \geq k$. In this case we can estimate $Es_{\overline{A \cup F}}(w)$ for $w$ "close to the origin" by using the fact that $\{\ell e_i : -k \leq \ell \leq k\} \subset A$. This, together with (2.8) gives for $|w| \leq k/2$

$$(2.59) \qquad Es_{\overline{A \cup F}}(w) \leq Es_{\{\ell e_i : -k \leq \ell \leq k\}}(w) = O((\log k)^{-1}).$$

We next state the (partial) analogue of Proposition 1 for $d = 2$.

PROPOSITION 7. *Let $d = 2$ and $F \subset \mathbb{Z}^2$ a fixed finite set.*

*(a) Assume that $A$ is a line segment*

$$(2.60) \qquad \{\ell e_i : -k(i, -) \leq \ell \leq k(i, +)\}$$

*with $k \leq k(i, -)$, $k(i, +) \leq 2k$ for some $k > 0$ and $i = 1$ or $2$. Then for some constants $C_1 > 0$, $K_1 < \infty$ and $a \in (0, 1)$, and sufficiently large $k$*

$$(2.61) \qquad \mu_{\partial(A \cup F)}(t_\pm) \sim C_1 \{k(i, +) + k(i, -)\}^{-1/2},$$

$$(2.62) \qquad |\mu_{\partial(A \cup F)}(t_+) - \mu_{\partial(A \cup F)}(t_-)| \leq K_1 \{k(i, +) + k(i, -)\}^{-5/2}$$

*and*

$$(2.63) \qquad \frac{\mu_{\partial(A \cup F)}(w)}{\mu_{\partial(A \cup F)}(t_\pm)} \leq 1 - a,$$

*where $t_\varepsilon$ stands for the tip $\varepsilon(k(i, \varepsilon) + 1)e_i$ ($\varepsilon = +$ or $-$) and $w$ is any point of $\partial(A \cup F)$ which is not a tip.*

*(b) Assume that $A$ is L-shaped, i.e., for some $\varepsilon_1, \varepsilon_2 \in \{-, +\}$, $k(1, \varepsilon_1) > 0$, $k(2, \varepsilon_2) > 0$*

$$(2.64) \qquad A = \left\{ \varepsilon_1 j e_1 : 0 \leq j \leq k(1, \varepsilon_1) \right\} \cup \left\{ \varepsilon_2 j e_2 : 0 \leq j \leq k(2, \varepsilon_2) \right\}.$$

*Write $t_i$ for the tip $\varepsilon_i(k(i, \varepsilon_i) + 1)e_i$. Then there exist $C_2(\beta) > 0$, $C_3(\beta) > 0$ such that*

$$(2.65) \qquad \lim \left[ k(1, \varepsilon_1) + k(2, \varepsilon_2) \right]^{1/2} \mu_{\partial A}(t_1) = C_2(\beta) > 0,$$

$$(2.66) \qquad \lim \left[ k(1, \varepsilon_1) + k(2, \varepsilon_2) \right]^{1/2} \mu_{\partial A}(t_2) = C_3(\beta) > 0$$

as $k(i, \varepsilon_i) \to \infty$, $i = 1, 2$ in such a way that

$$(2.67) \qquad \frac{k(1, \epsilon_1)}{k(2, \epsilon_2)} \to \beta \in [0, \infty].$$

There also exists a constant $C_4 > 0$ such that

$$(2.68) \qquad \mu_{\partial A}(t_1) - \mu_{\partial A}(t_2) = \left( C_4 + o(1) \right) \frac{k(1, \varepsilon_1) - k(2, \varepsilon_2)}{[k(1, \varepsilon_1) + k(2, \varepsilon_2)]^{3/2}}$$

as $k(i, \varepsilon_2) \to \infty$, $i = 1, 2$ in such a way that

$$(2.69) \qquad \frac{k(1, \varepsilon_1)}{k(2, \varepsilon_2)} \to 1.$$

There exists a constant $a \in (0, 1)$ such that

$$(2.70) \qquad \frac{\mu_{\partial A}(w)}{\mu_{\partial A}(t_i)} \leq 1 - a$$

for $t_i$ the tip with $k(i, \varepsilon_i) \geq k(3 - i, \varepsilon_{3-i})$, $w \in \partial A$ not a tip and $k(1, \varepsilon_1) \wedge k(2, \varepsilon_2)$ sufficiently large.

Finally, for all $\varepsilon > 0$ there exists an $a(\varepsilon) \in (0, 1)$ such that for all sufficiently large $k(1, \varepsilon_1)$ and $k(2, \varepsilon_2) \geq (1 + \varepsilon)k(1, \varepsilon_1)$ and all $w \in \partial A$ other than $w = t_2$,

$$(2.71) \qquad \frac{\mu_{\partial A}(w)}{\mu_{\partial A}(t_2)} \leq \left( 1 - a(\epsilon) \right).$$

Note that $w = t_1$ is included in (2.71), but that $w \neq t_1, t_2$ in (2.70). The proof of part (a) is very similar to that of Proposition 1. However, part (b) requires new techniques and has a very long proof. The proof of Proposition 2 will therefore be given elsewhere ([8], [9]). Here we only show how to obtain Theorems 1 and 2 from Propositions 1 and 7.

The next step shows that we can take $C_0$ in our procedures so large that if $A_n$ is a large generalized plus sign, then $y_n$ is essentially chosen uniformly from the tips of $A_n$, according to each of the procedures 1 and 2. $|S|$ will denote the cardinality of $S$.

PROPOSITION 8. *Let $A_n = A(k(i,\varepsilon,n))$ be a generalized plus sign for which (2.3) holds for some $k > 0$. For $d = 2$ assume that $A_n$ has the form (2.60) or (2.64). Let $a > 0$ be such that (2.5) holds (respectively (2.63) when $A_n$ has the form (2.60), and (2.70) when $A_n$ has the form (2.64)). Without loss of generality take $a \leq 1/2$. If*

$$(2.72) \qquad C_0 > 3\,|\log(1-a)|^{-1} \vee 24a^{-2},$$

*then if $y_n$ is chosen according to Procedure 1 or 2*

$$(2.73) \qquad P\{y_n = w \,|\, A_n\} \leq K_1 n^{-3}$$
$$\text{for all } w \in \partial A_n \text{ which are not a tip of } A_n.$$

*Also, with $S = S_n = \{(i,\varepsilon) : k(i,\varepsilon,n) > 0\}$,*

$$(2.74) \qquad \left| P\{y_n = t \,|\, A_n\} - \frac{1}{|S|} \right| \leq K_2(\log n)^{-1/2} \quad \text{if } t \text{ is any tip of } A_n,$$

*provided $d \geq 4$, or $A_n$ has the form (2.60) when $d = 2$, or that Procedure 1 is used when $d = 3$.*

In order not to interrupt the proof we first give a lemma about urn schemes which will be needed to prove (2.74) for Procedure 1.

LEMMA 9. *Let $U_1, U_2, \ldots, U_s$ be $s$ urns and assume balls are successively and independently put into these urns. The probability of any given ball being put into $U_i$ is $p_i$, $1 \leq i \leq s$. Let*

$$p_i = \frac{1}{s} + \varepsilon_i, \qquad \sum_1^s \varepsilon_i = 0.$$

Define

$$T(m) = \text{first time one of the urns contains } m \text{ balls},$$

$$(2.75)$$
$$\pi(\varepsilon, m) = P\{\text{the } T(m)\text{-th ball is added to } U_1\}$$
$$= P\{\text{at time } T(m), U_1 \text{ contains } m \text{ balls and for}$$
$$\qquad 2 \leq i \leq s, U_i \text{ contains fewer than } m \text{ balls}\}.$$

Then there exist constants $K_1 = K_1(s)$ and $M_0 = M_0(s)$ such that

(2.76)     $$\left| \pi(\varepsilon, m) - \frac{1}{s} \right| \leq K_1 \sqrt{m} \max_i |\varepsilon_i| \quad \text{for} \quad m \geq M_0.$$

If $s = 2$, $\varepsilon_1 > 0 = -\varepsilon_2$, then as $m \to \infty$ and $\varepsilon_1 \downarrow 0$ such that $\varepsilon_1 \sqrt{m} \to 0$

(2.77)     $$\pi(\varepsilon, m) - \frac{1}{2} \sim \frac{2}{\sqrt{\pi}} \varepsilon_1 \sqrt{m}.$$

On the other hand, if for some fixed $\delta > 0$

(2.78)     $$p_i \leq (1 - \delta)p_1, \quad 2 \leq i \leq s,$$

then for $\delta \leq 1/2$ one has

(2.79)     $$\pi(\varepsilon, m) \geq 1 - (s - 1) \exp\left( -\frac{m\delta^2}{8} \right).$$

**Proof:** We begin with (2.76). We set

$$\rho = \sqrt{m} \max_i |\varepsilon_i|.$$

It then suffices to prove (2.76) only for small $\rho$ and large $m$. Indeed it is easy to see that $|\pi(\varepsilon, m) - 1/s| = O(\rho)$ for fixed $m$, and once we have (2.76) for $\rho \leq \rho_0$ it follows for all $\rho \geq 0$ by taking $K_1 \geq \rho_0^{-1}$. Let $B = B(m)$ be the collection of $(s - 1)$ tuples $(k_2, \ldots, k_s)$ with $0 \leq k_i \leq m - 1$. Then

$$\pi(\varepsilon, m) = \sum_B \frac{(m - 1 + k_2 + \cdots + k_s)!}{(m - 1)! \, k_2! \cdots k_s!} \, p_1^m \, p_2^{k_2} \cdots p_s^{k_s}$$

(compare with the standard derivation for the negative binomial distribution when $s = 2$). By symmetry $\pi(s, m) = 1/s$ when $\varepsilon_i = 0$ for all $i$. Therefore, if we set

$$P(k_2, \ldots, k_s) = \frac{(m - 1 + k_2 + \cdots + k_s)!}{(m - 1)! \, k_2! \cdots k_3!} \left( \frac{1}{s} \right)^{m + k_2 + \cdots + k_s},$$

then

$$\pi(\varepsilon) - \frac{1}{s} = \sum_B P(k_2, \ldots, k_s) \Big\{ (sp_1)^m (sp_2)^{k_2} \cdots (sp_s)^{k_s} - 1 \Big\}.$$

Now

$$(sp_1)^m (sp_2)^{k_2} \cdots (sp_s)^{k_s} = \left\{ \prod_{j=1}^{s} (sp_j) \right\}^m \prod_{j=2}^{s} (sp_j)^{k_j - m}$$

and, since $sp_j = 1 + s\varepsilon_j$, we have for some $\theta_i \in [-1, +1]$

$$\log\{(sp_1)^m (sp_2)^{k_2} \cdots (sp_s)^{k_s}\}$$
$$= m \sum_{j=1}^{s} \{s\varepsilon_j + \theta_j s^2 \varepsilon_j^2\} + \sum_{j=2}^{s} (k_j - m)(s\varepsilon_j + \theta_{s+j} s^2 \varepsilon_j^2)$$
$$= \sum_{j=2}^{s} (k_j - m) s\varepsilon_j + 2\theta s^3 \rho^2$$

(recall $\sum_1^s \varepsilon_j = 0$ and $0 \le k_j < m$). Thus

(2.80) $$\left| \pi(\varepsilon, m) - \frac{1}{s} \right| \le \sum_B P(k_2, \ldots, k_s)$$
$$\left| \exp\left\{ \sum_{j=2}^{s} (k_j - m) s\varepsilon_j + 2\theta s^3 \rho^2 \right\} - 1 \right|$$

To estimate the right hand side of (2.80) we write the sum as an expectation. Let $Q$ be the probability measure governing the following process. Balls are put successively into $U_1, \ldots, U_s$, each ball being uniformly distributed over the $s$ urns. We continue until $\tau(m)$, the time when $U_1$ first contains $m$ balls, and define

(2.81) $$\ell_j = \text{number of balls in } U_j \text{ at time } \tau(m).$$

We write $I$ for the indicator function of the event that $\{\ell_j < m = \ell_1, 2 \le j \le s\}$. Then the right hand side of (2.80) equals

(2.82) $$\int I \left| \exp\left\{ \sum_{j=2}^{s} \frac{\ell_j - m}{\sqrt{m}} s\sqrt{m}\varepsilon_j + 2\theta s^3 \rho^2 \right\} - 1 \right| dQ.$$

Note that under $Q$

$$\left( \frac{\ell_2 - m}{\sqrt{m}}, \ldots, \frac{\ell_s - m}{\sqrt{m}} \right)$$

is the sum of $m$ i.i.d. vectors and hence converges in distribution (as $m \to \infty$) to an $(s-1)$-dimensional normal distribution with mean zero and some covariance matrix $\Sigma$. Thus we expect that, as $m \to \infty$ and $\rho \to 0$, (2.82) will behave asymptotically as

(2.83)
$$E\Big\{ J \, \Big| \sum_{j=2}^{s} s\sqrt{m}\varepsilon_j \xi_j \Big| \Big\},$$

where $(\xi_2, \ldots, \xi_s)$ has an $N(0, \Sigma)$ distribution and $J = \{\xi_j \leq 0, \, 2 \leq j \leq s\}$. Once we show that the difference of (2.82) and (2.83) is $o(\rho)$ as $\rho \downarrow 0$, uniformly for $m$ greater than or equal to some $M_1$, (2.75) is immediate.

To justify the above asymptotic equivalence we introduce the abbreviation $\rho_j = \sqrt{m}\,\varepsilon_j$ and the events

$$E(j, \eta) \;=\; \{\, |\rho_j \, (m - \ell_j)| \geq \eta\sqrt{m} \,\}.$$

It then suffices to prove that for some $M_1$ and for all $\eta > 0$ we can find $\rho_0$ (independent of $m$) such that

(2.84)
$$\int_{\cup_i E(i, \eta)} \Big[1 + \exp\Big\{\Big| \sum_{j=2}^{s} \frac{s\rho_j}{\sqrt{m}} (m - \ell_j)\Big|\Big\}\Big] dQ \;\leq\; \eta\rho$$

$$\text{for} \quad \rho \leq \rho_0 \text{ and } m \geq M_1.$$

But by Hölder's inequality

(2.85)
$$\int_{E(i, \eta)} \exp\Big\{\Big| \sum_{j=2}^{s} \frac{s\rho_j}{\sqrt{m}}(m - \ell_j)\Big|\Big\} dQ$$

$$\leq \Big[\prod_{j \neq i} \int \exp\Big\{\Big|\frac{\rho_j}{\sqrt{m}} s(s - 1)(m - \ell_j)\Big|\Big\} dQ\Big]^{1/(s-1)}$$

$$\cdot \Big[\int_{E(i, \eta)} \exp\Big\{\Big|\frac{\rho_i}{\sqrt{m}} s(s - 1)(m - \ell_i)\Big|\Big\} dQ\Big]^{1/(s-1)}$$

Under $Q$, $(m - \ell_j)$ has generating function

(2.86)
$$\int e^{\theta(m - \ell_j)} dQ \;=\; \Big[\frac{e^{\theta}}{2 - e^{-\theta}}\Big]^{m} \;=\; \Big[\frac{e^{2\theta}}{2e^{\theta} - 1}\Big]^{m}$$

(see [2, Ex. XI.2.d]). Taking $\theta = \pm\rho_j s(s - 1)m^{-1/2}$ we see that the product over $j \neq i$ in the right hand side of (2.85) is for small $\rho$ bounded by $2^{s-2} \exp(2\rho s(s - 1))^2$. The last integral in (2.85) is at most

$$e^{-\frac{\eta}{\rho}} \int \exp\Big\{ \big(|\rho_i|s(s - 1) + 1\big)\Big|\frac{m - \ell_i}{\sqrt{m}}\Big|\Big\} dQ$$

$$\leq 2e^{-\frac{\eta}{\rho}+5} \quad \text{for} \;\; \rho s(s - 1) \leq 1 \text{ and } m \geq M_1$$

(by (2.86)). By replacing $\rho_i$ by 0 in the last estimate we see that also

$$\int_{E(i,\eta)} 1 \, dQ \leq 2e^{-\frac{\eta}{\rho}+5}.$$

Thus (2.84) holds and the difference of (2.81) and (2.82) is $o(\rho)$ as $\rho \downarrow 0$, uniformly in $m \geq M_1$.

This proves (2.76). We leave it to the reader to derive (2.77) for $s = 2$ by dropping the absolute value signs in (2.80).

Lastly, (2.79) is a standard large deviation estimate. Indeed, for any $\theta \geq 0$

$$1 - \pi(\varepsilon, m) \leq \sum_{2}^{s} P\{U_j \text{ contains at least } m \text{ balls at time } \tau(m)\}$$

$$\leq \sum_{j=2}^{s} e^{-m\theta} E_\varepsilon \left\{ e^{\theta \ell_j} \right\},$$

where $\ell_j$ is again defined by (2.81), but $E_\varepsilon$ denotes expectation when $p_i = s^{-1} + \varepsilon_i$. Again by [2, Ex. XI.2.d] we find

$$E_\varepsilon \left\{ e^{\theta \ell_j} \right\} = \left\{ \frac{p_1}{p_1 + p_j} \left[ 1 - \frac{p_j}{p_1 + p_j} e^\theta \right]^{-1} \right\}^m = \left\{ \frac{p_1}{p_1 + p_j(1 - e^\theta)} \right\}^m.$$

Since, by assumption $p_j/p_1 \leq 1 - \delta$, it is easy to see that for $\delta \leq 1/2$ one has

$$e^{-\delta/2} \frac{p_1}{p_1 + p_j(1 - e^{\delta/2})} \leq 1 - \frac{\delta^2}{8} \leq e^{-\delta^2/8}.$$

Thus (2.79) holds.     ∎

**Proof of Proposition 8.** We restrict ourselves to the representative cases of Procedure 2 with $d \geq 4$ and Procedure 1 with $d = 3$. For these cases the proof is based on Proposition 1. We note that in all our constructions $A_n$ contains exactly $n$ vertices. Thus if $A_n$ satisfies (2.3) for a certain $k$ and $S = S_n$ is as defined before (2.73), then

(2.87) $$(k - 1)|S| \leq n = 1 + \sum_{(i,\varepsilon) \in S} (k(i, \varepsilon) - 1) \leq 2k|S|.$$

Thus $k$ and $n$ have to be of the same order and we may replace $k$ by $n$ in the right hand side of (2.4) and (2.6) for $A_n$.

We first consider Procedure 2 and $d \geq 4$, which is the easiest case. Now by (2.5), if $t$ is a tip of $A_n$ and $w \in \partial A_n$ is not a tip, then

$$(2.88) \qquad [\mu_{\partial A_n}(w)]^{\eta(n)} \leq (1-a)^{\eta(n)} \mu_{\partial A_n}(t) \leq (1-a)^{\eta(n)} Z_n(\eta),$$

where $Z_n(\eta)$ is as in (1.2). Thus (2.73) is immediate from (1.1) and $\eta(n) \sim C_0 \log n$ with $C_0 > 3|\log(1-a)|^{-1}$.

We also obtain from (2.88) that

$$Z_n(\eta) - \sum_{\substack{t \text{ a tip} \\ \text{of } A_n}} [\mu_{\partial A_n}(t)]^{\eta(n)} = \sum_{\substack{w \text{ not a tip} \\ \text{of } A_n}} [\mu_{\partial A_n}(w)]^{\eta(n)}$$

$$\leq n(1-a)^{\eta(n)} Z_n(\eta) \leq n^{-2} Z_n(\eta).$$

Thus if $t_1, \ldots, t_{|S|}$ are the tips of $A_n$, and $d \geq 4$, then by (2.6), for some $\theta_i \in [-1, +1]$

$$P\{y_n = t_\ell | A_n\} = [Z_n(\eta)]^{-1} [\mu_{\partial A_n}(t_\ell)]^{\eta(n)}$$

$$= (1 + \theta_1 n^{-2})^{-1} \left\{ \sum_{j=1}^{|S|} [\mu_{\partial A_n}(t_j)]^{\eta(n)} \right\}^{-1} [\mu_{\partial A_n}(t_\ell)]^{\eta(n)}$$

$$= (1 + \theta_1 n^{-2})^{-1} (1 + \theta_2 K_1 n^{3-d})^{\eta(n)}$$

$$\left\{ \sum_{j=1}^{|S|} [\mu_{\partial A_n}(t_\ell)]^{\eta(n)} \right\}^{-1} [\mu_{\partial A_n}(t_\ell)]^{\eta(n)}$$

$$= \frac{1}{|S|} (1 + \theta_1 n^{-2})^{-1} \exp(2\theta_3 K_1 n^{3-d} C_0 \log n).$$

Thus (2.74) holds if $d \geq 4$. This completes the proof for Procedure 2 when $d \geq 4$.

For Procedure 1 and $d = 3$ we again begin with proving (2.73). Let $t$ be a tip of $A_n$ and $w \in \partial A_n$ not a tip. Then by (2.5)

$$\mu_{\partial A_n}(w) \leq (1-a)\mu_{\partial A_n}(t).$$

Now, in the notation of Procedure 1,

$$(2.89) \qquad P\{y_n = w | A_n\} \leq P\{s(w) \text{ reaches the value } m(n) \sim C_0 \log n,$$
$$\text{before } s(t) \text{ reaches } m(n)\}.$$

In estimating the last probability we can ignore all particles which land at boundary sites other than $w$ or $t$. In other words, we can calculate the probability as if in each trial we either increase $s(t)$ or $s(w)$ with probabilities

$$\frac{\mu_{\partial A_n}(t)}{\mu_{\partial A_n}(t) + \mu_{\partial A_n}(w)} \quad \text{and} \quad \frac{\mu_{\partial A_n}(w)}{\mu_{\partial A_n}(t) + \mu_{\partial A_n}(w)} \,,$$

respectively. Thus, by (2.79) with $\delta = a$, $s = 2$ the right hand side of (2.89) is at most

$$\exp\left(-\frac{a^2}{8}m(n)\right) \leq n^{-3},$$

for $C_0 > 24a^{-2}$, and (2.73) follows.

For (2.74) we again denote the tips of $A_n$ by $t_1, \ldots, t_{|S|}$. We modify Procedure 1 a bit. Instead of taking $y_n$ equal to the first $z \in \partial A_n$ for which $s(z) = m(n)$, we ignore all sites which are not tips. We simply take $y_n$ equal to the first *tip* whose score reaches $m(n)$. By what we proved already this change affects $y_n$ with a probability at most $|\partial A_n| n^{-3} \leq 2dn^{-2}$. This can be ignored since we allow a much larger error in (2.74).

Now in the modified procedure we can again ignore all sites of $\partial A_n$ which are not tips and assume that at each trial the score $s(t_j)$ of the tip $t_j$ is increased by 1 with probability

$$p_j := \left[\sum_{i=1}^{|S|} \mu_{\partial A_n}(t_i)\right]^{-1} \mu_{\partial A_n}(t_j).$$

By (2.4) and (2.87)

$$\left| p_j - \frac{1}{|S|} \right| \leq K_1 (\log k)^{-1} \leq 2K_1 (\log n)^{-1}$$

and (2.76) now shows

$$\left| P\{y_n = t_j \mid A_n\} - \frac{1}{|S|} \right| \leq K_2 (C_0 \log n)^{1/2} (\log n)^{-1}$$

for large $n$, and $1 \leq j \leq |S|$. Thus (2.74) is proven. ∎

Once again no change of proof is necessary when $A_n$ is replaced by $A_n \cup F$ for a fixed finite set $F$ and $w \in \partial(A_n \cup F)$, $w$ not a tip of $A_n$, in the cases

$d = 3$ with Procedure 1 when $\{(i,-),(i,+)\} \subset S$ for some $i$ (cf. (1.9)) or $d = 2$ with $A_n$ of the form (2.60) (and either procedure).

We are finally ready for the

**Proof of Theorem 1.**

We fix $C_0 > 3|\log(1 - a)|^{-1} \vee 24a^{-2}$ and take $m(n) \sim C_0 \log n$ in Procedure 1, respectively $\eta(n) \sim C_0 \log n$ in Procedure 2. Also fix $S \subset \{1,\ldots,d\} \times \{-,+\}$ with at least two elements; when $d = 2$ take $S = \{(i,-),(i,+)\}$ for $i = 1$ or 2. Define the following events, random variables and stopping times:

$$F(n,\eta) := \Big\{A_n = A_n(k(i,\varepsilon,n)) \text{ is a generalized plus sign}$$
$$\text{with } k(i,\varepsilon,n) = 0 \text{ for } (i,\varepsilon) \notin S \text{ and}$$
$$\big|k(i,\varepsilon,n) - n/|S|\big| \leq \eta n \text{ for all } (i,\varepsilon) \in S\Big\}.$$

It is necessary to extend the definition of $k(i,\varepsilon,n)$ to a general $A_n$ which is not necessarily a generalized plus sign. In general we put

$$k(i,\varepsilon,n) = \max\{k : \varepsilon k e_i \in A_n\}, \quad (i,\varepsilon) \in \{1,\ldots,d\} \times \{+,-\}.$$

Let $\mathcal{F}_r$ be the $\sigma$-field generated by $A_1,\ldots,A_r$ and

$$Z_n = Z_n(i,\varepsilon,\ell) = k(i,\varepsilon,n)-k(i,\varepsilon,\ell)-\sum_{r=\ell}^{n-1} E\Big\{k(i,\varepsilon,r+1)-k(i,\varepsilon,r)\,|\,\mathcal{F}_r\Big\},$$

$$G = G(\ell) = \Big\{|Z_r(i,\varepsilon,\ell)| \leq \frac{\eta}{2}\ell+\frac{\eta}{4}(r-\ell) \text{ for all } r \geq \ell$$
$$\text{and all } (i,\varepsilon) \in S\Big\},$$

$$G(\ell,n) = \Big\{|Z_r(i,\varepsilon,\ell)| \leq \frac{\eta}{2}\ell+\frac{\eta}{4}(r-\ell) \text{ for all } \ell \leq r \leq n$$
$$\text{and all } (i,\varepsilon) \in S\Big\},$$

$$\tau = \inf\{r : A_r \text{ is not a generalized plus sign or } k(i,\varepsilon,r) > 0$$
$$\text{for some } (i,\varepsilon) \notin S\},$$

$$\sigma_\ell = \inf\left\{r \geq \ell : \left|P\{y_r = \varepsilon(k(i,\varepsilon,r)+1)e_r | \mathcal{F}_r\} - 1/|S|\right| \geq \frac{\eta}{4}\right.$$
$$\left. \text{for some } (i,\varepsilon) \in S\right\}.$$

It should be noted that for $r < \tau$, $k(i,\varepsilon,r+1) - k(i,\varepsilon,r) = 1$ or $0$, according as $y_r$ equals $\varepsilon(k(i,\varepsilon,r)+1)e_i$ or not. Thus, for $r < \tau$

$$E\left\{k(i,\varepsilon,r+1) - k(i,\varepsilon,r) | \mathcal{F}_r\right\} = P\left\{y_r = \varepsilon(k(i,\varepsilon,r)+1)e_i | \mathcal{F}_r\right\},$$

and for $n \leq \tau$

$$Z_n(i,\varepsilon,\ell) = k(i,\varepsilon,n) - k(i,\varepsilon,\ell)$$
$$(2.90) \qquad\qquad -\sum_{r=\ell}^{n-1} P\left\{y_r = \varepsilon(k(i,\varepsilon,r)+1)e_i | \mathcal{F}_r\right\}.$$

Assume now that for some $\ell$, $n \geq \ell$ and $0 < \eta < (6|S|)^{-1}$ (with the $K_2$ of (2.74))

$$(2.91) \qquad K_2(\log \ell)^{-1/2} < \frac{\eta}{4} \quad \text{and } F(\ell,\eta) \text{ and } G(\ell,n) \text{ occur.}$$

Then for $\ell \leq r < \tau$, $\ell \leq r \leq \sigma_\ell \wedge n$, $A_r$ is still a generalized plus sign. Moreover, for $(i,\varepsilon) \in S$

$$(2.92)$$
$$\left|k(i,\varepsilon,r) - \frac{r}{|S|}\right| \leq |Z_r(i,\varepsilon,\ell)| + \left|k(i,\varepsilon,\ell) - \frac{\ell}{|S|}\right|$$
$$+ \left|\sum_{q=\ell}^{r-1} P\{y_q = \varepsilon(k(i,\varepsilon,q)+1)e_i | \mathcal{F}_q\} - \frac{r-\ell}{|S|}\right|$$
$$\leq \frac{\eta}{2}\ell + \frac{\eta}{4}(r-\ell) + \eta\ell + \frac{\eta}{4}(r-\ell) = \eta\ell + \frac{\eta}{2}r.$$

The first inequality here is immediate from (2.90), while for the second inequality we used that

$$|Z_r(i,\varepsilon,\ell)| \leq \frac{\eta}{2}\ell + \frac{\eta}{4}(r-\ell) \quad \text{on} \quad G(\ell,r),$$

$$\left|k(i,\varepsilon,\ell) - \frac{\ell}{|S|}\right| \leq \eta\ell \quad \text{on} \quad F(\ell,\eta),$$

and for $q < \sigma_\ell$, $(i, \varepsilon) \in S$,

$$(2.93) \qquad \left| P\{y_q = \varepsilon(k(i, \varepsilon, q) + 1)e_i \mid \mathcal{F}_q\} - \frac{1}{|S|} \right| < \frac{\eta}{4}.$$

Since we assumed $r < \tau$, we also have

$$k(i, \varepsilon, r) = 0 \quad \text{for} \quad (i, \varepsilon) \notin S.$$

Thus $F(r, 2\eta)$ occurs. By virtue of (2.74) and (2.91) we see that (2.93) also holds for $j = r$, or $\sigma_\ell \geq r + 1$. Note that (2.93) automatically holds for $q = \ell$ on $F(\ell, \eta)$ by this same argument. It follows that if (2.91) occurs, then

$$(2.94) \qquad \sigma_\ell \geq \tau \wedge (n + 1).$$

For the remainder fix $0 < \eta < (6|S|)^{-1}$ and $\ell$ such that $K_2(\log \ell)^{-1/2} < \frac{\eta}{4}$. Now $G(\ell) = \bigcap_{n \geq \ell} G(\ell, n)$, and also $\tau > \ell$ on $F(\ell, \eta)$. Therefore

$$P\Big\{ G(\ell) \text{ occurs and } \tau < \infty \mid F(\ell, \eta) \Big\}$$

$$\leq \sum_{n=\ell}^{\infty} P\Big\{ G(\ell, n) \text{ occurs and } \tau = n + 1 \mid F(\ell, \eta) \Big\}$$

$$\leq \sum_{n=\ell}^{\infty} P\Big\{ \tau = n + 1 \mid F(\ell, \eta) \cap G(\ell, n) \cap \{\tau > n\} \Big\}.$$

However, we just saw that on

$$(2.95) \qquad F(\ell, \eta) \cap G(\ell, n) \cap \{\tau > n\}$$

we also have $F(n, 2\eta)$. By (2.73) we therefore have on (2.95)

$$P\{\tau = n + 1 \mid \mathcal{F}_n\} \leq |\partial A_n| K_1 n^{-3} \leq 2dK_1 n^{-2}.$$

Thus

$$(2.96) \qquad P\{F(\ell, \eta) \text{ and } G(\ell) \text{ occur, but } \tau < \infty\} \leq \sum_{n=\ell}^{\infty} 2dK_1 n^{-2}$$

$$\leq K_3 \ell^{-1}.$$

We further note that from (2.94), on $F(\ell, \eta) \cap G(\ell) \cap \{\tau = \infty\}$ also $\sigma_\ell = \infty$. Thus outside a set of probability at most $K_3 \ell^{-1}$ we have $\tau = \sigma_\ell = \infty$

on $F(\ell,\eta) \cap G(\ell)$. In addition, (2.92) shows that $F(r, 3\eta/4)$ occurs on $F(\ell,\eta) \cap G(\ell) \cap \{\tau = \sigma_\ell = \infty\}$ for all $r \geq 4\ell$.

Finally we estimate

$$P\{\,G(\ell)\ \text{fails}\mid F(\ell,\eta)\,\}.$$

This is done by means of standard exponential bounds for martingales. Indeed $\{Z_n(i,\varepsilon,\ell)\}_{n\geq\ell}$ is a martingale. Since $k(i,\varepsilon,n) - k(i,\varepsilon,n-1)$ equals 0 or 1 the increments of $Z_n$ are at most 1 in absolute value, and we conclude from [15, pp. 154, 155]

$$P\{G(\ell)\ \text{fails}\mid F(\ell,\eta)\}$$

$$\leq \sum_{(i,\varepsilon)\in S} P\left\{|Z_r(i,\varepsilon,\ell)| > \frac{\eta}{2}\ell + \frac{\eta}{4}(r-\ell)\ \text{for some}\ r \geq \ell\right\}$$

$$\leq 4d \exp\left(-\lambda\frac{\eta}{2}\ell\right),$$

where $\lambda = \lambda(\eta) > 0$ is such that

$$\frac{1}{\lambda}\left[e^\lambda - 1 - \lambda\right] = \frac{\eta}{4}.$$

Thus, for $0 < \eta < (6|S|)^{-1}$ and all sufficiently large $\ell$

$$P\left\{G(\ell)\ \text{occurs},\ \tau = \infty\ \text{and}\ F(r, 3\eta/4)\ \text{occurs for all}\ r \geq 4\ell\mid F(\ell,\eta)\right\}$$

$$\geq 1 - 2K_3\ell^{-1}.$$

It follows that for large $\ell$

$$P\left\{\text{For all}\ q, F\left(r,\left(\frac{3}{4}\right)^q\eta\right)\ \text{for}\ r \geq 4^q\ell\mid F(\ell,\eta)\right\}$$

$$\geq 1 - 2K_3\sum_{q=0}^{\infty}(4^q\ell)^{-1} \geq 1 - \frac{8}{3}K_3\ell^{-1} \geq \frac{1}{2}.$$

Since clearly $P\{F(\ell,\eta)\} > 0$ we have

$$P\left\{\bigcap_{q\geq 0}\left\{F(r,(\tfrac{3}{4})^q\ell)\ \text{for}\ r \geq 4^q\ell\right\}\right\} > 0.$$

This proves (1.5) since

$$\bigcap_{q\geq 0}\left\{F(r,(\tfrac{3}{4})^q\ell)\ \text{for}\ r \geq 4^q\ell\right\} \subset E(S).$$

∎

We leave it to the reader to check that one also obtains (1.8) in the cases (1.9) and (1.10). The necessary minor modifications have already been indicated in comments immediately following the proofs of Lemmas 4, 6 and Proposition 8.

## 3. Proof of Theorem 2.

To simplify notation we take $\varepsilon_1 = \varepsilon_2 = +$ so that $S = \{(1,+),(2,+)\}$. We also write $k(i,n)$ instead of $k(i,+,n)$ so that we are dealing with $A_n$ of the form

$$(3.1) \qquad A_n \;=\; \{je_1 : 0 \le j \le k(1,n)\} \;\cup\; \{je_2 : 0 \le j \le k(2,n)\}$$

with

$$(3.2) \qquad\qquad\qquad k(1,n) + k(2,n) \;=\; n-1.$$

The tips of such an $A_n$ will be written as

$$t_i \;=\; t_i(n) \;=\; \big(k(i,n)+1\big)e_i.$$

Even though it is not needed for Theorem 2, we point out that if $a$ is as in (2.70), and $C_0$ satisfies (2.72), then we may assume that for all $n$, $A_n$ is of the form (3.1) and $y_n$ is one of the tips $t_i(n)$, $i = 1,2$. In fact, the proof of (2.73) shows that if for some $\ell$, $A_\ell$ is of the form (3.1), then

$$(3.3) \qquad P\{A_n \text{ is of the form (3.1) for all } n \ge \ell \,|\, A_\ell\} \;\ge\; 1 - K_3\ell^{-1}$$

(compare (2.96)).

The preceding observation is not needed for (1.6), for $E(S)$ fails in any case as soon as (3.1) fails for some $n$. If there is a first time when $y_n$ is not a tip of $A_n$, then $E(S)$ can no longer occur. We may therefore stop the construction of $A_m$, $m \ge n$, once $A_n$ is no longer of the form (3.1). If $A_n$ is of the form (3.1) we write $\gamma(i, k(\cdot,n))$ for the conditional probability that $y_n = t_i(n)$, given $y_n \in \{t_1(n), t_2(n)\}$. Thus

$$\gamma(i, k(\cdot,n)) \; \frac{P\{y_n = t_i(n) \,|\, A_n\}}{P\{y_n = t_1(n) \,|\, A_n\} + P\{y_n = t_2(n) \,|\, A_n\}}$$

on the event $\{A_n$ is of the form $(3.1)\}$. Define an auxiliary Markov chain $Y(n) = (Y_1(n), Y_2(n))$, $n \geq 1$, with state space $\mathbb{Z}_+^2$ and transition probabilities

$$P\{Y_i(n+1) = Y_i(n) + 1, Y_{3-i}(n+1) = Y_{3-i}(n) \,\big|\, Y_j(n)$$
$$(3.5) \qquad\qquad = k(j,n),\ j=1,2\} = \gamma(i, k(\cdot, n)), \quad i = 1, 2.$$

We may view the vector process $(k_1(n), k_2(n))$ as a realization of the $Y(n)$ process which is stopped at the (random) time when $A_n$ is of the form (3.1) for the last time (this time may be infinite). In order to prove (1.6) it therefore suffices to prove

$$(3.6) \qquad\qquad P\Big\{ \lim \frac{Y_1(n)}{Y_2(n)} = 0 \ \text{ or } \ \infty \Big\} = 1.$$

For (3.6) will imply that w.p.1 either $A_n$ is not of the form (3.1) for some $n < \infty$ or $k(1,n)/k(2,n)$ converges to 0 or $\infty$.

To attack (3.6) we first need to extract some information about the $\gamma(i, n)$ from Proposition 7 b). Write

$$\pi_i = \pi_i\big(k(\cdot, n)\big) = \mu_{\partial A_n}(t_i)$$

if $A_n$ has the form (3.1). Then for Procedure 2

$$(3.7) \qquad\qquad \gamma(i, k(\cdot, n)) = \frac{[\pi_i]^{\eta(n)}}{[\pi_i]^{\eta(n)} + [\pi_2]^{\eta(n)}}.$$

It is easy to see in this case (from (2.68), (2.65) and (2.66) and symmetry considerations and (3.2)) that for each constant $D > 0$ there exists an $0 < \varepsilon_0 \leq 1$ and an $n_0 = n_0(D, \varepsilon_0)$ such th at

$$(3.8) \qquad \gamma(i, k(\cdot, n)) \geq \frac{1}{2} + \frac{D}{n}\big[k(i,n) - k(3-i, n)\big]$$

on the set

$$(3.9) \qquad k(3-i, n) \leq k(i, n) \leq (1 + \varepsilon_0)k(3 - i, n), \quad n \geq n_0.$$

Note that this relies only on $\eta(n) \to \infty$, not on the precise behavior of $\eta(n)$. In addition, by (2.71), on

$$(3.10) \qquad k(i, n) \geq \Big(1 + \frac{\varepsilon_0}{16}\Big)k(3 - i, n), \quad n \geq n_0.$$

we have

$$(3.11) \qquad \pi_{3-i} \leq \left(1 - a\left(\tfrac{\varepsilon_0}{16}\right)\right)\pi_i,$$

which implies for some $\kappa = \kappa(C_0) > 0$

$$(3.12) \qquad \gamma\big(i, k(\cdot, n)\big) \geq 1 - n^{-\kappa}.$$

Here we do use the fact that $\eta(n) \sim C_0 \log n$. We also see that we can make $\kappa > 1$ by taking $C_0$ large.

For Procedure 1 we do not have as explicit an expression as (3.7) for the $\gamma\big(i, k(\cdot, n)\big)$. Now $\gamma\big(i, k(\cdot, n)\big)$ is the probability that the urn $U_i$ receives $m(n)$ balls before the urn $U_{3-i}$, in the setup of Lemma 9 with $s = 2$ and

$$p_i = \frac{\pi_i\big(k(j, n)\big)}{\pi_1\big(k(j, n)\big) + \pi_2\big(k(j, n)\big)}.$$

We again conclude from (2.68), (2.65), (2.66) and symmetry and monotonicity considerations that on (3.9)

$$p_i \geq \frac{1}{2} + \frac{K_1}{n}\big(k(i, n) - k(3 - i, n)\big)$$

for some $K_1 > 0$. We now conclude from (2.77) that (3.8) holds also for Procedure 1 on the set (3.9). Similarly, by virtue of (3.11)

$$p_{3-i} \leq (1 - a(\varepsilon_0/16))p_i \quad \text{on (3.10)}.$$

(2.79) now shows that also for Procedure 1 (3.12) holds under condition (3.10).

From our previous remarks we see that (1.6) is a consequence of the following proposition, which has some independent interest. We are happy to acknowledge some help from R. Durrett with its proof.

PROPOSITION 10. *Let* $\{Y(n)\}_{n \geq 1}$ *be a Markov chain on* $\mathbb{Z}^2$ *in which at each time exactly one of the coordinates increases by 1. Let* $\gamma(i, k(\cdot, n))$ *be the transition probabilities as in (3.5). Assume that for some* $D > \frac{1}{2}$ *and* $0 < \varepsilon_0 \leq 1$, $\kappa > 0$, *(3.8) holds on (3.9) and (3.12) holds on (3.10). Then*

$$P\left\{\lim \frac{Y_1(n)}{Y_2(n)} = 0 \text{ or } \infty\right\} = 1.$$

*If $\kappa > 1$ in (3.12) then even*

(3.13) $\qquad P\{Y_1(n) \text{ or } Y_2(n) \text{ remains bounded}\} = 1.$

**Proof:** We break the proof up into several steps. We define

$$X_n = \max_{i=1,2} Y_i(n) - \min_{i=1,2} Y_i(n)$$

$$= |Y_1(n) - Y_2(n)|,$$

$$\Delta_{n+1} = X_{n+1} - X_n.$$

Then $\Delta_{n+1}$ takes on the values $+1$ or $-1$. In step (i) we show that $\limsup X_n/\sqrt{n} = \infty$ w.p.1. This comes about by more or less random fluctuations. We then show that once $X_n/\sqrt{n}$ is large, (3.8) provides enough upwards drift to give $\sup X_n/n \geq \varepsilon_0(2 + e\varepsilon_0)^{-1}$ w.p.1 (steps (ii) and (iii)). Finally then (3.12) takes over to give (3.6).

**Step (i).** Let $\mathcal{F}_n$ denote the $\sigma$-fields generated by $Y(1), \ldots, Y(n)$. Then (3.8) shows that for sufficiently large $n$

(3.14) $\qquad P\{\Delta_{n+1} = 1 \mid \mathcal{F}_n\} \geq \dfrac{1}{2} + \dfrac{D}{n} X_n \quad \text{on} \ \left\{X_n \leq \dfrac{\varepsilon_0}{2(1 + \varepsilon_0)} n\right\}$

(recall that $Y_1(n) + Y_2(n) = (n-1) + Y_1(1) + Y_2(1)$). In particular, $X_n$, which is positive, is stochastically larger than the absolute value of a one dimensional simple symmetric random walk, $\{U_n\}$ say. Since for a simple random walk $\limsup |U_n|/\sqrt{n} = \infty$ w.p.1 we also have

(3.15) $\qquad \limsup \dfrac{X_n}{\sqrt{n}} = \infty \quad w.p.1.$

**Step (ii).** Let $n_0$ be so large that (3.14) holds for $n \geq n_0$ and let

(3.16) $\qquad \sigma = \inf\left\{n \geq n_0 : \dfrac{X_n}{n} \geq \varepsilon_0(2 + 2\varepsilon_0)^{-1}\right\}.$

In this step we prove

(3.17) $\qquad P\left\{\sigma < \infty \ \text{ or } \ \lim \dfrac{X_n}{\sqrt{n}} = \infty\right\} = 1.$

Let $A$ be a (large) fixed number and $\varepsilon, \gamma, n_1$ be strictly positive numbers such that

$$(3.18) \qquad 2D(1-\varepsilon) > 1, \qquad (1-\varepsilon)(1+\gamma)^{1/2} \le 1 - \frac{\varepsilon}{2},$$

$$2D(1-\varepsilon)(1+\gamma)^{-1/2}\gamma \ge (1+\frac{1}{4}\gamma)(1+\gamma)^{1/2} - 1 + \frac{1}{4}\gamma.$$

$$\varepsilon A \le \gamma n_1^{1/2}, \qquad A \le 8n_1^{1/2},$$

Assume that also

$$(3.19) \qquad n_1^{-1/2}X_{n_1} \ge A \quad \text{but} \quad n_1 < \sigma.$$

Now consider the martingale

$$V_n := X_{n \wedge \sigma} - X_{n_1} - \sum_{n_1+1}^{n \wedge \sigma} E\{\Delta_j \,|\, \mathcal{F}_{j-1}\}, \quad n \ge n_1.$$

Note that for $j \le \sigma$,

$$(3.20) \; E\{\Delta_j \,|\, \mathcal{F}_{j-1}\} \ge \left(\frac{1}{2} + \frac{D}{j}X_j\right) - \left(\frac{1}{2} - \frac{D}{j}X_j\right) = \frac{2D}{j}X_j \ge 0,$$

by virtue of (3.14). Therefore, on the event (3.19),

$$(3.21) \qquad X_n \le (1-\varepsilon)An^{1/2} \quad \text{for some } n \in [n_1, n_1(1+\gamma) \wedge \sigma]$$

implies that for this $n$ also

$$V_n \le X_{n \wedge \sigma} - X_{n_1} \le (1-\varepsilon)A\sqrt{n} - A\sqrt{n_1}$$
$$\le A\sqrt{n_1}[(1-\varepsilon)(1+\gamma)^{1/2} - 1] \le -\frac{\varepsilon A}{2}\sqrt{n_1}.$$

Since also

$$E\{\Delta_j^2 \,|\, \mathcal{F}_{j-1}\} \le 1,$$

we have on (3.19)

$$P\{(3.21) \text{ occurs} \,|\, \mathcal{F}_{n_1}\}$$

$$\le P\left\{-V_n \ge \frac{\varepsilon A}{4}n_1^{1/2} + \frac{\varepsilon A}{4\gamma}n_1^{-1/2}\sum_{n_1+1}^{n} E\{\Delta_j^2 \,|\, \mathcal{F}_{j-1}\}\right.$$

$$\left. \text{for some } n_1 \le n \le n_1(1+\gamma) \,|\, \mathcal{F}_{n_1}\right\}$$

The right hand side here is at most

$$\exp\left(-\frac{\varepsilon^2 A^2}{16\gamma}\right),$$

by exponential bounds for martingales with bounded increments; for instance one may apply [15, pp. 154, 155] with $c = 1$, $a = (\varepsilon A/4)n_1^{1/2}$, $\lambda = (\varepsilon A/4)\gamma^{-1}n_1^{-1/2} \leq 1$.

Next consider a sample path for which (3.19) holds, but (3.21) fails. We then have from (3.20) that for $n_1 \leq n \leq n_1(1+\gamma)$

$$\sum_{n_1+1}^{n\wedge\sigma} E\{\Delta_j \mid \mathcal{F}_{j-1}\} \geq \sum_{n_1+1}^{n\wedge\sigma} 2D(1-\varepsilon)Aj^{-1/2}$$

$$\geq 2D(1-\varepsilon)A(1+\gamma)^{-1/2}\,n_1^{-1/2}(n\wedge\sigma-n_1).$$

In particular, if $\sigma \geq n_1(1+\gamma)$ then (cf. (3.18))

$$\sum_{n_1+1}^{(1+\gamma)n_1} E\{\Delta_j \mid \mathcal{F}_{j-1}\} \geq An_1^{1/2}\left\{(1+\frac{1}{4}\gamma)(1+\gamma)^{1/2} - 1 + \frac{1}{4}\gamma\right\}.$$

Therefore, on (3.19)

$$P\Big\{(3.21) \text{ fails, } \sigma \geq n_1(1+\gamma) \text{ but}$$

$$X_{(1+\gamma)n_1} \leq (1+\frac{1}{4}\gamma)A[(1+\gamma)n_1]^{1/2} \,\Big|\, \mathcal{F}_{n_1}\Big\}$$

$$\leq P\Big\{-V_{n_1(1+\gamma)} \geq \frac{\gamma}{4}An_1^{1/2} \,\Big|\, \mathcal{F}_{n_1}\Big\}$$

$$\leq P\Big\{-V_{n_1(1+\gamma)} \geq \frac{\gamma}{8}An_1^{1/2} + \frac{A}{8}n_1^{-1/2}\sum_{n_1+1}^{n_1(1+\gamma)} E\{\Delta_j^2 \mid \mathcal{F}_{j-1}\} \,\Big|\, \mathcal{F}_{n_1}\Big\}$$

$$\leq \exp\left(-\frac{\gamma}{64}A^2\right).$$

In the last step we again used [15, pp. 154, 155]. It follows from the above estimates that on (3.19).

$$P\Big\{\sigma \geq n_1(1+\gamma) \text{ and } \{X_n \leq (1-\varepsilon)An^{1/2} \text{ for some } n \in [n_1, n_1(1+\gamma)]$$

$$\text{or } X_{n_1(1+\gamma)} \leq (1+\frac{1}{4}\gamma)A[(1+\gamma)n_1]^{1/2}\} \,\Big|\, \mathcal{F}_{n_1}\Big\}$$

$$\leq 2\exp(-C_1 A^2),$$

with

$$C_1 = \frac{\varepsilon^2}{16\gamma} \wedge \frac{\gamma}{64}.$$

We can now repeat this estimate with $A$ and $n_1$ replaced by $(1 + \frac{1}{4}\gamma)^j A$ and $(1 + \gamma)^j n_1$, respectively, for $j = 1, 2, \ldots$. This results in

(3.22)     $P\{\sigma = \infty \text{ and } X_n \leq (1 - \varepsilon)An^{1/2} \text{ for some } n \geq n_1 \,|\, \mathcal{F}_{n_1}\}$

$$\leq \sum_{j=0}^{\infty} 2\exp\{-C_1(1 + \frac{1}{4}\gamma)^{2j} A^2\}$$

on the event (3.19). Since, for every $A$, $X_{n_1} \geq An_1^{1/2}$ for infinitely many $n_1$ (by step (i)), (3.17) follows.

**Step (iii).** We now improve (3.17) to

(3.23)                $P\{\sigma < \infty\} = 1.$

To prove (3.23) we introduce a further submartingale. Let

$$W_n = \log\left\{\frac{1}{n \wedge \sigma} X_{n \wedge \sigma}\right\}.$$

By definition of $\sigma$, and the inequality $X_{k+1} - X_k \leq 1$, $W_n$ is bounded above. Again assume that (3.19) occurs, and let

$$\tau = \tau(A) = \inf\{n \geq n_1 : X_n \leq (1 - \varepsilon)An^{1/2}\},$$

with $\varepsilon$ as in step (ii). We saw in (3.22) that

(3.24)     $P\{\sigma = \infty \text{ and } \tau < \infty \,|\, \mathcal{F}_{n_1}\} \to 0 \text{ as } A \to \infty.$

On the other hand, on $n < \sigma \wedge \tau$ we have

$$\begin{aligned}
E\{W_{n+1} \,|\, \mathcal{F}_n\} &= E\{\log(X_n + \Delta_{n+1}) \,|\, \mathcal{F}_n\} - \log(n + 1) \\
&= \log\{\frac{1}{n} X_n\} + E\{\log(1 + X_n^{-1}\Delta_{n+1}) \,|\, \mathcal{F}_n\} - \log(1 + \frac{1}{n}) \\
&\geq W_n + X_n^{-1} E\{\Delta_{n+1} \,|\, \mathcal{F}_n\} - \frac{1}{2} X_n^{-2} E\{\Delta_{n+1}^2 \,|\, \mathcal{F}_n\} - \frac{1}{n} \\
&\geq W_n + \frac{2D}{n+1} - \frac{1}{2n}(1 - \varepsilon)^{-2}A^{-2} - \frac{1}{n}
\end{aligned}$$

(by (3.20) and $\Delta_{n+1}^2 = 1$). If $A$ and $n_1$ are chosen large enough, then with $\delta = \frac{1}{2}(2D - 1) > 0$ (recall $D > \frac{1}{2}$ by assumption)

$$\frac{2D}{n+1} - \frac{1}{2n}(1 - 2\varepsilon)^{-2}A^{-2} - \frac{1}{n} \geq \frac{\delta}{n+1}$$

for $n \geq n_1$. Therefore

$$W_{n \wedge \tau} - \sum_0^{n \wedge \sigma \wedge \tau - 1} \frac{\delta}{j+1}$$

is a submartingale, which is bounded above, and has a finite limit with probability 1. However, on $\sigma \wedge \tau = \infty$, the limit is (by (3.16)) at most

$$\log \varepsilon_0 (2 + 2\varepsilon_0)^{-1} - \sum_0^\infty \frac{\delta}{n+1} = -\infty.$$

Thus we must have $\sigma \wedge \tau$ finite with probability 1. In view of (3.24) this proves (3.23).

**Step (iv).** Finally, we prove in this step that (3.6) holds, as well as (3.13) when $\kappa > 1$ in (3.12). Choose $n_0$ so large that steps (ii) and (iii) apply and let $\sigma$ be as in (3.16). Assume that $\sigma < \infty$, and for the sake of argument, let

$$Y_1(\sigma) \geq Y_2(\sigma) + \varepsilon_0(2 + 2\varepsilon_0)^{-1}\sigma.$$

Without loss of generality assume that $n_0^{\kappa/2} \geq 8$ and

$$n_0 \geq (\varepsilon_0/16)(1 + |Y_1(1)| + |Y_2(1)|).$$

Define

$$\nu = \inf\{n \geq \sigma : Y_1(n) \leq Y_2(n) + \frac{1}{2}\varepsilon_0(2 + 2\varepsilon_0)^{-1}\sigma + (n - \sigma)(1 - \sigma^{-\kappa/2})\}.$$

Then $\nu > \sigma$ and for $\sigma \leq n < \nu$, $X_n = Y_1(n) - Y_2(n) \geq \frac{1}{2}\varepsilon_0(2 + 2\varepsilon_0)^{-1}n$ so that

$$E\{\Delta_{n+1}|\mathcal{F}_n\} \geq 1 - 2n^{-\kappa}, \quad \sigma^2\{\Delta_{n+1}|\mathcal{F}_n\} \leq 4n^{-\kappa}$$

(by virtue of (3.12) and $Y_1(n) + Y_2(n) = n - 1 + Y_1(1) + Y_2(1)$. Thus, if $\sigma < \nu < \infty$, then

$$[Y_1(\nu) - Y_2(\nu)] - [Y_1(\sigma) - Y_2(\sigma)] - \sum_\sigma^{\nu-1} E\{\Delta_{n+1}|\mathcal{F}_n\}$$

$$\leq -\frac{1}{2}\varepsilon_0(2 + 2\varepsilon_0)^{-1}\sigma - (\nu - \sigma)(1 - 2\sigma^{-\kappa} - 1 + \sigma^{-\kappa/2})$$

$$\leq -\frac{1}{2}\varepsilon_0(2 + 2\varepsilon_0)^{-1}n_0 - \frac{1}{8}\sigma^{\kappa/2}\sum_\sigma^{\eta-1}\sigma^2\{\Delta_{n+1}|\mathcal{F}_n\}.$$

Once more by the exponential bounds of [15, pp. 154, 155] we obtain on the event $\{\sigma < \infty\}$

$$P\{\nu < \infty \,|\, \mathcal{F}_\sigma\} \leq \exp\{-\tfrac{\varepsilon_0}{2}(2 + 2\varepsilon_0)^{-1} n_0\}.$$

Since this estimate holds for every large $n_0$, there exists with probability 1 for every $n_1 < \infty$ an $n_2 < \infty$ and an $i \in \{1, 2\}$ such that

$$Y_i(n) \geq Y_{3-i}(n) + (n - n_2)(1 - n_2^{-\kappa/2}) \quad \text{for all } n \geq n_2.$$

Since $Y_1(n) + Y_2(n) = n + 0(1)$, this implies

$$\liminf \frac{Y_i(n)}{Y_{3-i}(n)} \geq \frac{2 - n_2^{-\kappa/2}}{n_2^{-\kappa/2}} \geq 2n_1^{\kappa/2} - 1.$$

(3.6) follows because $n_1$ can be taken as large as desired.

If $\kappa > 1$ in (3.12), then on $\{Y_i(\sigma) \geq \varepsilon_0(2 + 2\varepsilon_0)^{-1} Y_{3-i}(\sigma)\}$ we have

$$P\{Y_i(n+1) = Y_i(n) + 1, \, Y_{3-i}(n+1) = Y_{3-i}(n) \quad \text{for all } n \geq \sigma \,|\, \mathcal{F}_\sigma\}$$

$$\geq \prod_{n \geq n_0} (1 - n^{-\kappa}) \geq \exp\{-K_1 n_0^{1-\kappa}\}.$$

(3.13) follows immediately.                                              ■

To conclude we note that (1.11) is proven by (3.3) and (3.13).

## REFERENCES

1. J.-P. Eckman, P. Meakin, I. Procaccia, R. Zeitak, *Growth and form of noise-reduced diffusion-limited aggregation*, Phys. Rev. A **39** (1989), 3185–3195.

2. W. Feller, "An introduction to probability theory and its applications," Vol. I, 3rd ed., John Wiley and Sons, 1968.

3. W. Feller, "An introduction to probability theory and its applications," Vol. II, 2nd ed., John Wiley and Sons, 1971.

4. Y. Hayakawa, H. Kondo, M. Mathshushita, *Monte Carlo simulations of generalized diffusion limited aggregation*, J. Phys. Soc. Japan **55** (1986), 2479–2482.

5. J. Kertész, J., T. Vicsek, *Diffusion-limited aggregation and regular patterns: fluctuations versus anisotropy*, J. Phys. A **19** (1986), L257–L262.

6. H. Kesten, *How long are the arms in DLA?*, J. Phys. A **20** (1987), L29–L33.

7. H. Kesten, *Hitting probabilities of random walks on* $\mathbb{Z}^d$, Stoch. Proc. and their Appl. **25** (1987), 165–184.

8. H. Kesten, *Relations between solutions to a discrete and continuous Dirichlet problem*, in "Festschrift in honor of Frank Spitzer," (R. Durrett, H. Kesten, eds. ) Birkhäuser-Boston, 1991.

9. H. Kesten, *Relations between solutions to a discrete and continuous Dirichlet problem II*, preprint.

10. G. Lawler, "Intersections of random walks," Birkhäuser-Boston, 1991.

11. M. Matsushita, K. Honda, H. Toyoki, Y. Hayakawa, H. Kondo, *Generalization and the fractal dimensionality of diffusion-limited aggregation*, J. Phys. Soc. Japan **55** (1986), 2618–2626.

12. P. Meakin, *Noise-reduced diffusion-limited aggregation*, Phys. Rev. A **36** (1987), 332–339.

13. P. Meakin, *The growth of fractal aggregates and their fractal measures*, in "Phase transitions," (C. Domb, J. L. Lebowitz, eds.), Academic Press, 1988, pp. 335–489.

14. P. Meakin, S. Tolman, *Fractals' physical origin and properties*, (L. Pietronero, ed.), Plenum.

15. J. Neveu, *Martingales a temps discrete*, Masson and Cie.

16. L. Niemeyer, L. Pietronero, H. J. Wiesmann, *Fractal dimension of dielectric breakdown*, Phys. Rev. Lett. **52** (1984), 1033–1036.

17. J. Nittman, H. E. Stanley, *Tip splitting without interfacial tension and dendritic growth patterns arising from molecular anisotropy*, Nature **321** (1986), 663–668.

18. W. Rudin, "Real and complex analysis," 3rd ed., McGraw Hill Book Co., 1987.

19. F. Spitzer, "Principles of random walk," D. van Nostrand Co, 1984.

20. C. Tang, *Diffusion-limited aggregation and the Saffman-Taylor problem*, Phys. Rev. A **31** (1985), 1977–1979.

21. T. Vicsek, "Fractal growth phenomena," World Scientific, 1989.

22. T. A. Witten, L. M. Sander, *Diffusion-limited aggregation, a kinetic phenomenon*, Phys. Rev. Lett **47** (1981), 1400–1403.

# Limits on Random Measures and Stochastic Difference Equations Related to Mixing Array of Random Variables

## H. Kunita

Department of Applied Science, Faculty of Engineering, Kyushu University, Fukuoka 812 Japan

## 0. Introduction

Let $\{\xi_j^n, j \in \mathbf{N}\}, n = 1, 2, \ldots$ be a sequence of stationary stochastic processes with values in $\mathbf{R}^m$ converging in law to a stationary process $\{\xi_j, j \in \mathbf{N}\}$. Denote the laws of $\xi_1^n$ and $\xi_1$ by $\pi^n$ and $\pi$ respectively. Suppose that there exists an increasing sequence of positive numbers $\{c_n\}$ tending to infinity such that the sequence of measures $\{\mu^n\}$ defined by $\mu^n(F) = n\pi^n(c_n F)$ converges vaguely to a measure $\mu$ on $\mathbf{R}^m - \{0\}$. Define two families of random measures by

$$B^n(t, E) = \frac{1}{\sqrt{n}} \sum_{j=1}^{[nt]} (\chi_E(\xi_j^n) - \pi_n(E)), \qquad (0.1)$$

$$N^n(t, F) = \sum_{j=1}^{[nt]} \chi_F\left(\frac{\xi_j^n}{c_n}\right), \qquad (0.2)$$

where $E$, $F$ are Borel sets in $\mathbf{R}^m$ such that $\bar{F} \subset \mathbf{R}^m - \{0\}$ and $\chi_E$, $\chi_F$ are indicator functions of the sets $E$, $F$. Then for each $n$, $((B^n(t, .), N^n(t, .))$, $t \in [0, \infty))$ may be regarded as a stochastic process with values in the space of (signed) measures on $\mathbf{R}^m \times (\mathbf{R}^m - \{0\})$, cadlag (right continuous with left hand limits) with respect to time t. In the previous paper [7], the author has shown the weak convergence of the sequence $\{(B^n(t, .), N^n(t, .))\}_n$ in the space $\mathbf{D}$ with respect to Skorohod's $J_1$-topology, assuming that $\xi_j^n, j \in \mathbf{N}$ are independent random variables for any n. He proved that (1) the limit $B(t, E)$ is characterized as a Brownian random measure, i.e., it is a Gaussian random measure for each fixed $t$ and is a Brownian motion for each fixed $E$, (2) the limit $N(t, F)$ is a Poisson random measure with the intensity $t\mu(F)$, and (3) these two are independent processes.

The first object of this paper is to extend the above limit theorem to the case where $\xi_j^n, j \in \mathbf{N}$ are not necessarily independent but have some mixing propery. Let $\{\phi_k^n, k \in \mathbf{N}\}$ be the uniform mixing rate of the stationary process $\{\xi_j^n; j \in \mathbf{N}\}$. (For the definition, see (1.2) in Section 1). We show that if it satisfies $\sum_{k=1}^{\infty}(\sup_n \phi_k^n)^{1/2} < \infty$, then the sequence $\{(B^n(t), N^n(t))\}_n$ converges in law in the space $\mathbf{D}$ and the limit $(B(t), N(t))$ has properties (1),(2),(3) mentioned above. See Theorem 1.1.

Another object will be to discuss the weak convergence of the solutions of stochastic difference equations represented by

$$\psi_j = \psi_{j-1} + \frac{1}{\sqrt{n}} f^n(\psi_{j-1}, \xi_j^n) + \frac{1}{n} g^n(\psi_{j-1}), \qquad j = 1, 2, \ldots \qquad (0.3)$$

where $f^n(x, \lambda)$ and $g^n(x)$ ; $x \in \mathbf{R}^d$, $\lambda \in \mathbf{R}^m$ are continuous functions with values in $\mathbf{R}^d$ converging to $f(x, \lambda)$ and $g(x)$ respectively as $n \to \infty$. Let $\psi_j^n$ be the solution of (0.3) with the initial condition $\psi_0 = x_0$ and set $\varphi_t^n = \psi_{[nt]}^n$. Then for any $n$ the process $\varphi_t^n$ satisfies

$$\begin{aligned} \varphi_t^n &= x_0 + \int_0^t \int_{|\lambda| \leq c_n M} f^n(\varphi_{s-}^n, \lambda) B^n(ds, d\lambda) \\ &\quad + \int_0^t \int_{|\lambda| > c_n M} \hat{f}^n(\varphi_{s-}^n, \lambda) N^n(ds, d\lambda) \\ &\quad + \int_0^{[nt]/n} \{g^n(\varphi_{s-}^n) + b_M^n(\varphi_{s-}^n)\} ds, \end{aligned} \qquad (0.4)$$

where $M > 0$ and

$$\hat{f}^n(x, \lambda) = \frac{1}{\sqrt{n}} f^n(x, c_n \lambda), \qquad (0.5)$$

$$b_M^n(x) = \sqrt{n} \int_{|\lambda| \leq c_n M} f^n(x, \lambda) \pi^n(d\lambda). \qquad (0.6)$$

Assume that the sequences $\{\hat{f}^n(x, \lambda)\}_n$ and $\{b_M^n(x)\}_n$ converge to $\hat{f}(x, \lambda)$ and $b_M(x)$, respectively and $\lim_{M \to 0} b_M(x) = b_0(x)$ exists. A question is that whether the limit $\varphi_t$ exists or not and if it exists, it satisfies

$$\begin{aligned} \varphi_t &= x_0 + \int_0^t \int_{\mathbf{R}^m} f(\varphi_{s-}, \lambda) B(ds, d\lambda) \\ &\quad + \int_0^t \int_{\mathbf{R}^m - \{0\}} \hat{f}(\varphi_{s-}, \lambda) N(ds, d\lambda) \\ &\quad + \int_0^t \{g + b_0\}(\varphi_{s-}) ds. \end{aligned} \qquad (0.7)$$

We will prove that the sequence $\{\varphi_t^n\}_n$ converges in law. However equation (0.7) is false in many cases. In general the limit process $\varphi_t$ can not be represented by the above two random measures $B(t, E)$ and $N(t, F)$. We need another Brownian random measure $K_0(t, G)$, which is concentrated at the boundary of $\mathbf{R}^m$ (homeomorphic to $m - 1$ dimensional sphere $S^{m-1}$). We show in Theorem 2.1 that the stochastic differential equation governing the limit process $\varphi_t$ is given by

$$\varphi_t = x_0 + \int_0^t \int_{\mathbf{R}^m} f(\varphi_{s-}, \lambda) B(ds, d\lambda)$$

$$+ \int_0^t \int_{S^{m-1}} \bar{h}(\varphi_{s-}, \infty, \theta) K_0(ds, d\theta)$$

$$+ \int_0^t \int_{0 < |\lambda| \leq 1} \hat{f}(\varphi_{s-}, \lambda)(N(ds, d\lambda) - ds\,\mu(d\lambda))$$

$$+ \int_0^t \int_{|\lambda| > 1} \hat{f}(\varphi_{s-}, \lambda) N(ds, d\lambda)$$

$$+ \int_0^t \{g + b_1 + c\}(\varphi_{s-}) ds. \tag{0.8}$$

Here the function $\bar{h}$ is defined by $\bar{h}(x, \infty, \theta) = \lim_{r \to \infty} f(x, r, \theta)/\rho(r)$, where $(r, \theta)$ is the polar coordinate of $\lambda$ and $\rho(r)$ is a positive nondecreasing function such that $\lim_{n \to \infty} \rho(c_n r)/\sqrt{n}$ exists. Further, $c(x)$ is a certain correction term.

Limit theorems for stochastic difference equation similar to this paper have been studied by Fujiwara [1], [2] and Kunita [8]. In the papers [1], [2] he discusses the case where the driving processes $\{\xi_j^n\}$ appear linearly in equation (0.3), i.e., the coefficients $f^n(x, \lambda)$ are of the forms $f^n(x, \lambda) = f^n(x)\lambda$. On the other hand, in the paper [8], the case of nonlinear coefficients $f^n(x, \lambda)$ are studied but the limit process is restricted to a diffusion process.

## 1.  Convergence of random measures

Let $\mathbf{N}$ be the set of all positive integers. Let $\{\xi_j^n; j \in \mathbf{N}\}, n = 1, 2, \cdots$ and $\{\xi_j; j \in \mathbf{N}\}$ be stationary stochastic processes with values in the common space $\mathbf{R}^m$ defined on the probability space $(\Omega, \mathcal{F}, P)$. We set

$$\pi^n(E) = P(\xi_j^n \in E), \quad \pi(E) = P(\xi_j \in E). \tag{1.1}$$

These measures do not depend on $j$ since they are stationary. We will introduce three conditions to $\{\xi_j^n\}$.

**Condition $(\xi.1)$** *Any finite dimensional distributions of $\{\xi_j^n\}, n = 1, 2, \cdots$ converge weakly to the corresponding finite dimensional distribution of $\{\xi_j\}$.*

Then the sequence $\{\pi^n\}$ converges weakly to $\pi$.

The second condition is concerned with the mixing property. Let $h, k(h < k)$ be elements of $\mathbf{N}$. We denote by $\mathcal{F}_{h,k}^n$ the least $\sigma$ field for which $\xi_j^n, h \leq j \leq k$ are measureble. We often write $\mathcal{F}_{1,k}^n$ by $\mathcal{F}_k^n$ and $\mathcal{F}_{h,\infty}^n = \sigma(\cup_{k=h+1}^\infty \mathcal{F}_{h,k}^n)$. The uniform mixing rate ($\phi$-mixing rate) of the process $\{\xi_j^n; j \in \mathbf{N}\}$ is a sequence $\{\phi_k^n, k \in \mathbf{N}\}$ of nonnegative numbers defined by

$$\phi_k^n = \sup_{h \in \mathbf{N}} \sup\{|P(F|E) - P(F)|; E \in \mathcal{F}_h^n, P(E) > 0, F \in \mathcal{F}_{h+k,\infty}^n\}. \tag{1.2}$$

**Condition $(\xi.2)$** *The sequence $\{\phi_k^n\}$ satisfies $\sum_{k=1}^\infty (\sup_n \phi_k^n)^{\frac{1}{2}} < \infty$ .*

Let $\{\phi_k\}$ be the uniform mixing rate of the process $\{\xi_j\}$. Then the inequality $\phi_k \leq \sup_n \phi_k^n$ is easily verified for any $k \in \mathbf{N}$. Therefore $\sum_{k=1}^{\infty} \phi_k^{1/2} < \infty$ is satisfied. Then using the mixing inequality, we can show that the infinite sum

$$V(E_1, E_2) = \sum_{j=2}^{\infty} \{E[\chi_{E_1}(\xi_1)\chi_{E_2}(\xi_j)] - \pi(E_1)\pi(E_2)\} \tag{1.3}$$

converges. Further if $f, g \in L^2(\pi)$, then

$$V(f, g) = \int \int_{\mathbf{R}^m \times \mathbf{R}^m} f(\lambda)g(\lambda')V(d\lambda, d\lambda') \tag{1.4}$$

is well defined and satisfies

$$|V(f, g)| \leq 2(\sum_{j=1}^{\infty} \phi_j^{\frac{1}{2}})(\int |f|\pi)(\int g^2\pi)^{\frac{1}{2}}. \tag{1.5}$$

Let $\{c_n\}$ be an increasing sequence of positive numbers diverging to infinity. The third condition is concerned with the vague convergence of measures $\{\mu^n\}$ on $\mathbf{R}^m - \{0\}$ defined by

$$\mu^n(F) = n\pi^n(c_n F). \tag{1.6}$$

The sequence $\{c_n\}$ will be fixed throughout this paper.

**Condition ($\xi$.3)** *The sequence $\{\mu^n\}$ converges vaguely to a Radon measure $\mu$ on $\mathbf{R}^m - \{0\}$, i.e.,*

$$\lim_{n \to \infty} \int_{\mathbf{R}^m - \{0\}} h(\lambda)\mu^n(d\lambda) = \int_{\mathbf{R}^m - \{0\}} h(\lambda)\mu(d\lambda) \tag{1.7}$$

*holds for any bounded continuous function $h(\lambda)$ such that $h(\lambda) = 0$ holds on a certain neighborhood of 0.*

We will give the definition of the weak convergence rigorously. Let $\mathcal{G}(\mathbf{R}^m)$ $= \{E_k\}$ (or $\mathcal{G}(\mathbf{R}^m - \{0\}) = \{F_k\}$) be a ring of Borel sets of $\mathbf{R}^m$ (or $\mathbf{R}^m - \{0\}$) such that $E_k \subset \mathbf{R}^m$ (or $\bar{F}_k \subset \mathbf{R}^m - \{0\}$) for any $k$ and it generates the Borel field of $\mathbf{R}^m$ (or $\mathbf{R}^m - \{0\}$). We assume that any element $E$ of $\mathcal{G}(\mathbf{R}^m)$ (or $F$ of $\mathcal{G}(\mathbf{R}^m - \{0\})$) satisfies $\pi^n(E) \to \pi(E)$ (or $\mu^n(F) \to \mu(F)$). Let $\mathcal{M}(\mathbf{R}^m)$ (or $\mathcal{M}(\mathbf{R}^m - \{0\})$) be the set of all additive set functions on $\mathcal{G}(\mathbf{R}^m)$ (or $\mathcal{G}(\mathbf{R}^m - \{0\})$). Then we may define the law of $(B^n, N^n)$ on the space $\mathbf{D} = \mathbf{D}([0, \infty); \mathcal{M}(\mathbf{R}^m) \times \mathcal{M}(\mathbf{R}^m - \{0\}))$. (For the topology of the space $\mathbf{D}$, see Kunita [7].) We denote it by $P^n$. If the sequence $\{P^n\}$ converges weakly as $n \to \infty$, the sequence $\{(B^n, N^n)\}_n$ is said to converge in law.

**Theorem 1.1.** *Assume Conditions ($\xi$.1) $-$ ($\xi$.3) for $\{\xi_j^n, j \in \mathbf{N}\}, n = 1, 2, \cdots$. Then the sequence $\{(B^n, N^n)\}_n$ converges in law. Let $(B, N, P^\infty)$ be its limit law. Then the followings hold.*

*(1) For any $E_1, ..., E_q$ of $\mathcal{G}(\mathbf{R}^m)$, $(B(t, E_1), ..., B(t, E_q))$ is a $q$ dimensional Brownian motion with mean $0$ and covariance*

$$E[B(t, E_i)B(t, E_j)] = t(\pi(E_i \cap E_j) + V(E_i, E_j) + V(E_j, E_i) - \pi(E_i)\pi(E_j)). \quad (1.8)$$

*Further for any $t$, $B(t, E)$ can be extended continuously to any $E \in \mathcal{B}(\mathbf{R}^m)$, which is a Gaussian random measure.*

*(2) For any $F$ of $\mathcal{G}(\mathbf{R}^m - \{0\})$, $N(t, F)$ is a Poisson process with intensity $\mu(F)$. Further for any $t$, $N(t, F)$ can be extended to any $F$ of $\mathcal{B}(\mathbf{R}^m - \{0\})$, which is a Poisson random measure.*

*(3) The processes $B$ and $N$ are mutually independent.*

Now we will represent points of $\mathbf{R}^m - \{0\}$ by polar coordinate $(r, \theta)$, where $r \in (0, \infty)$ and $\theta \in S^{m-1}$ ($m - 1$ dimensional unit sphere with center $0$). Then $\mathbf{R}^m - \{0\}$ is homeomorphic to $(0, \infty) \times S^{m-1}$ and $(0, \infty] \times S^{m-1} \cup \{0\}$ is a compactification of $\mathbf{R}^m$. We can regard $\{\infty\} \times S^{m-1}$ as the boundary of $\mathbf{R}^m$ and denote it by $\partial\mathbf{R}^m$. We set $\bar{\mathbf{R}}^m = \mathbf{R}^m \cup \partial\mathbf{R}^m$.

Let $\rho(r)$, $r \geq 0$ be a strictly positive continuous nondecreasing function such that

$$\lim_{n \to \infty} \frac{\rho(c_n r)}{\sqrt{n}} = \eta(r) \quad (1.9)$$

exists and is a continuous function of r. Then $\lim_{r \to 0} \eta(r) = 0$ holds.

Associated with the function $\rho(r)$ and the sequence $\{c_n\}$, we will define another sequence of random measures on $\bar{\mathbf{R}}^m$ by

$$K_\varepsilon^n(t, G) = \int_G \rho(|\lambda|)\chi_{[0, c_n\varepsilon]}(|\lambda|)B^n(t, d\lambda), \quad (1.10)$$

and discuss the weak convergence of the sequence $\{(B^n, N^n, K_\varepsilon^n)\}_n$. For this purpose, we assume two more conditions. For a positive constant $M$, we define the truncated random variable of $\xi_j^n$ by $\xi_j^{n,M} = \xi_j^n \chi_{[0, c_n M]}(|\xi_j^n|)$.

**Condition $(\xi.4)$** *For any $j$ and $M$, the sequence $\left\{ E[\rho(|\xi_j^{n,M}|)^2 | \mathcal{F}_{j-1}^n] \right\}_n$ is uniformly integrable, i.e.,*

$$\lim_{c \to \infty} \sup_n E\left[ \rho(|\xi_j^{n,M}|)^2 ; E[\rho(|\xi_j^{n,M}|)^2 | \mathcal{F}_{j-1}^n] > c \right] = 0. \quad (1.11)$$

Given $\varepsilon > 0$ and $n \geq 1$, define a measure on $\bar{\mathbf{R}}^m$ by

$$\nu_\varepsilon^n(G) = \int_G \rho(|\lambda|)^2 \chi_{[0, c_n\varepsilon]}(|\lambda|)\pi^n(d\lambda). \quad (1.12)$$

Then $\sup_n \nu_\epsilon^n(\bar{\mathbf{R}}^m) < \infty$ holds by Condition ($\xi$.4).

**Condition ($\xi$.5)** *For any $\epsilon > 0$, the sequence of measures $\{\nu_\epsilon^n\}_n$ converges weakly to a measure $\nu_\epsilon$ on $\bar{\mathbf{R}}^m$.*

The family of the measures $\{\nu_\epsilon\}_\epsilon$ decreases as $\epsilon$ decreases. We set

$$\nu = \lim_{\epsilon \to 0} \nu_\epsilon. \qquad (1.13)$$

Now we will define the law of $(B^n, N^n, K_\epsilon^n)$. Let $\mathcal{G}(\bar{\mathbf{R}}^m) = \{G_k\}$ be a ring of Borel sets of $\bar{\mathbf{R}}^m$ such that it generates the Borel field of $\bar{\mathbf{R}}^m$. We assume that any element $G$ of $\mathcal{G}(\bar{\mathbf{R}}^m)$ satisfies $\nu_\epsilon^n(G) \to \nu_\epsilon(G)$ for any $\epsilon > 0$. Let $\mathcal{M}(\bar{\mathbf{R}}^m)$ be the set of all additive set functions on $\mathcal{G}(\bar{\mathbf{R}}^m)$. Then we may define the law of $(B^n, N^n, K_\epsilon^n)$ on the space $\mathbf{D} = \mathbf{D}([0,\infty); \mathcal{M}(\mathbf{R}^m) \times \mathcal{M}(\mathbf{R}^m - \{0\}) \times \mathcal{M}(\bar{\mathbf{R}}^m))$. We denote it by $\bar{P}_\epsilon^n$. If the sequence $\{\bar{P}_\epsilon^n\}_n$ converges weakly as $n \to \infty$, the sequence $\{(B^n, N^n, K_\epsilon^n)\}_n$ is said to converge in law.

**Theorem 1.2.** *Assume Conditions ($\xi$.1) $-$ ($\xi$.5) for $\{\xi_j^n, j \in \mathbf{N}\}, n = 1, 2, \cdots$. Let $\epsilon$ be a positive number such that $\{|\lambda| \leq \epsilon\}$ is a $\mu$-continuity set. Then the sequence $\{(B^n, N^n, K_\epsilon^n)\}_n$ converges in law. Let $(B, N, K_\epsilon, P_\epsilon^\infty)$ be its limit law. Then the followings hold.*

*(1) $(B, N)$ has the same property as in Theorem 1.1.*

*(2) For any $G_1, ..., G_s$, $(K_\epsilon(t, G_1), ..., K_\epsilon(t, G_s))$ is a Lévy process. Further, set*

$$K_0(t, G \cap \partial \mathbf{R}^m) \equiv K_\epsilon(t, G) - \int_{G \cap \mathbf{R}^m} \rho(|\lambda|) B(t, d\lambda)$$

$$- \int_{G \cap (\mathbf{R}^m - \{0\})} \eta(|\lambda|) \chi_{(0,\epsilon]}(|\lambda|) \{N(t, d\lambda) - t\mu(d\lambda)\}. \qquad (1.14)$$

*Then $(K_0(t, G_1 \cap \partial \mathbf{R}^m), ..., K_0(t, G_s \cap \partial \mathbf{R}^m))$ is a Brownian motion with mean $0$ and covariance*

$$E[K_0(t, G_i \cap \partial \mathbf{R}^m) K_0(t, G_j \cap \partial \mathbf{R}^m)] = t\nu(G_i \cap G_j \cap \partial \mathbf{R}^m). \qquad (1.15)$$

*For any $t$, $K_0(t,.)$ can be extended continuously to a Gaussian random measure on $\partial \mathbf{R}^m$.*

*(3) $B, N, K_0$ are mutually independent.*

The proof of these theorems will be given at Section 4.

*Remark.* Instead of Condition ($\xi$.4), we introduce the following.

**Condition ($\xi$.4)'** *For any $j$ and $N$, the sequence $\{\rho(|\xi_j^{n,M}|)^2\}_n$ is uniformly integrable.*

Clearly Condition ($\xi$.4)' implies ($\xi$.4). Further, under ($\xi$.1) and ($\xi$.4), the sequence of measures $\{\nu_\epsilon^n\}_n$ converges weakly. The limit measure $\nu_\epsilon$ is supported by $\mathbf{R}^m$ and is represented by

$$\nu(G) = \nu_e(G) = \int_G \rho(|\lambda|)^2 \pi(d\lambda).$$

Therfore the Brownian random measure $K_0$ of Theorem 1.2 is identically 0. As an example, consider the case where $\{\xi_j^n\}$ does not depend on $n$, which we denote by $\{\xi_j\}$. If $E[\rho(\xi_1^M)^2] < \infty$ holds for any $M > 0$, Condition $(\xi.4)'$ is satisfied. Hence the Brownian random measure $K_0$ is 0.

*Remark.* Let $\hat{\mathbf{R}}^m$ be another $m$-dimensional Euclidean space and let $\hat{S}^{m-1}$ be the unit sphere in $\hat{\mathbf{R}}^m$, i.e., $\hat{S}^{m-1} = \{x \in \hat{\mathbf{R}}^m; |x| = 1\}$. Let $\Gamma_1 = \{x \in \hat{\mathbf{R}}^m; |x| < 1\}$ and $\Gamma_2 = \{x \in \hat{\mathbf{R}}^m; |x| > 1\}$. Then $\hat{\mathbf{R}}^m$ is splitted into two domains $\Gamma_1$ and $\Gamma_2$ and $\hat{S}^{m-1}$ is the splitting boundary of the inner domain $\Gamma_1$ and the outer domain $\Gamma_2$. Further $\Gamma_1$ is homeomorphic to $\mathbf{R}^m$ and $\Gamma_2$ is homeomorphic to $\mathbf{R}^m - \{0\}$. Therefore we may regard that the Brownian random measure $B(t, E)$ of Theorem 1.1 is defined on the inner domain $\Gamma_1$ and the Poisson random measure $N(t, F)$ is defined on the outer domain $\Gamma_2$. Further the Brownian random measure $K_0(t, G)$ is defined on the splitting boundary $\hat{S}^{m-1}$.

## 2.   Convergence of solutions of stochastic difference equations.

Let us return to the stochastic difference equations (0.3) or (0.4). We will introduce assumptions to coefficients $\{f^n(x, \lambda)\}$ and $\{g^n(x)\}$ of the equations. Let $\mathbf{C}^k = \mathbf{C}^k(\mathbf{R}^d; \mathbf{R}^d)$ be the space of $k$-times continuously differentiable maps from $\mathbf{R}^d$ into itself. For $f \in \mathbf{C}^k$ we define seminorms $\| \ \|_{k,N}$, $k = 1, 2$ by

$$\|f\|_{k,N} = \sup_{|x| \le N} \frac{|f(x)|}{1+|x|} + \sum_{1 \le |\alpha| \le k} \sup_{|x| \le N} |D^\alpha f(x)|, \qquad (2.1)$$

and $\| \ \|_k^* = \lim_{N \to \infty} \| \ \|_{k,N}$. We denote by $\mathbf{C}_{b*}^k$ the set of all $f \in \mathbf{C}^k$ such taht $\|f\|_k^* < \infty$. We assume that $f^n(\lambda) \equiv f^n(x, \lambda)$ in equation (0.3) is a $\mathbf{C}_{b*}^2$-valued function which is continuous with respect to $\lambda$ or is a step function of $\lambda$.

**Condition (f.1)** *(1) The sequence $\{f^n(\lambda)\}_n$ satisfies*

$$\sup_{n, \lambda} \|f^n(\lambda)\|_{2,N} / \rho(|\lambda|) < \infty$$

*for any $N$. There exists a $\mathbf{C}_{b*}^2$-valued function $f(\lambda)$ such that $\|f^n(\lambda) - f(\lambda)\|_{1,N}$ converges to 0 uniformly on compact sets of $\mathbf{R}^m$ for any $N$. Furthermore, the limit $f(x, \lambda)$ is integrable with respect to $\pi$ for any $x$ and satisfies*

$$\int_{\mathbf{R}^m} f(x, \lambda) \pi(d\lambda) = 0. \qquad (2.2)$$

(2) The sequence $\{b^n_M(x)\}_n$ defined by (0.6) is in $\mathbf{C}^2_{b*}$, $\sup_n \|b^n_M\|_{2,N} < \infty$ for any $N$ and converges to $b_M(x)$ of $\mathbf{C}^1_{b*}$ with respect to $\|\ \|_{1,N}$ for any $N$.

**Condition (f.2)** Set $h^n(\lambda) \equiv f^n(\lambda)/\rho(|\lambda|)$. Each $h^n(\lambda)$ is continuous and is extended to a continuous $\mathbf{C}^1_{b*}$-valued function $\bar{h}^n(\lambda)$ on $\bar{\mathbf{R}}^m$. Further there exists a continuous $\mathbf{C}^1_{b*}$-valued function $\bar{h}(\lambda)$ such that $\|\bar{h}^n(\lambda) - \bar{h}(\lambda)\|_{1,N}$ converges to 0 uniformly on $\bar{\mathbf{R}}^m$ for any $N$.

**Condition (g.1)** The sesquence $\{g^n(x)\}$ is in $\mathbf{C}^1_{b*}$ and converges to $g(x)$ of $\mathbf{C}^1_{b*}$ with respect to $\|\ \|_{1,N}$ for any $N$.

By Condition (f.1), the sequence $\{f^n(\lambda)\}$ satisfies $\|f^n(\lambda)\|_{1,N} \leq C_N \rho(|\lambda|)$ with some positive constant $C_N$. Then $\|f(\lambda)\|_{1,N} \leq C_N \rho(|\lambda|)$ holds. Set

$$\hat{f}^n(\lambda) \equiv \frac{1}{\sqrt{n}} f^n(c_n \lambda). \qquad (2.3)$$

These functions satisfy $\|\hat{f}^n(\lambda)\|_{1,N} \leq C_N \rho(c_n|\lambda|)/\sqrt{n}$. Further since

$$\hat{f}^n(\lambda) = \frac{\rho(c_n|\lambda|)}{\sqrt{n}} h^n(c_n \lambda),$$

the sequence $\{\hat{f}^n(\lambda)\}$ converges uniformly to a continuous $\mathbf{C}^1_{b*}$ valued function $\hat{f}(\lambda) = \hat{f}(x, \lambda)$ with respect to $\|\ \|_{1,N}$ seminorms by Condition (f.2). It satisfies

$$\hat{f}(\lambda) = \eta(|\lambda|)\bar{h}(\infty, \frac{\lambda}{|\lambda|}) \qquad \text{if} \quad |\lambda| > 0. \qquad (2.4)$$

Now the law of the 4-ple $(B^n, N^n, K^n_\varepsilon, \varphi^n)$ is defined on $\mathbf{D} = \mathbf{D}([0, \infty) : \mathcal{M}(\mathbf{R}^m) \times \mathcal{M}(\mathbf{R}^m - \{0\}) \times \mathcal{M}(\bar{\mathbf{R}}^m) \times \mathbf{R}^d)$. We denote it by $\bar{P}^n$. We show the weak convergence of $\{\bar{P}^n\}$ and characterize the limit $(B, N, K_\varepsilon, \varphi, \bar{P}^\infty)$ as a stochastic differential equation driven by random measures. For this purpose we define stochastic integrals based on random measures.

Set

$$\mathcal{B}_t = \bigcap_{\varepsilon > 0} \sigma(B(s, \cdot), N(s, \cdot), K_0(s, \cdot) : s \leq t + \varepsilon). \qquad (2.5)$$

Let $u(x, \lambda)$, $x \in \mathbf{R}^d$, $\lambda \in \mathbf{R}^m$ be a measurable function, continuous in $x$ such that $u(x, \cdot) \in L^2(\pi)$ for any x. Let $\psi_t$ be an $\mathbf{R}^d$-valued $(\mathcal{B}_t)$ adapted cadlag process. Then the stochastic integral of $u(\psi_s, \lambda)$ based on $B(ds, d\lambda)$ is defined by

$$\int_0^t \int_{\mathbf{R}^m} u(\psi_{s-}, \lambda) B(ds, d\lambda)$$

$$= \lim_{|\Delta| \to 0} \sum_{k=0}^n \left( \int_{\mathbf{R}^m} u(\psi_{t_{k-1}}, \lambda) B(t_k, d\lambda) - \int_{\mathbf{R}^m} u(\psi_{t_{k-1}}, \lambda) B(t_{k-1}, d\lambda) \right),$$

$$(2.6)$$

where $\Delta = \{0 = t_0 < \cdots < t_n = t\}$ and $|\Delta| = \max_k t_k - t_{k-1}$. It is a continuous local martingale. Next let $v(x, \lambda)$ be another function with the same property as $u(x, \lambda)$ and let $\eta_t$ be a cadlag process adapted to $(\mathcal{B}_t)$. Then the stochastic integral $\int_0^t \int v(\eta_{s-}, \lambda) B(ds, d\lambda)$ is well defined. We have

$$
< \int_0^t \int_{\mathbf{R}^m} u(\psi_{s-}, \lambda) B(ds, d\lambda), \int_0^t \int_{\mathbf{R}^m} v(\eta_{s-}, \lambda) B(ds, d\lambda) >
$$

$$
= \int_0^t \left( \int_{\mathbf{R}^m} u(\psi_{s-}, \lambda) v(\eta_{s-}, \lambda) \pi(d\lambda) \right) ds
$$

$$
+ \int_0^t \left( \int \int_{\mathbf{R}^m \times \mathbf{R}^m} u(\psi_{s-}, \lambda) v(\eta_{s-}, \lambda') V(d\lambda, d\lambda') \right) ds
$$

$$
+ \int_0^t \left( \int \int_{\mathbf{R}^m \times \mathbf{R}^m} u(\psi_{s-}, \lambda') v(\eta_{s-}, \lambda) V(d\lambda, d\lambda') \right) ds
$$

$$
- \int_0^t \left( \int_{\mathbf{R}^m} u(\psi_{s-}, \lambda) \pi(d\lambda) \right) ds \int_0^t \left( \int_{\mathbf{R}^m} v(\eta_{s-}, \lambda) \pi(d\lambda) \right) ds. \quad (2.7)
$$

Now set

$$
\tilde{N}(t, F) = N(t, F) - t\mu(F). \quad (2.8)
$$

We shall define stochastic integrals based on $\tilde{N}$. If $u \in L^2(\mu)$, the integral $\int u(\lambda) \tilde{N}(t, d\lambda)$ is well defined. It is a Lévy process with mean 0 and covariance $t \int u(\lambda)^2 \mu(d\lambda)$. Let $u(x, \lambda)$ and $v(x, \lambda)$ be measurable functions continuous in $x$ such that $u(x, \lambda)$ and $v(x, \lambda)$ belong to $L^2(\mu)$ for any x. Then the stochastic integrals

$$
\int_0^t \int_{\mathbf{R}^m - \{0\}} u(\psi_{s-}, \lambda) \tilde{N}(ds, d\lambda), \quad \int_0^t \int_{\mathbf{R}^m - \{0\}} v(\eta_{s-}, \lambda) \tilde{N}(ds, d\lambda)
$$

are well defined similarly to (2.6). These are local martingales and their bracket process is

$$
< \int_0^t \int_{\mathbf{R}^m - \{0\}} u(\psi_{s-}, \lambda) \tilde{N}(ds, d\lambda), \int_0^t \int_{\mathbf{R}^m - \{0\}} v(\eta_{s-}, \lambda) \tilde{N}(ds, d\lambda) >
$$

$$
= \int_0^t \left( \int_{\mathbf{R}^m - \{0\}} u(\psi_{s-}, \lambda) v(\eta_{s-}, \lambda) \mu(d\lambda) \right) ds. \quad (2.9)
$$

Similarly we can define the integrals of the form

$$
\int_{S^{m-1}} u(\theta) K_0(t, d\theta), \quad \int_0^t \int_{S^{m-1}} u(\psi_{s-}, \theta) K_0(ds, d\theta).
$$

The former is a Brownian motion with mean 0 and variance $t \int u(\theta)^2 \nu(d\theta)$. The latter is a continuous local martingale. We have

$$
< \int_0^t \int_{S^{m-1}} u(\psi_{s-}, \theta) K_0(ds, d\theta), \int_0^t \int_{S^{m-1}} v(\eta_{s-}, \theta) K_0(ds, d\theta) >
$$

$$
= \int_0^t \left( \int_{S^{m-1}} u(\psi_{s-}, \theta) v(\eta_{s-}, \theta) \nu(d\theta) \right) ds. \quad (2.10)
$$

**Theorem 2.1.** *Assume Conditions* $(\xi.1) - (\xi.5)$, $(f.1),(f.2)$ *and* $(g.1)$ *for equations* $(0.4)$. *Then the sequence* $\{(B^n, N^n, K^n_\epsilon, \varphi^n)\}_n$ *converges in law. Let* $(B, N, K_\epsilon, \varphi, \bar{P}^\infty_\epsilon)$ *be its limit law.*

*(1)* $(B, N, K_\epsilon, \bar{P}^\infty_\epsilon)$ *has the same property as in Theorem 1.2.*

*(2)* $\varphi_t$ *is represented by*

$$
\begin{aligned}
\varphi_t = \; & x_0 + \int_0^t \int_{\mathbf{R}^m} f(\varphi_{s-}, \lambda) B(ds, d\lambda) \\
& + \int_0^t \int_{S^{m-1}} \bar{h}(\varphi_{s-}, \infty, \theta) K_0(ds, d\theta) \\
& + \int_0^t \int_{0<|\lambda|\le 1} \hat{f}(\varphi_{s-}, \lambda) \tilde{N}(ds, d\lambda) \\
& + \int_0^t \int_{|\lambda|>1} \hat{f}(\varphi_{s-}, \lambda) N(ds, d\lambda) \\
& + \int_0^t \{g + b_1 + c\}(\varphi_{s-}) ds.
\end{aligned}
\tag{2.11}
$$

*Here the function* $c(x) = (c_1(x), ..., c_d(x))$ *is defined by*

$$
c_i(x) = \int \int_{\mathbf{R}^m \times \mathbf{R}^m} f(x, \lambda) \cdot \nabla f_i(x, \lambda') V(d\lambda, d\lambda'),
\tag{2.12}
$$

*where* $f(x, \lambda) = (f_1(x, \lambda), ..., f_d(x, \lambda))$ *and* $a \cdot b$ *denotes the inner product of two vectors.*

The case where the limit process $\varphi_t$ becomes a diffusion process is studied in [7]. In the present context, it is stated as follows.

**Theorem 2.2.** *Assume Conditions* $(\xi.1) - (\xi.3)$, $(\xi.4)'$ *and* $(f.1),(g.1)$ *for equations* $(0.4)$. *Then the sequence* $\{(B^n(t), \varphi^n_t)\}_n$ *converges in law. Let* $(B, \varphi)$ *be its limit law. Then*

$$
\varphi_t = x_0 + \int_0^t \int_{\mathbf{R}^m} f(\varphi_{s-}, \lambda) B(ds, d\lambda) + \int_0^t \{g + b_1 + c\}(\varphi_{s-}) ds.
\tag{2.13}
$$

Finally we will discuss the case where the coefficients $f^n$ and $g^n$ of equation $(0.3)$ does not depend on n. Consider random difference equations

$$
\psi_j = \psi_{j-1} + \frac{1}{\sqrt{n}} f(\psi_{j-1}, \xi^n_j) + \frac{1}{n} g(\psi_{j-1}),
\tag{2.14}
$$

where $f(\lambda) = f(x, \lambda)$ is a continuous $\mathbf{C}^2_{b*}$-valued function satisfying $(2.2)$, $g$ is an element of $\mathbf{C}^1_{b*}$ and $\{\xi^n_j\}_n$ is a sequence of stationary processes satisfying

Conditions $(\xi.1) - (\xi.3)$. Let $\psi_j^n$ be the solution of the above equation with the initial condition $\psi_0 = x_0$ and set $\varphi_t^n = \psi_{[nt]}^n$ as before.

**Corollary 1.** *Assume* $\sup_\lambda \|f(\lambda)\|_{1,N} < \infty$ *for any* $N > 0$. *Then the sequence* $\{(B^n(t), \varphi_t^n)\}_n$ *converges in law. The limit* $(B(t), \varphi_t)$ *satisfies* (2.13).

We next consider the case where $\sup_\lambda \|f(\lambda)\|_{1,N} = \infty$ for some N.

**Corollary 2.** *Assume that Conditions* $(\xi.4)$ *and* $(\xi.5)$ *are satisfied for* $\rho(r) = (1 + r)^\alpha$ *and* $c_n = n^\beta$ *where* $\alpha, \beta > 0$. *Assume further*

$$h(x, \infty, \theta) = \lim_{|\lambda| \to \infty} \frac{f(x, |\lambda|, \theta)}{(1 + |\lambda|)^\alpha} \tag{2.15}$$

*exists and is a continuous function on* $S^{m-1}$.
*(1) If* $\beta < 1/2\alpha$, *the sequence* $\{(B^n(t), K_\varepsilon^n(t), \varphi_t^n)\}_n$ *converges in law. The limit* $(B(t), K_\varepsilon(t), \varphi_t)$ *satisfies*

$$\begin{aligned}
\varphi_t = {} & x_0 + \int_0^t \int_{\mathbf{R}^m} f(\varphi_s, \lambda) B(ds, d\lambda) \\
& + \int_0^t \int_{S^{m-1}} h(\varphi_s, \infty, \theta) K_0(ds, d\theta) \\
& + \int_0^t \{g + b_1 + c\}(\varphi_s) ds.
\end{aligned} \tag{2.16}$$

*(2) If* $\beta = 1/2\alpha$, *then the sequence* $\{(B^n(t), N^n(t), K_\varepsilon^n(t), \varphi_t^n)\}_n$ *converges in law. The limit satisfies equation* (2.11) *with*

$$\hat{f}(x, \lambda) = |\lambda|^\alpha \bar{h}(x, \infty, \frac{\lambda}{|\lambda|}). \tag{2.17}$$

The proof of Theorem 2.1 will be given at Section 4. In the next section we will formulate Theorem 2.1 in a different way and establish Theorem 3.1, which will be applied in Section 4 for the proof of Theorem 2.1.

## 3.  Martingale problem

Let $\varphi_t^n$ be the solution of equation (0.4). Its law $Q^n$ can be defined on the space $\mathbf{D} = \mathbf{D}([0, \infty); \mathbf{R}^d)$. In this section we show the weak convergence of the sequence of the laws $\{Q^n\}$ assuming the same conditions as in Theorem 2.1. Our goal is to characterize the limit law $Q^\infty$ as a solution of a certain martingale problem.

We first define two integro-differential operators $\mathcal{L}^{(1)}$ and $\mathcal{L}^{(2)}$. Let $M$ be a positive constant. For a $\mathbf{C}^2$-function $F(x)$, set

$$
\mathcal{L}^{(1)}F(x) = \frac{1}{2}\sum_{i,j} a_{ij}(x)\frac{\partial^2 F}{\partial x^i \partial x^j}(x)
$$
$$
+ \sum_i \{b_{M,i}(x) + c_i(x) + g_i(x)\}\frac{\partial F}{\partial x^i}(x)
$$
$$
+ \int_{0<|\lambda|\leq M}\Big\{F(x + \hat{f}(x,\lambda)) - F(x) - \sum_i \hat{f}_i(x,\lambda)\frac{\partial F}{\partial x^i}(x)\Big\}\mu(d\lambda), \quad (3.1)
$$

where

$$
a_{ij}(x) = \int_{\mathbf{R}^m} f_i(x,\lambda)f_j(x,\lambda)\pi(d\lambda)
$$
$$
+ \int\int_{\mathbf{R}^m \times \mathbf{R}^m}\Big\{f_i(x,\lambda)f_j(x,\lambda') + f_j(x,\lambda)f_i(x,\lambda')\Big\}V(d\lambda, d\lambda')
$$
$$
+ \int_{S^{m-1}} \bar{h}_i(x,\infty,\theta)\bar{h}_j(x,\infty,\theta)\nu(d\theta) \quad (3.2)
$$

and for a $\mathbf{C}_b^2$-function $F$,

$$
\mathcal{L}^{(2)}F(x) = \int_{|\lambda|>M}\Big\{F(x + \hat{f}(x,\lambda)) - F(x)\Big\}\mu(d\lambda) \quad (3.3)
$$

and define

$$
\mathcal{L} = \mathcal{L}^{(1)} + \mathcal{L}^{(2)}. \quad (3.4)
$$

**Theorem 3.1.** *Assume Conditions* $(\xi.1)-(\xi.5)$, $(f.1)$, $(f.2)$ *and* $(g.1)$ *for equations (0.4). Then the sequence* $\{\varphi_t^n\}$ *converges in law. Let* $(\varphi_t, Q^\infty)$ *be its limit law. Then for any* $\mathbf{C}^2$ *function* $F$,

$$
F(\varphi_t) - \int_0^t \mathcal{L}^{(1)}F(\varphi_{s-})ds
$$
$$
- \sum_{0\leq s\leq t}\{F(\varphi_s) - F(\varphi_{s-})\}\Delta N(s, \{\lambda : |\lambda| > M\}) \quad (3.5)
$$

*is a locally square integrable martingale. Further for any* $\mathbf{C}_b^2$-*function* $F$,

$$
F(\varphi_t) - \int_0^t \mathcal{L}F(\varphi_{s-})ds \quad (3.6)
$$

*is a square integrable martingale, where* $\Delta N(s, F) = N(s, F) - N(s-, F)$.

In the rest of this section, we give the proof of Theorem 3.1. We will apply a criterion of the weak convergence obtained by Fujiwara- Kunita [3]. It tells us that Theorem 3.1 is valid if the following Conditions (A.I)-(A.IV) are satisfied.

For a positive constant $M$, we set

$$\bar{f}^{n,M}(x) = E[f^n(x, \xi_j^{n,M})], \quad \tilde{f}^{n,M}(x, \lambda) = f^{n,M}(x, \lambda) - \bar{f}^{n,M}(x).$$

Then by Condition (f.1) (2), $\bar{f}^{n,M} \to 0$ and $\tilde{f}^{n,M}(\cdot, \lambda) \to f(\cdot, \lambda)$ (uniformly on compact sets for the latter convergence) with respect to $\| \ \|_{1,N}$ seminorms.

**Condition (A.I)** (1) For every $M > 0$ and $N > 0$ there exists a sequence of non-decreasing $(\mathcal{F}^n_{[nt]})$ adapted cadlag processes $\{D^n_t\}$ such that

$$\frac{1}{n} \sum_{k=[ns]+1}^{[nt]} \sum_{\ell=k+1}^{[nt]} \|E[\tilde{f}^{n,M}(\xi_\ell^{n,M})|\mathcal{F}^n_k]\|_{2,N} \|\tilde{f}^{n,M}_k(\xi_k^{n,M})\|_{1,N}$$

$$+ \frac{1}{n} \sum_{k=[ns]+1}^{[nt]} (\|f^{n,M}(\xi_k^{n,M})\|_{1,N})^2 \leq D^n_t - D^n_s, \qquad (3.7)$$

for any $0 \leq s < t$ and that its compensators $\{D^{n,p}\}$ are $C$-tight.

(2) $\{\|\bar{f}^{n,M}\|_{1,N}\}_n$ is bounded for every $M > 0$ and $N > 0$.

**Condition (A.II)** (1) For any bounded continuous function $h$, $N > 0$, $0 < \delta < M$ and $0 \leq s < t$,

$$\lim_{n \to \infty} E\left[ \sup_{|x| \leq N} \left| E\left[ \sum_{k=[ns]+1}^{[nt]} h\left(\hat{f}^n(x, \frac{\xi_k^n}{c_n})\right) \chi_{(\delta,M]}\left( \|\hat{f}^n(\frac{\xi_k^n}{c_n})\|_{1,N} \right) \Big| \mathcal{F}^n_{[ns]} \right] \right. \right.$$

$$\left. \left. - (t-s) \int_{\mathbf{R}^m - \{0\}} h\left(\hat{f}(x, \lambda)\right) \chi_{(\delta,M]}\left( \|\hat{f}\|_{1,N} \right) \mu(d\lambda) \right| \right] = 0.$$

(2) For any $N > 0$ and $0 \leq s < t$,

$$\lim_{\varepsilon \to 0} \lim_{n \to \infty} E\left[ \sup_{|x| \leq N} \left| \frac{1}{n} E\left[ \sum_{k=[ns]+1}^{[nt]} \tilde{f}_i^{n,\varepsilon}(x, \xi_k^{n,\varepsilon}) \tilde{f}_j^{n,\varepsilon}(x, \xi_k^{n,\varepsilon}) \Big| \mathcal{F}^n_{[ns]} \right] \right. \right.$$

$$- (t-s) \left\{ \int_{\mathbf{R}^m} f_i(x, \lambda) f_j(x, \lambda) \pi(d\lambda) \right.$$

$$\left. \left. + \int_{S^{m-1}} \bar{h}_i(x, \infty, \theta) \bar{h}_j(x, \infty, \theta) \nu(d\theta) \right\} \right| \right] = 0.$$

(3) For any $N > 0$ and $0 \le s < t$,

$$\lim_{\varepsilon \to 0} \lim_{n \to \infty} E\left[ \sup_{|x| \le N} \left| \frac{1}{n} E\left[ \sum_{k=[ns]+1}^{[nt]} \sum_{\ell=k+1}^{[nt]} \tilde{f}_i^{n,\varepsilon}(x, \xi_k^{n,\varepsilon}) \tilde{f}_j^{n,\varepsilon}(x, \xi_\ell^{n,\varepsilon}) \Big| \mathcal{F}_{[ns]}^n \right] \right. \right.$$
$$\left. \left. -(t-s) \int\int_{\mathbf{R}^m \times \mathbf{R}^m} f_i(x, \lambda) f_j(x, \lambda') V(d\lambda, d\lambda') \right| \right] = 0.$$

(4) For any $N > 0$ and $0 \le s < t$,

$$\lim_{\varepsilon \to 0} \lim_{n \to \infty} E\left[ \sup_{|x| \le N} \left| \frac{1}{n} E\left[ \sum_{k=[ns]+1}^{[nt]} \sum_{\ell=k+1}^{[nt]} \tilde{f}^{n,\varepsilon}(x, \xi_k^{n,M}) \cdot \nabla \tilde{f}_i^{n,\varepsilon}(x, \xi_\ell^{n,M}) \Big| \mathcal{F}_{[ns]}^n \right] \right. \right.$$
$$\left. \left. -(t-s) \int\int_{\mathbf{R}^m \times \mathbf{R}^m} f(x, \lambda) \cdot \nabla f_i(x, \lambda') V(d\lambda, d\lambda') \right| \right] = 0.$$

(5) $\lim_{n \to \infty} g^n(x) = g(x)$ uniformly on compact sets.

**Condition (A.III)** For any $\alpha$ with $|\alpha| \le 1$, $M > 0$, $N > 0$ and $0 \le s < t$,

$$\lim_{n \to \infty} E\left[ \sup_{|x| \le N} \left| \frac{1}{\sqrt{n}} \sum_{k=[ns]+1}^{[nt]} E\left[ D^\alpha \tilde{f}^{n,M}(x, \xi_k^{n,M}) \Big| \mathcal{F}_{[ns]}^n \right] \right| \right] = 0, \quad (3.8)$$

$$\lim_{n \to \infty} E\left[ \sup_{|x| \le N} \left( \frac{1}{\sqrt{n}} \right)^3 \sum_{k=[ns]+1}^{[nt]} \sum_{\ell=k+1}^{[nt]} \left| E\left[ D^\alpha \tilde{f}^{n,M}(x, \xi_\ell^{n,M}) \Big| \mathcal{F}_k^n \right] \right| \right.$$
$$\left. \times \left| \tilde{f}^{n,M}(x, \xi_k^{n,M}) \right|^2 \right] = 0. \quad (3.9)$$

**Condition (A.IV)** For any $t > 0$ and $N > 0$,

$$\lim_{M \to \infty} \limsup_{n \to \infty} P\left( \sup_{k \le [nt]} \|\hat{f}^n(\frac{\xi_k^n}{c_n})\|_{1,N} > M \right) = 0.$$

In the remainder of this section, we will examine these conditions one by one.

**Lemma 3.2.** *Condition (A.I) is satisfied.*

Proof can be carried out similarly as in the proof of Lemma 2.2 in [8]. It is omitted.

**Lemma 3.3.** *Condition (A.II)(1) is satisfied.*

*Proof.* Using the mixing inequality,

$$
E\left[\sup_{|x|\le N}\left|E\left[\sum_{k=[ns]+1}^{[nt]} h\left(\hat{f}^n\left(x,\frac{\xi_k^n}{c_n}\right)\right)\chi_{(\delta,M]}\left(\|\hat{f}^n\left(\frac{\xi_k^n}{c_n}\right)\|_{1,N}\right)\Big|\mathcal{F}_{[ns]}^n\right]\right.\right.
$$

$$
\left.\left.-E\left[\sum_{k=[ns]+1}^{[nt]} h\left(\hat{f}^n\left(x,\frac{\xi_k^n}{c_n}\right)\right)\chi_{(\delta,M]}\left(\|\hat{f}^n\left(\frac{\xi_k^n}{c_n}\right)\|_{1,N}\right)\right]\right|\right]
$$

$$
\le \sum_{k=[ns]+1}^{[nt]} 2(\phi_{k-[ns]}^n)^{\frac{1}{2}} E\left[\sup_{|x|\le N} h\left(\hat{f}^n\left(x,\frac{\xi_k^n}{c_n}\right)\right)^2\right]^{\frac{1}{2}}
$$

$$
\le \frac{1}{n}\left(\sum_{k=1}^{\infty} 2(\phi_k^n)^{\frac{1}{2}}\right)\int_{\mathbf{R}^m-\{0\}}\sup_{|x|\le N} h\left(\hat{f}^n(x,\lambda)\right)\mu_n(d\lambda) \le \frac{C'}{n},
$$

where $C'$ is a positive constant independent of n. Therefore the above converges to 0 as $n \to \infty$. On the other hand, we have by Conditions $(\xi.3)$ and (f.2),

$$
\lim_{n\to\infty}\sup_{|x|\le N}\left|E\left[\sum_{k=[ns]+1}^{[nt]} h\left(\hat{f}^n\left(x,\frac{\xi_k^n}{c_n}\right)\right)\chi_{(\delta,M]}\left(\|\hat{f}^n\left(\frac{\xi_k^n}{c_n}\right)\|_{1,N}\right)\right]\right.
$$

$$
\left.-(t-s)\int_{\mathbf{R}^m-\{0\}} h\left(\hat{f}(x,\lambda)\right)\chi_{(\delta,M]}\left(\|\hat{f}(\lambda)\|_{1,N}\right)\mu(d\lambda)\right| = 0.
$$

Therefore Condition (A.II)(1) is satisfied. $\square$

**Lemma 3.4.** *Condition (A.II)(2) is satisfied.*

*Proof.* For the convenience of notations, we denote $\tilde{f}_i^{n,\epsilon}$ ( the i-th component of $\tilde{f}^{n,\epsilon}$) by $f_i^{n,\epsilon}$. We have

$$
E\left[\sup_{|x|\le N}\frac{1}{n}\left|\sum_{k=[ns]+1}^{[nt]}\left(E\left[f_i^{n,\epsilon}(x,\xi_k^{n,\epsilon})f_j^{n,\epsilon}(x,\xi_k^{n,\epsilon})\Big|\mathcal{F}_{[ns]}^n\right]\right.\right.\right.
$$

$$
\left.\left.\left.-E\left[f_i^{n,\epsilon}(x,\xi_k^{n,\epsilon})f_j^{n,\epsilon}(x,\xi_k^{n,\epsilon})\right]\right)\right|\right]
$$

$$
\le \sum_{k=[ns]+1}^{[nt]} 2(\phi_{k-[ns]}^n)^{\frac{1}{2}} E\left[\sup_{|x|\le N}\hat{f}_i^{n,\epsilon}\left(x,\frac{\xi_k^{n,\epsilon}}{c_n}\right)^2\hat{f}_j^{n,\epsilon}\left(x,\frac{\xi_k^{n,\epsilon}}{c_n}\right)^2\right]^{\frac{1}{2}}.
$$

There exists a positive constant $C_N$ such that $\|\hat{f}^n(\lambda)\|_{1,N} \le C_N\rho(c_n|\lambda|)/\sqrt{n}$ holds for all n and $\lambda$. Since $|\xi_k^{n,\epsilon}/c_n| \le \epsilon$, the last term is bounded by $2C_N^2(\rho(c_n\epsilon)/\sqrt{n})^2\sum_{k=1}^{\infty}(\phi_k^n)^{1/2}$. It converges to 0 as $n \to \infty$ and $\epsilon \to 0$.

Now define $h^{n,\varepsilon}(x, \lambda) = f^{n,\varepsilon}(x, \lambda)/\rho(|\lambda|)$. Then $h^{n,\varepsilon}(\cdot, \lambda)$ can be extended to a $\mathbf{C}^1_{b*}$-valued function $\bar{h}^{n,\varepsilon}(\lambda)$. The sequence $\{\bar{h}^{n,\varepsilon}(x, \lambda)\}_n$ converges to $\bar{h}(x, \lambda)$ of Condition (f.2) with respect to $\| \ \|_{1,N}$ uniformly in $\lambda$ for any fixed $\varepsilon$. Note the equality

$$\frac{1}{n} \sum_{k=[ns]+1}^{[nt]} E\left[f_i^{n,\varepsilon}(x, \xi_k^{n,\varepsilon}) f_j^{n,\varepsilon}(x, \xi_k^{n,\varepsilon})\right]$$

$$= \frac{[nt] - [ns]}{n} \int_{\mathbf{R}^m} h_i^{n,\varepsilon}(x, \lambda) h_j^{n,\varepsilon}(x, \lambda) \nu_\varepsilon^n(d\lambda).$$

Then the right hand side converges to

$$(t - s) \int_{\mathbf{R}^m} \bar{h}_i(x, \lambda) \bar{h}_j(x, \lambda) \nu_\varepsilon(d\lambda)$$

uniformly on compact sets with respect to x. Further as $\varepsilon \to 0$, the above converges to

$$(t - s) \int_{\mathbf{R}^m} \bar{h}_i(x, \lambda) \bar{h}_j(x, \lambda) \nu(d\lambda)$$

$$= (t - s) \left( \int_{\mathbf{R}^m} f_i(x, \lambda) f_j(x, \lambda) \pi(d\lambda) \right.$$

$$\left. + \int_{S^{m-1}} \bar{h}_i(x, \infty, \theta) \bar{h}_j(x, \infty, \theta) \nu(d\theta) \right). \qquad \square$$

Before we proceed to the proof of (A.II)(3) and (4), we need a lemma.

**Lemma 3.5.** *Let $k < \ell$. Then*

$$\lim_{c \to \infty} \sup_n E\left[\rho(|\xi_k^{n,\varepsilon}|)\rho(|\xi_\ell^{n,\varepsilon}|); \ \rho(|\xi_k^{n,\varepsilon}|) > c \ \text{ or } \ \rho(|\xi_\ell^{n,\varepsilon}|) > c\right] = 0. \quad (3.10)$$

*Proof.* We will first prove

$$\lim_{c \to \infty} \sup_n E\left[\rho(|\xi_k^{n,\varepsilon}|)\rho(|\xi_\ell^{n,\varepsilon}|); \ \rho(|\xi_k^{n,\varepsilon}|) > c\right] = 0. \quad (3.11)$$

We have

$$E\left[\rho(|\xi_k^{n,\varepsilon}|)\rho(|\xi_\ell^{n,\varepsilon}|); \ \rho(|\xi_k^{n,\varepsilon}|) > c\right]$$

$$\leq E\left[\rho(|\xi_k^{n,\varepsilon}|) E[\rho(|\xi_\ell^{n,\varepsilon}|)^2 | \mathcal{F}_{\ell-1}^n]^{\frac{1}{2}}; \ \rho(|\xi_k^{n,\varepsilon}|) > c\right]$$

$$\leq E\left[\rho(|\xi_k^{n,\varepsilon}|) E[\rho(|\xi_\ell^{n,\varepsilon}|)^2 | \mathcal{F}_{\ell-1}^n]^{\frac{1}{2}}; \ E[\rho(|\xi_\ell^{n,\varepsilon}|)^2 | \mathcal{F}_{\ell-1}^n]^{\frac{1}{2}} > c'\right]$$

$$+c'E\left[\rho(|\xi_k^{n,\epsilon}|);\ \rho(|\xi_k^{n,\epsilon}|) > c\right]$$

$$\leq E[\rho(|\xi_k^{n,\epsilon}|)^2|]^{\frac{1}{2}}E\left[E[\rho(|\xi_\ell^{n,\epsilon}|)^2|\mathcal{F}_{\ell-1}^n];\ E[\rho(|\xi_\ell^{n,\epsilon}|)^2|\mathcal{F}_{\ell-1}^n] > c'^2\right]^{\frac{1}{2}}$$

$$+\frac{c'}{c^\gamma}E\left[\rho(|\xi_k^{n,\epsilon}|)^{1+\gamma};\ \rho(|\xi_k^{n,\epsilon}|) > c\right],$$

where $0 < \gamma < 1$ and $c' > 0$. Note that $\{E[\rho(|\xi_k^{n,\epsilon}|)^2]\}_n$ is bounded in $n$ and both $\left\{E\left[\rho(|\xi_\ell^{n,\epsilon}|)^2\big|\mathcal{F}_{\ell-1}^n\right]\right\}_n$ and $\{\rho(|\xi_k^{n,\epsilon}|)^{1+\gamma}\}_n$ are uniformly integrable by Condition $(\xi.5)$. Then the supremum of the above with rspect to $n$ converges to 0 as $c \to \infty$ and $c' \to \infty$. Therefore (3.11) is established.

Now for any $\delta > 0$, there exists $c > 0$ such that

$$\sup_n E\left[\rho(|\xi_\ell^{n,\epsilon}|)\rho(|\xi_k^{n,\epsilon}|);\ \rho(|\xi_k^{n,\epsilon}|) > c\right] < \delta \qquad (3.12)$$

by (3.11). We have

$$\sup_n E\left[\rho(|\xi_\ell^{n,\epsilon}|)\rho(|\xi_k^{n,\epsilon}|);\ \rho(|\xi_k^{n,\epsilon}|) \leq c, \rho(|\xi_\ell^{n,\epsilon}|) > c'\right]$$

$$< c \cdot \sup_n E\left[\rho(|\xi_\ell^{n,\epsilon}|);\ \rho(|\xi_\ell^{n,\epsilon}|) > c'\right].$$

Since $\{\rho(|\xi_\ell^{n,\epsilon}|)\}_n$ is uniformly integrable, there exists $c' > 0$ such that

$$c \cdot \sup_n E\left[\rho(|\xi_\ell^{n,\epsilon}|);\ \rho(|\xi_\ell^{n,\epsilon}|) > c'\right] < \delta. \qquad (3.13)$$

Then (3.12) and (3.13) imply

$$\sup_n E\left[\rho(|\xi_\ell^{n,\epsilon}|)\rho(|\xi_k^{n,\epsilon}|);\ \rho(|\xi_k^{n,\epsilon}|) > c \text{ or } \rho(|\xi_\ell^{n,\epsilon}|) > c'\right] < 2\delta.$$

Therefore (3.10) is satisfied. $\square$

**Lemma 3.6.** *Condition (A.II),(3) is satisfied.*

*Proof.* Set

$$L_{k,\ell}^n = \tilde{f}_i^{n,M}(x,\xi_k^{n,M})\tilde{f}_j^{n,M}(x,\xi_\ell^{n,M}) \quad \text{and} \quad \tilde{L}_{k,\ell}^n(x) = L_{k,\ell}^n(x) - E[L_{k,\ell}^n(x)].$$

Then we can prove similarly as in the proof of Theorem 4.2 in [3] that

$$\lim_{n\to\infty}\sup_{|x|\leq N}\frac{1}{n}\left|\sum_{k=[ns]+1}^{[nt]}\sum_{\ell=k+1}^{[nt]}\left(E[L_{k,\ell}^n(x)|\mathcal{F}_{[ns]}^n] - \bar{L}_{k,\ell}^n(x)\right)\right| = 0. \qquad (3.14)$$

Hence it is sufficient to prove

$$\lim_{n \to \infty} \sup_{|x| \leq N} \left| \frac{1}{n} \sum_{k=[ns]+1}^{[nt]} \sum_{\ell=k+1}^{[nt]} \bar{L}_{k,\ell}^n(x) - (t-s)V(f_i(x), f_j(x)) \right| = 0. \quad (3.15)$$

By Lemma 3.5 for any $\delta' > 0$ there exists a continuous function $\psi_c(x)$ such that $0 \leq \psi_c(x) \leq 1$, $\psi_c(x) = 1$ if $|x| \leq c$ and $\psi_c(x) = 0$ if $|x| \geq c+1$, and the inequality

$$\sup_{|x| \leq N} \left| \frac{1}{n} \sum_{k=[ns]+1}^{[nt]} \sum_{\ell=k+1}^{k+k_0} E\left[ L_{k,\ell}^k(x)\left(1 - \psi_c(\xi_k^{n,M})\psi_c(\xi_\ell^{n,M})\right) \right] \right| < \delta' \quad (3.16)$$

holds for all n. Then use this inequality instead of (58) in the proof of Lemma 2.3 in [8]. Then we obtain (3.15). □

The proof of (A.II)(4) can be carried out similarly. Condition (A.II)(5) is immediate from Condition ($\xi$.5) and (g.1).

**Lemma 3.7.** *Conditions (A.III) and (A.IV) are satisfied.*

*Proof.* Note that

$$\frac{1}{\sqrt{n}} \sum_{k=[ns]+1}^{[nt]} E\left[ \left| E\left[ D^\alpha \tilde{f}_i^{n,M}(x, \xi_k^{n,M}) \Big| \mathcal{F}_{[ns]}^n \right] \right| \right]$$

$$\leq \frac{1}{\sqrt{n}} \sum_{k=[ns]+1}^{[nt]} 2(\varphi_{k-[ns]}^n)^{\frac{1}{2}} E\left[ D^\alpha \tilde{f}_i^{n,M}(x, \xi_k^{n,M})^2 \right]^{\frac{1}{2}}.$$

Since $\left\{ \sup_{|x| \leq N} E\left[ D^\alpha \tilde{f}_i^{n,M}(x, \xi_k^{n,M})^2 \right] \right\}_n$ is a bounded sequence, the last term of the above converges to 0 uniformly on compact sets as $n \to \infty$. This proves (3.8). The convergence (3.9) can be proved similarly.

Now, since $\| \hat{f}^n(\lambda) \|_{1,N} \leq C_N \rho(c_n|\lambda|)/\sqrt{n}$, we have

$$P\left( \sup_{k \leq [nt]} \| \hat{f}^n(\frac{\xi_k^n}{c_n}) \|_{1,N} > M \right) \leq \mu_n\left( \left\{ \lambda : \frac{\rho(c_n|\lambda|)}{\sqrt{n}} > \frac{M}{C_N} \right\} \right).$$

The right hand side converges to $\mu(\{\lambda : \eta(|\lambda|) > M/C_N\})$ if the set $\{...\}$ is the continuity point of $\mu$. It converges to 0 as $M \to \infty$. Therefore Condition (A.IV) is satisfied. □

Finally we remark that Condition (f.2) can be relaxed in Theroem 3.1 in the following way.

**Condition (f.2)'** *(1) There exists $\mathbf{C}_{b*}$-valued functions $\bar{h}^n(\lambda) = \bar{h}^n(x, \lambda)$*

$(n \in \mathbf{N})$ and $\bar{h}(\lambda) = \bar{h}(x, \lambda)$ on $\bar{\mathbf{R}}^m$ such that

$$\bar{h}^n(x, \lambda) = \frac{f^n(x, \lambda)}{\rho(|\lambda|)}, \qquad \bar{h}(x, \lambda) = \frac{f(x, \lambda)}{\rho(|\lambda|)} \qquad if \qquad \lambda \in \mathbf{R}^m$$

and satisfies

$$\lim_{n \to \infty} \sup_{|x| \leq N} \left| \int \bar{h}_i^n(x, \lambda) \bar{h}_j^n(x, \lambda) \nu_\epsilon^n(d\lambda) - \int \bar{h}_i(x, \lambda) \bar{h}_j(x, \lambda) \nu_\epsilon(d\lambda) \right| = 0.$$

(2) Define $\hat{f}^n(x, \lambda)$ by (2.3). Then $\hat{f}(x, \lambda) = \lim_{n \to \infty} \hat{f}^n(x, \lambda)$ exists for all $x$ and $\lambda$ and satisfies

$$\lim_{n \to \infty} \sup_{|x| \leq N} \left| \int h\left( \hat{f}^n(x, \lambda) \right) \chi_{(\delta, M]}\left( \|\hat{f}^n(\lambda)\|_{1, N} \right) \mu^n(d\lambda) \right.$$
$$\left. - \int h\left( \hat{f}(x, \lambda) \right) \chi_{(\delta, M]}\left( \|\hat{f}(\lambda)\|_{1, N} \right) \mu(d\lambda) \right| = 0$$

for any bounded continuous function $h$.

**Theorem 3.1'.** *Assume* $(\xi.1) - (\xi.5), (f.1), (f.2)'$ *and* $(g.1)$ *for equations* $(0.4)$. *Then the assertion of Theorem 3.1 is valid.*

Indeed, the proof of the theorem can be carried out without any essential change as in the proof of Theorem 3.1.

## 4.   Proofs of Theorems

Let us first observe that Theorems 1.1 and 2.1 can be reduced to Theorem 3.1 or Theorem 3.1'. Let $E_1, .., E_q \in \mathcal{G}(\mathbf{R}^m)$, $F_1, ..., F_r \in \mathcal{G}(\mathbf{R}^m - \{0\})$ be arbitrarily fixed sets. Set for any $n = 1, 2, ...$

$$\chi^n(\lambda) = (\chi_{E_1}(\lambda) - \pi^n(E_1), ..., \chi_{E_q}(\lambda) - \pi^n(E_q)), \tag{4.1}$$

$$\kappa^n(\lambda) = \sqrt{n}\left( \chi_{F_1}(\frac{\lambda}{c_n}) - \frac{1}{n}\mu^n(F_1), ..., \chi_{F_r}(\frac{\lambda}{c_n}) - \frac{1}{n}\mu^n(F_r) \right), \tag{4.2}$$

and for $y = (u, v)$, $u \in \mathbf{R}^q, v \in \mathbf{R}^r$ define $F^n(y, \lambda)$ and $G^n(y, \lambda)$ by

$$F^n(y, \lambda) = (\chi^n(\lambda), \kappa^n(\lambda)), \tag{4.3}$$

$$G^n(y) = (0, ..., 0, \mu^n(F_1), ..., \mu^n(F_r)), \tag{4.4}$$

and consider a stochastic difference equation on $\mathbf{R}^{q+r}$:

$$\Psi_k^n = \Psi_{k-1}^n + \frac{1}{\sqrt{n}} F_n(\Psi_{k-1}^n, \xi_k^n) + \frac{1}{n} G^n(\Psi_{k-1}^n), \tag{4.5}$$

$$\Psi_0^n = (0, x). \tag{4.6}$$

Set $\Phi_t^n = \Psi_{[nt]}^n$. Then $\Phi_t^n$ is represented by $\Phi_t^n = (B^n(t), N^n(t))$.

Next let $G_1, ..., G_s \in \mathcal{G}(\tilde{\mathbf{R}}^m)$ be arbitrary fixed sets. Set

$$\zeta^n(\lambda) = \left( \rho(|\lambda|)\chi_{G_1}(\lambda)\chi_{[0,c_n\epsilon]}(|\lambda|) - \int_{G_1} \rho(|\lambda|)\mu_\epsilon^n(d\lambda), \cdots, \right.$$
$$\left. \rho(|\lambda|)\chi_{G_s}(\lambda)\chi_{[0,c_n\epsilon]}(|\lambda|) - \int_{G_s} \rho(|\lambda|)\mu_\epsilon^n(d\lambda) \right), \qquad (4.7)$$

and define for $y = (u, v, w, x) \in \mathbf{R}^{q+r+s+d}$

$$F^n(y, \lambda) = (\chi^n(\lambda), \kappa^n(\lambda), \zeta^n(\lambda), f^n(x, \lambda)), \qquad (4.8)$$
$$G^n(y) = (0, ..., 0, \mu^n(F_1), ..., \mu^n(F_r), 0, ..., 0, g^n(x)), \qquad (4.9)$$

and consider a stochastic difference equation (4.5), (4.6) on $\mathbf{R}^{q+r+s+d}$. Set $\Phi_t^n = \Psi_{[nt]}^n$. Then $\Phi_t^n$ is represented by $\Phi_t^n = (B^n(t), N^n(t), K_\epsilon^n(t), \varphi_t^n)$. Therefore we can apply Theorems 3.1 or 3.1' for the weak convergence of the above sequence $\{\Phi_t^n\}_n$, by checking Conditions (f.1),(f.2)' and (g.1) in these cases.

*Proof of Theorem 1.1.* Define $F^n$ and $G^n$ by (4.3) and (4.4), and consider equations (4.5), (4.6). Conditions ($\xi$.4) and ($\xi$.5) are trivially satisfied by setting $\rho(r) \equiv 1$. Further conditions (f.1),(f.2)'and (g.1) are satisfied for these $F^n$ and $G^n$ in place of $f^n$ and $g^n$. Indeed, the sequences $\{F^n(y, \lambda)\}$ and $\{G^n(y)\}$ converge to $F(y, \lambda) = (\chi(\lambda), 0)$ and $G(y) = (0, \mu(F_1), ..., \mu(F_r))$, respectively in the sense of Conditions (f.1) and (g.1), where $\chi(\lambda)$ is defined by (4.1) replacing $\pi^n$ by $\pi$. Further, $H^n(y, \lambda) \equiv F^n(y, \lambda)(\rho = 1)$ satisfies Condition (f.2). Let $Q^n$ be the law of $(B^n(t, E_1), ..., B^n(t, E_q), N^n(t, F_1), ..., N^n(t, F_r))$ in the space $\mathbf{D} = \mathbf{D}([0; \infty), \mathbf{R}^{q+r})$. Then the sequence $\{Q^n\}$ converges weakly by Theorem 3.1. Let $Q^\infty$ be its limit law.

We will apply Theorem 3.1 to the $C_b^2$-function $F(u, v)$, where we assume $q = r = 1$ for simplicity of notation. Then the operator $\mathcal{L}$ of (3.4) is represented by

$$\mathcal{L}F(u, v) = \frac{1}{2}\left( \pi(E) + 2V(E, E) - \pi(E)^2 \right)\frac{\partial^2 F}{\partial u^2}(u, v)$$
$$+ \int_{0 < |\lambda|} \left\{ F(u, v + \chi_F(\lambda)) - F(u, v) \right\}\mu(d\lambda). \qquad (4.10)$$

Therefore,

$$F(B(t), \tilde{N}(t)) - \int_0^t \mathcal{L}F(B(s-), \tilde{N}(s-))ds \qquad (4.11)$$

is a martingale with respect to $Q^\infty$. The fact implies that $(B(t), N(t))$ is a Lévy process and satisfies all assertions of Theorem 1.1. See Kunita [7].

Finally let $P^n$ be the law of $(B^n(t), N^n(t))$ on the space $\mathbf{D}([0, \infty), \mathcal{M}(\mathbf{R}^m) \times \mathcal{M}(\mathbf{R}^m - \{0\}))$. Then the sequence $\{P^n\}$ is tight, which follows from the tightness of $\{Q^n\}$ for any choice of $\{E_1, ..., E_q, F_1, ..., F_r\}$, $q, r = 1, 2, ...$ Let $P^\infty$ be any weak limit. Then the law of $(B(t, E_1), ..., B(t, E_q), N(t, F_1), ..., N(t, F_r))$ with respect to $P^\infty$ coincides with $Q^\infty$. Hence $P^\infty$ is unique. $\square$

*Proof of Theorems 1.2 and 2.1.* Consider equations (4.5), (4.6), where $F^n$ and $G^n$ are defined by (4.8) and (4.9). Then the sequences $\{F^n(y, \lambda)\}_n$ and $\{G^n(y)\}_n$ converge to

$$F(y, \lambda) = (\chi(\lambda), 0, \zeta(\lambda), f(x, \lambda)), \quad G(y) = (0, \mu(F), 0, g(x)), \quad (4.12)$$

where $\chi(\lambda)$ and $\zeta(\lambda)$ are defined by (4.1) and (4.7) replacing $\pi^n$ and $\nu_\varepsilon^n$ by $\pi$ and $\nu_\varepsilon$, and $\chi_{[0,c_n\varepsilon]}(|\lambda|)$ by 1, resepectively. Hence Conditions (f.1) and (g.1) are satisfied. Set

$$H^n(y, \lambda) = \frac{F^n(y, \lambda)}{\rho(|\lambda|)}. \quad (4.13)$$

Then $H^n(\cdot, \lambda)$ can be extended to a $\mathbf{C}_{b*}^1$-valued function $\bar{H}^n(\cdot, \lambda)$ on $\bar{\mathbf{R}}^m$. Further the sequence $\{\bar{H}^n(\cdot, \lambda)\}$ converges to $\bar{H}(\cdot, \lambda)$ with respect to $\| \ \|_{1,N}$ seminorms. It is represented by

$$\bar{H}(y, \lambda) = \frac{F(y, \lambda)}{\rho(|\lambda|)} \quad \text{if} \quad \lambda \in \mathbf{R}^m, \quad (4.14)$$

$$\bar{H}(y, \infty, \frac{\lambda}{|\lambda|}) = \left(0, 0, \chi_G\left(\infty, \frac{\lambda}{|\lambda|}\right)\chi_{[0,\varepsilon]}(|\lambda|), \hat{h}\left(x, \infty, \frac{\lambda}{|\lambda|}\right)\right) \quad (4.15)$$

if $\lambda = (\infty, \lambda/|\lambda|) \in \partial\mathbf{R}^m$. Therefore Condition (f.2)' is satisfied for the sequence $\{F^n\}$. Further, the sequence $\hat{F}^n(y, \lambda) = F^n(y, c_n\lambda)/\sqrt{n}, \ n = 1, 2, ...$ converges to

$$\hat{F}(y, \lambda) = \left(0, \chi_F(\lambda), \eta(|\lambda|)\chi_G\left(\infty, \frac{\lambda}{|\lambda|}\right)\chi_{[0,\varepsilon]}(|\lambda|), \hat{f}(x, \lambda)\right), \quad (4.16)$$

if $|\lambda| > 0$.

Let $\bar{Q}^n$ be the law of $(B^n(t), N^n(t), K_\varepsilon^n(t), \varphi_t^n)$ on the space $\mathbf{R}^{q+r+s+d}$. Then $\{\bar{Q}^n\}$ converges weakly by Theorem 3.1'. Let $\bar{Q}^\infty$ be its limit law. We will apply Theorem 3.1 to the function $F = F(u, v, w)$ assuming $q = r = s = 1$. The operator $\mathcal{L}$ of (3.4) is represented by

$$\mathcal{L}F(u, v, w) = \frac{1}{2}\left(\pi(E) + 2V(E, E) - \pi(E)^2\right)\frac{\partial^2 F}{\partial u^2}$$

$$+ \frac{1}{2}\left(\int_G \rho(|\lambda|)^2\pi(d\lambda) - \left(\int_G \rho(|\lambda|)\pi(d\lambda)\right)^2\right.$$

$$+ 2\int\int_{G\times G} \rho(|\lambda|)\rho(|\lambda'|)V(d\lambda, d\lambda') + \nu(G)\left.\right)\frac{\partial^2 F}{\partial w^2}$$

$$+ \left(\int_{G\cap E} \rho(|\lambda|)\pi(d\lambda) - \pi(E)\int_G \rho(|\lambda|)\pi(d\lambda)\right)$$

$$+ \int \int \left( \chi_{E \times G}(\lambda, \lambda') \rho(|\lambda'|) + \chi_{G \times E}(\lambda, \lambda') \rho(|\lambda|) \right) V(d\lambda, d\lambda') \right) \frac{\partial^2 F}{\partial u \partial w}$$

$$+ \int_{0 < |\lambda| \leq M} \left\{ F\left( u, v + \chi_F(\lambda), w + \eta(|\lambda|) \chi_{(0,\epsilon]}(|\lambda|) \chi_G\left(\infty, \frac{\lambda}{|\lambda|}\right) \right) \right.$$

$$\left. - F(u, v, w) - \eta(|\lambda|) \chi_{(0,\epsilon]}(|\lambda|) \chi_G\left(\infty, \frac{\lambda}{|\lambda|}\right) \frac{\partial F}{\partial w} \right\} \mu(d\lambda)$$

$$+ \int_{|\lambda| > M} \left\{ F\left( u, v + \chi_F(\lambda), w + \eta(|\lambda|) \chi_{(0,\epsilon]}(|\lambda|) \chi_G\left(\infty, \frac{\lambda}{|\lambda|}\right) \right) \right.$$

$$\left. - F(u, v, w) \right\} \mu(d\lambda). \tag{4.17}$$

Therefore

$$F(B(t), \tilde{N}(t), K_\epsilon(t)) - \int_0^t \mathcal{L} F(B(s-), \tilde{N}(s-), K_\epsilon(s-)) ds$$

is a martingale by Theorem 3.1. The fact implies that $(B(t), N(t), K_\epsilon(t))$ is a Lévy process and satisfies all assertions of Theorem 2.2. See [7].

Next we apply Theorem 3.1' to the function $F = F(u, v, w, x)$. We will not represent the operator $\mathcal{L}$ explicitly since it is long. However if $F = x$, the theorem tells us that

$$M_t \equiv \varphi_t - \int_0^t (b_M + c + g)(\varphi_s) ds - \sum_{s \leq t} \Delta \varphi_s \Delta N(s, \{|\lambda| > M\}) \tag{4.18}$$

is a locally square integrable martingale. Let $M_t^c$ be its continuous part and $M_t^d$ be its discontinuous part. Their bracket processes are computed from coefficients of $\partial^2 / \partial x^2$ of the operator $\mathcal{L}$. Indeed, we have

$$< M_t^c >= \int_0^t a_1(\varphi_s) ds, \tag{4.19}$$

where

$$a_1(x) = \int_{\mathbf{R}^m} f(x, \lambda)^2 \pi(d\lambda)$$

$$+ 2 \int \int_{\mathbf{R}^m \times \mathbf{R}^m} f(x, \lambda) f(x, \lambda') V(d\lambda, d\lambda')$$

$$+ \int_{S^{m-1}} \bar{h}(x, \infty, \theta)^2 \nu(d\theta). \tag{4.20}$$

Similarly the joint quadratic variation of $M_t^c$ and $B(t, E)$ is obtained from the coefficient of $\partial^2 / \partial u \partial x$ of the operator $\mathcal{L}$ i.e.,

$$< M_t^c, B(t, E) >= \int_0^t a_2(\varphi_s, E) ds, \tag{4.21}$$

where

$$a_2(x, E) = \int_E f(x, \lambda)\pi(d\lambda)$$
$$+ \int\int_{\mathbf{R}^m \times \mathbf{R}^m} \{\chi_E(\lambda)f(x, \lambda') + \chi_E(\lambda')f(x, \lambda)\}V(d\lambda, d\lambda'). \quad (4.22)$$

We have further

$$< M_t^c, K_\varepsilon(t, G) >= \int_0^t a_3(\varphi_{s-}, G)ds, \quad (4.23)$$

where

$$a_3(x, G) = \int_{G \cap \mathbf{R}^m} \rho(|\lambda|)f(x, \lambda)\pi(d\lambda)$$
$$+ \int\int_{\mathbf{R}^m \times \mathbf{R}^m} \{\chi_{G \cap \mathbf{R}^m}(\lambda)\rho(|\lambda|)f(x, \lambda')$$
$$+ \chi_{G \cap \mathbf{R}^m}(\lambda')\rho(|\lambda'|)f(x, \lambda)\}V(d\lambda, d\lambda')$$
$$+ \int_{G \cap \partial \mathbf{R}^m} \bar{h}(x, \infty, \theta)\nu(d\theta). \quad (4.24)$$

We will next consider the bracket processes of discontinuous martingales. It can be computed from the part

$$\int_{0 < |\lambda| \leq M} \left\{ F(y + \hat{F}(y, \lambda)) - F(y) - \sum_i \hat{F}_i(y, \lambda)\frac{\partial F}{\partial y^i}(y) \right\} \mu(d\lambda) \quad (4.25)$$

of the operator $\mathcal{L}$ where $y = (u, v, w, x)$. Set $F = x^2$. Then the above is equal to $\int_{0 < |\lambda| \leq M} |\hat{f}(x, \lambda)|^2\mu(d\lambda)$. Then $(M_t^d)^2 - \int_0^t \left( \int_{0 < |\lambda| \leq M} |\hat{f}(\varphi_{s-}, \lambda)|^2\mu(d\lambda) \right) ds$ is a local martingale, proving that

$$< M_t^d > = \int_0^t \left( \int_{0 < |\lambda| \leq M} |\hat{f}(\varphi_{s-}, \lambda)|^2\mu(d\lambda) \right) ds. \quad (4.26)$$

Next set $F = vx$. Then (4.25) is equal to $\int_{0 < |\lambda| \leq M} \hat{f}(x, \lambda)\chi_F(\lambda)\mu(d\lambda)$. This implies

$$< M_t^d, \tilde{N}(t, F) >= \int_0^t \left( \int_{0 < |\lambda| \leq M} \hat{f}(\varphi_{s-}, \lambda)\chi_F(\lambda)\mu(d\lambda) \right) ds. \quad (4.27)$$

Similarly we have

$$< M_t^d, K_\varepsilon(t, G) >=< M_t^d, K_\varepsilon^d(t, G) >$$
$$= \int_0^t \left( \int_{0 < |\lambda| \leq M} \hat{f}(\varphi_{s-}, \lambda)\eta(|\lambda|)\chi_G(\lambda)\chi_{(0,\varepsilon]}(|\lambda|)\mu(d\lambda) \right) ds. \quad (4.28)$$

Now define $\tilde{M}_t$ by

$$\tilde{M}_t = \int_0^t \int_{\mathbf{R}^m} f(\varphi_{s-}, \lambda) B(ds, d\lambda)$$
$$+ \int_0^t \int_{S^{m-1}} \bar{h}(\varphi_{s-}, \infty, \theta) K_0(ds, d\theta)$$
$$+ \int_0^t \int_{0 < |\lambda| \le M} \hat{f}(\varphi_{s-}, \lambda) \tilde{N}(ds, d\lambda). \tag{4.29}$$

Three terms of the right hand side are orthogonal local martingales. We can compute $< \tilde{M}_t >$ by (2.7), (2.9) and (2.10). Then we get $< M >_t = < \tilde{M} >_t$. We have by (4.21),

$$< M_t, \int_0^t \int_{\mathbf{R}^m} f(\varphi_{s-}, \lambda) B(ds, d\lambda) >$$
$$= < M_t^c, \int_0^t \int_{\mathbf{R}^m} f(\varphi_{s-}, \lambda) B(ds, d\lambda) >$$
$$= \int_0^t \left( \int_{\mathbf{R}^m} f(\varphi_{s-}, \lambda) a_2(\varphi_{s-}, d\lambda) \right) ds$$
$$= \int_0^t \left( \int_{\mathbf{R}^m} f(\varphi_{s-}, \lambda)^2 \pi(d\lambda) \right) ds$$
$$+ 2 \int_0^t \left( \int \int_{\mathbf{R}^m \times \mathbf{R}^m} f(\varphi_{s-}, \lambda) f(\varphi_{s-}, \lambda') V(d\lambda, d\lambda') \right) ds. \tag{4.30}$$

Further by (4.23)

$$< M_t^c, \int_0^t \int_{\mathbf{R}^m} \bar{h}(\varphi_{s-}, \lambda) K_\epsilon(ds, d\lambda) >$$
$$= \int_0^t \int_{\mathbf{R}^m} \bar{h}(\varphi_{s-}, \lambda) d_s < M_s^c, K_\epsilon(s, d\lambda) >$$
$$= \int_0^t \left( \int_{\mathbf{R}^m} \bar{h}(\varphi_{s-}, \lambda) a_3(\varphi_{s-}, d\lambda) \right) ds$$
$$= \int_0^t \left( \int_{\mathbf{R}^m} f(\varphi_{s-}, \lambda)^2 \pi(d\lambda) \right) ds$$
$$+ \int_0^t \left( \int \int_{\mathbf{R}^m \times \mathbf{R}^m} f(\varphi_{s-}, \lambda) f(\varphi_{s-}, \lambda') V(d\lambda, d\lambda') \right) ds$$
$$+ \int_0^t \left( \int_{S^{m-1}} \bar{h}(\varphi_{s-}, \infty, \theta)^2 \nu(d\theta) \right) ds. \tag{4.31}$$

Since $K_0(ds, d\lambda)$ is defined by (1.14), we obtain from (4.30) and (4.31),

$$< M_t^c, \int_0^t \int_{S^{m-1}} \bar{h}(\varphi_{s-}, \infty, \theta) K_0(ds, d\theta) >$$
$$= \int_0^t \left( \int_{S^{m-1}} \bar{h}(\varphi_{s-}, \infty, \theta)^2 \nu(d\theta) \right) ds. \tag{4.32}$$

Also we have by (4.27)

$$< M_t^d, \int_0^t \int_{0<|\lambda|\le M} \hat{f}(\varphi_{s-}, \lambda)\tilde{N}(ds, d\lambda) >$$

$$= \int_0^t \left( \int_{0<|\lambda|\le M} \hat{f}(\varphi_{s-}, \lambda)^2 \mu(d\lambda) \right) ds. \tag{4.33}$$

Consequently from (4.30), (4.32) and (4.33)

$$< M_t, \tilde{M}_t >=< M_t^c, \int_0^t \int_{\mathbf{R}^m} f(\varphi_{s-}, \lambda)B(ds, d\lambda) >$$

$$+ < M_t^c, \int_0^t \int_{S^{m-1}} \bar{h}(\varphi_{s-}, \infty, \theta)K_0(ds, d\theta) >$$

$$+ < M_t^d, \int_0^t \int_{\mathbf{R}^m - \{0\}} \hat{f}(\varphi_{s-}, \lambda)\tilde{N}(ds, d\lambda) >=< M_t > . \tag{4.34}$$

Therefore we get $< M_t - \tilde{M}_t >=< M_t > -2 < M_t, \tilde{M}_t > + < \tilde{M}_t >= 0$, proving $M_t = \tilde{M}_t$. Then (4.18) and (4.29) imply

$$\varphi_t = x + \int_0^t \int_{\mathbf{R}^m} f(\varphi_{s-}, \lambda)B(ds, d\lambda)$$

$$+ \int_0^t \int_{S^{m-1}} \bar{h}(\varphi_{s-}, \infty, \theta)K_0(ds, d\lambda)$$

$$+ \int_0^t (b_M + c + g)(\varphi_{s-})ds$$

$$+ \int_0^t \int_{0<|\lambda|\le M} \hat{f}(\varphi_{s-}, \lambda)\tilde{N}(ds, d\lambda)$$

$$+ \sum_{s\le t} \Delta\varphi_s \Delta N(s, \{|\lambda| > M\}). \tag{4.35}$$

The functions $b_M$ and $b_1$ are related by

$$b_M(x) = b_1(x) + \int_{1<|\lambda|\le M} \hat{f}(x, \lambda)\mu(d\lambda).$$

Therefore

$$\int_0^t b_M(\varphi_{s-})ds + \int_0^t \int_{0<|\lambda|\le M} \hat{f}(\varphi_{s-}, \lambda)\tilde{N}(ds, d\lambda)$$

$$= \int_0^t b_1(\varphi_{s-})ds + \int_0^t \int_{0<|\lambda|\le 1} \hat{f}(\varphi_{s-}, \lambda)\tilde{N}(ds, d\lambda)$$

$$+ \int_0^t \int_{1<|\lambda|\le M} \hat{f}(\varphi_{s-}, \lambda)N(ds, d\lambda). \tag{4.36}$$

Substitute the above to the right hand side of (4.35) and let $M$ tend to infinity. The last term tends to 0. Then we obtain the representation (2.11).

Finally the weak convergence of $\{\bar{P}^n\}_n$ can be shown similarly as in the proof of Theroem 1.1. $\square$

## Bibliography

[1] T. Fujiwara, On the jump-diffusion approximation of stochastic difference equations driven by a mixing sequence, J. Math. Soc. Japan, 42(1990), 353-376.

[2] T. Fujiwara, Limit theorems for random difference equations driven by mixing processes, preprint.

[3] T. Fujiwara and H. Kunita, Limit theorems for stochastic difference-differential equations, Nagoya Math. J., submitted

[4] B.V. Gnedenko and A.N. Kolmogorov, *Limit distributions for some of independent random variables*, Addison-Wesley, 1954.

[5] J. Jacod and A.N. Shiryaev, *Limit theorems for stochastic processes*, Springer-Verlag, 1987.

[6] H. Kunita, *Stochastic flows and stochastic differential equations*, Cambridge Univ. Press, 1990.

[7] H. Kunita, Limits of random measures induced by an array of independent random variables, to appear in Constantin Caratheodory: An international tribute ed. T.M. Rassias.

[8] H. Kunita, Central limit theorems on random measures and stochastic difference equations, to appear in Proc. Conf. Gaussian random fields ed. T. Hida.

[9] H.J. Kushner and Hai-Huang, On the weak convergence of a sequence of general stochastic difference equations, SIAM J. Appl. Math., 40(1981),528-541.

[10] J.D. Samur, On the invariance principle for stationary $\varphi$-mixing triangular array with infinitely divisible limits, Prob. Th. and Rel. Fields, 75(1987), 245-259.

# CHARACTERIZING THE WEAK
## CONVERGENCE OF STOCHASTIC INTEGRALS

by

Thomas G. Kurtz
Dept. of Math & Statistics
University of Wisconsin-Madison
Madison, WI 53706

Philip Protter
Dept. of Math & Statistics
Purdue University
West Lafayette, IN 47907-1399

For $n = 1, 2, \ldots$, let $\Xi_n = (\Omega^n, \mathcal{F}^n, (\mathcal{F}^n_t)_{t \geq 0}, P^n)$ be a filtered probability space, let $H^n$ be càdlàg and adapted, and let $X^n$ be a càdlàg semimartingale. A fundamental question is: Under what conditions does the convergence in distribution of $(H^n, X^n)$ to $(H, X)$ imply that $X$ is a semimartingale and that $\int_0^t H^n_{s-} \, dX^n_s$ converges in distribution to $\int_0^t H_{s-} \, dX_s$? A slightly more general formulation would put conditions on the sequence $X^n$ alone such that the convergence above holds for all such sequences $H^n$. A sequence with this property will be called *good*. To be precise, let $\mathbf{M}^{km}$ denote the real-valued, $k \times m$ matrices, and let $\mathbf{D}_E[0, \infty)$ denote the space of càdlàg, $E$-valued functions with Skorohod topology.

**Definition:** *For $n = 1, 2, \ldots$, let $X^n$ be an $\mathbf{R}^k$-valued, $(\mathcal{F}^n_t)$-semimartingale, and let the sequence $(X^n)_{n \geq 1}$ converge in distribution in the Skorohod topology to a process $X$. The sequence $(X^n)_{n \geq 1}$ is said to be good if for any sequence $(H^n)_{n \geq 1}$ of $\mathbf{M}^{km}$-valued, càdlàg processes, $H^n$ $(\mathcal{F}^n_t)$-adapted, such that $(H^n, X^n)$ converges in distribution in the Skorohod topology on $\mathbf{D}_{\mathbf{M}^{km} \times \mathbf{R}^m}[0, \infty)$ to a process $(H, X)$, there exists a filtration $(\mathcal{F}_t)$ such that $H$ is $(\mathcal{F}_t)$-adapted, $X$ is an $(\mathcal{F}_t)$-semimartingale, and*

$$\int_0^t H^n_{s-} \, dX^n_s \quad \Rightarrow \quad \int_0^t H_{s-} \, dX_s.$$

Jakubowski, Mémin and Pagès [1] give a sufficient condition for a sequence $(X^n)_{n \geq 1}$ to be good called *uniform tightness* or *UT*. This condition uses the characterization of a semimartingale as a good integrator (see e.g., Protter [4]), and requires that it hold uniformly in $n$. On $\Xi_n$, let $\mathcal{H}^n$

---

[1] Supported in part by NSF grant #
[2] Supported in part by NSF grant #DMS-8805595

denote the set of elementary predictable processes bounded by 1: that is,

$$\mathcal{H}^n = \{H^n : H^n \text{ has the representation } H_t^n = H_0^n 1_{\{0\}}(t) + \sum_{i=1}^{p-1} H_i^n 1_{[t_i, t_{i+1})}(t),$$

with $H_i^n \in \mathcal{F}_{t_i}^n$, $p \in \mathbb{N}$, and $0 = t_0 < t_1 < \ldots < t_p < \infty$, $|H_i^n| \leq 1\}$.

**Definition:** *A sequence of semimartingales* $(X^n)_{n \geq 1}$, $X^n$ *defined on* $\Xi_n$, *satisfies the condition UT if for each* $t > 0$ *the set* $\{\int_0^t H_s^n dX_s^n, H^n \in \mathcal{H}^n, n \in \mathbb{N}\}$ *is stochastically bounded.*

**Theorem 1.** *(Jakubowski–Mémin–Pagès). If* $(H^n, X^n)$ *on* $\Xi_n$ *converges in distribution to* $(H, X)$ *in the Skorohod topology and if* $(X^n)_{n \geq 1}$ *satisfies UT, then there exists a filtration* $(\mathcal{F}_t)$ *such that* $X$ *is an* $(\mathcal{F}_t)$-*semimartingale and* $\int H_{s-}^n dX_s^n$ *converges in distribution in the Skorohod topology to* $\int H_{s-} dX_s$. *That is, the sequence* $(X^n)_{n \geq 1}$ *is good.*

The condition UT is sometimes difficult to verify in practice. An alternative condition is given in Kurtz and Protter [2]. To *subtract off* the large jumps in a Skorohod continuous manner, define $h_\delta : [0, \infty) \to [0, \infty)$ by $h_\delta(r) = (1 - \delta/r)^+$, and $J_\delta : \mathbf{D}_{\mathbf{R}^m}[0, \infty) \to \mathbf{D}_{\mathbf{R}^m}[0, \infty)$ by

$$J_\delta(x)(t) = \sum_{0 \leq s \leq t} h_\delta(|\Delta x_s|) \Delta x_s,$$

where $\Delta x_s = x(s) - x(s-)$. Let $\int_0^t |dA_s|$ denote the total variation of the process A from 0 to $t$ ($\omega$ by $\omega$).

**Theorem 2.** *(Kurtz–Protter). Let* $(H^n, X^n)$ *on* $\Xi_n$ *converge in distribution to* $(H, X)$ *on* $\Xi$ *in the Skorohod topology on* $\mathbf{D}_{\mathbf{M}^{km} \times \mathbf{R}^m}[0, \infty)$. *Fix* $\delta > 0$ *(allowing* $\delta = \infty$*) and let* $X^{n,\delta} = X^n - J_\delta(X^n)$. *Then* $X^{n,\delta}$ *is a semimartingale and let* $X^{n,\delta} = M^{n,\delta} + A^{n,\delta}$ *be a decomposition of* $X^{n,\delta}$ *into an* $(\mathcal{F}_t^n)$-*local martingale and an adapted process of finite variation on compacts. Suppose*

(∗) *For each* $\alpha > 0$, *there exist stopping times* $T^{n,\alpha}$ *such that* $P(T^{n,\alpha} \leq \alpha) \leq \frac{1}{\alpha}$ *and* $\sup_n E\{[M^{n,\delta}, M^{n,\delta}]_{t \wedge T^{n,\alpha}} + \int_0^{t \wedge T^{n,\alpha}} |dA_s^{n,\delta}|\} < \infty$.

*Then there exists a filtration* $(\mathcal{F}_t)$ *on* $\Xi$ *such that* $H$ *is* $(\mathcal{F}_t)$-*adapted and* $X$ *is an* $(\mathcal{F}_t)$-*semimartingale, and* $(H^n, X^n, \int H_{s-}^n dX_s^n)$ *converges in distribution to* $(H, X, \int H_{s-} dX_s)$ *in the Skorohod topology on* $\mathbf{D}_{\mathbf{M}^{km} \times \mathbf{R}^m \times \mathbf{R}^k}[0, \infty)$. *That is,* $(X^n)_{n \geq 1}$ *is good.*

It is shown in Kurtz and Protter [2] and also in Mémin and Slominski

[3] that UT and (*) are equivalent sufficient conditions for the sequence $(X^n)$ of (vector-valued) semimartingales to be good.

The next theorem, which is the principal result of this note, shows that the sufficient conditions of Jakubowski–Mémin–Pagès and Kurtz–Protter are also necessary.

**Theorem 3.** *Let $X^n$ be a sequence of vector-valued semimartingales on filtered probability spaces $\Xi_n$. If $X^n$ is a good sequence, then $X^n$ satisfies the condition UT and the condition (*) of Theorem 2.*

**Proof:** Since UT and (*) are equivalent, it suffices to show that UT holds. We treat the case $k = m = 1$ for notational simplicity.

Suppose that $(X^n)_{n \geq 1}$ is a good sequence but that UT does not hold. Then there exists a sequence $(H^n)_{n \geq 1}$, $H^n \in \mathcal{H}_n$, and a sequence $c_n$ tending to $\infty$ such that for some $\varepsilon > 0$,

$$\liminf_{n \to \infty} P^n \{ \int H^n_{s-} \, dX^n_s \geq c_n \} \geq \varepsilon.$$

But this implies

$$(1) \qquad \liminf_{n \to \infty} P^n \{ \int \frac{1}{c_n} H^n_{s-} \, dX^n_s \geq 1 \} \geq \varepsilon$$

as well. Since $|H^n| \leq 1$, we have that $\frac{1}{c_n} H^n$ converges in distribution (uniformly) to the zero process. Since $X^n$ is good by hypothesis, then $\int \frac{1}{c_n} H^n_{s-} \, dX^n_s$ converges in distribution to $\int 0 dX_s = 0$. This contradicts (1), and we have the result. $\qquad \square$

We can employ the argument in the previous proof to show that the property of goodness is inherited through stochastic integration.

**Theorem 4.** *Let $(X^n)_{n \geq 1}$ be a sequence of $\mathbf{R}^m$-valued semimartingales, $X^n$ defined on $\Xi_n$, with $(X^n)_{n \geq 1}$ being good. If $H^n$ defined on $\Xi_n$ are càdlàg, adapted, $\mathbf{M}^{km}$-valued processes, and $(H^n, X^n)$ converges in distribution in the Skorohod topology on $\mathbf{D}_{\mathbf{M}^{km} \times \mathbf{R}^m}[0, \infty)$, then $Y^n_t = \int_0^t H^n_{s-} \, dX^n_s$ is also a good sequence of semimartingales.*

**Proof:** Let $k = m = 1$. By Theorem 1, it is sufficient to show that $(Y^n)$ satisfies UT. Suppose not. Then, as in the proof of Theorem 3, there exists a sequence $(\tilde{H}^n)$ with $\tilde{H}^n \in \mathcal{H}^n$, a sequence $c_n$ tending to $\infty$, and $\varepsilon > 0$ such that

$$\liminf_{n \to \infty} P^n\{\int \tilde{H}^n_{s-} \, dY^n_s \geq c_n\} \geq \varepsilon.$$

and equivalently

$$\liminf_{n \to \infty} P^n\{\int \tilde{H}^n_{s-} H^n_{s-} \, dX^n_s \geq c_n\} \geq \varepsilon.$$

which implies

(2) $$\liminf_{n \to \infty} P^n\{\int \frac{1}{c_n} \tilde{H}^n_{s-} H^n_{s-} \, dX^n_s \geq 1\} \geq \varepsilon$$

But, as before, the goodness of $(X^n)$ implies that the stochastic integral in (2) converges to zero contradicting (2) and verifying UT for $(Y^n)$.     □

As an application of the preceding, we consider stochastic differential equations. Suppose that for each $n$, $U^n$ is adapted, càdlàg and $X^n$ is a semimartingale on $\Xi_n$, $(U^n, X^n) \Rightarrow (U, X)$, and $X^n$ is a good sequence. Let $F^n$, $F$ be, for example, Lipschitz continuous such that $F^n \to F$ uniformly on compacts, and let $Z^n$, $Z$ be the unique solutions of

$$Z^n_t = U^n_t + \int_0^t F^n(Z^n_{s-}) dX^n_s$$

$$Z_t = U_t + \int_0^t F(Z_{s-}) dX_s.$$

Then combining Theorem 4 with Theorem 5.4 of [2] (or similar results in [3] or [5]) yields that $Z^n$ is also a good sequence converging of course to $Z$. The preceding holds as well for much more general $F^n$, $F$; see [2]. In particular by taking $F^n(x) = F(x) = x$, we have that $X^n$ a good sequence implies that $Z^n = \mathcal{E}(X^n)$ is a good sequence, where $\mathcal{E}(Y)$ denotes the stochastic exponential of a semimartingale $Y$.

## REFERENCES

1. Jakubowski, A., Mémin, J., and Pagès, G.: Convergence en loi des suites d'intégrales stochastiques sur l'espace $\mathbf{D}^1$ de Skorokhod. *Probab. Th. Rel. Fields* **81**, 111–137 (1989).

2. Kurtz, T.G. and Protter, P.: Weak limit theorems for stochastic integrals and stochastic differential equations. To appear in *Ann. Probability*.

3. Mémin, J. and Slominski, L.: Condition UT et stabilité en loi des solutions d'équations différentielles stochastiques. Preprint (1990).

4. Protter, P.: *Stochastic Integration and Differential Equations: A New Approach*. Springer–Verlag, New York (1990).

5. Slominski, L.: Stability of strong solutions of stochastic differential equations. *Stoch. Processes Appl.* **31**, 173–202 (1989).

# STOCHASTIC DIFFERENTIAL EQUATIONS INVOLVING POSITIVE NOISE

Tom Lindstrøm, Bernt Øksendal and Jan Ubøe

Dept. of Mathematics, University of Oslo, Box 1053, Blindern,
N-0316 Oslo 3, Norway

## Contents

## §1. Introduction and motivation

An ordinary stochastic differential equation may be viewed as a differential equation where some of the coefficients are subject to white noise perturbations. Perhaps the most celebrated example is the equation

$$(1.1) \qquad \frac{dX_t}{dt} = (r + \alpha \cdot W_t)X_t; \quad X_0 = x,$$

which is the mathematical model for, say, a population $X_t$ whose relative growth rate has the form $r + \alpha W_t$, where $r, \alpha$ are constants and $W_t$ denotes white noise, which represents random fluctuations due to changes in the environment of the population.

Quite often, however, the nature of the noise is not "white" but biased in some sense. For example, if $X_t$ represents the size/weight of a tree in a random environment then a suitable stochastic differential equation for $X_t$ would be

(1.2)                    $$\frac{dX_t}{dt} = (r + \alpha \cdot N_t)X_t; \quad X_0 = x$$

where $N_t$ is a stochastic process representing "positive noise" in some sense.

As a second example, consider incompressible fluid flow in a porous medium. Combining Darcy's law and the continuity equation we end up with the partial differential equation (in $x$ for each $t$)

(1.3)            $$\begin{cases} div(k(x) \triangledown p(x,t)) = -f(x,t) & \text{for } x \in D_t \\ p(x,t) = 0 & \text{for } x \in \partial D_t \end{cases}$$

$k(x) \geq 0$ is the permeability of the medium at the point $x \in \mathbf{R}^3, f(x,t)$ is the given source/sink rate of the fluid at the point $x$ and at time $t$ and $p(x,t)$ is the (unknown) pressure of the fluid, while $D_t$ denotes the set of points $x$ where the fluid has obtained the maximal degree of saturation at time $t$. (See e.g. Øksendal (1990) for details). Because the permeability $k(x)$ is hard to measure and varies rapidly from point to point we find it natural to propose a stochastic mathematical model where $k(x)$ is regarded as a 3-parameter positive noise process.

There are of course several ways to interpret "positive noise process". For example, many papers have been written about (1.3) with the assumption that $k(x)$ is some ordinary stochastic process parametrized by $x$, i.e. $k(x) = k(x,w); w \in \Omega$, assuming only nonnegative values.

In particular, in Dikow and Hornung (1987), the following situation is studied:

(1.4)                    $$k(x) = k_0(x) \cdot exp\xi(x,w),$$

where $k_0(\cdot), \xi(\cdot, w)$ are (uniformly) Hölder $\alpha$-continuous for each $w, \xi$ is a bounded process and $k_0$ is bounded and bounded away from 0. Actually, under these assumptions the operator in (1.3) becomes uniformly elliptic and one can approach the corresponding boundary value problem pathwise (i.e. for each $w$) by known deterministic methods.

In Øksendal (1990), it is shown how to solve (1.3) for more general types of $k$: It suffices that $k(x)$ is an $A_2$-weight in the sense of Muckenhoupt, i.e. that

$$(1.5) \qquad \sup_B (\frac{1}{|B|} \int_B k(x)dx)(\frac{1}{|B|} \int_B \frac{1}{k(x)}dx) < \infty$$

the sup being taken over all balls $B \subset \mathbf{R}^3$, $dx$ denoting Lebesgue measure and $|B| = \int_B dx$ being the volume of $B$. In particular, this allows $k$ to have zeroes.

Here we propose a stochastic approach which is radically different from the methods mentioned above:

We represent the stochastic quantities involved by a suitable *functional of multi-parameter white noise* $W_x$; $x \in \mathbf{R}^n$. For example, in equation (1.3) we put $k(x)$ equal to some positive functional white noise (see §7). Then we look for a solution of the corresponding stochastic partial differential equation in some generalized/distribution sense. Finally explicit information about the solution can then be obtained by taking averages of the distribution solution. The philosophy behind this approach is in a sense similar to the philosophy behind ordinary stochastic differential equations involving white noise: It is better to calculate/solve the equation first and then take averages rather than take the average first and then calculate.

In this report we explain the first step in such a program: Here we will consider only 1-parameter equations, the multi-parameter case will be discussed in future papers.

To summarize, the purpose of this paper is to construct a (1-parameter) noise concept which is general enough to include good models for positive noise from real life situations, yet at the same time allows us to use calculus to solve differential equations involving this noise.

The key concept in our approach is the *functional process*, a distribution valued process which we construct in §4, after giving some background in §2-3. In §5 we introduce the Hermite transform $\mathcal{H}$, which associates to each functional process a (deterministic) complex valued function on

$\mathbf{C}_0^{\mathbf{N}} = \{(z_1, z_2, \cdots); z_j \in \mathbf{C}, \exists N \text{ with } z_j = 0 \text{ for } j > N\}$. This transform is fundamental for our subsequent calculus on functional processes.

Even though a functional process is distribution valued we show in §6 that one may define a functional calculus on such processes, in the sense that one can compose them with analytic functions which do not grow too fast at $\infty$.

In §7 we introduce the concept of a *positive* functional process and we prove that the (Wick) product of two positive processes is again a positive process. This justifies the use of this concept to model non-negative growth in a random environment.

Finally, in §8 we illustrate our method by solving the 1-dimensional version of (1.3). We also give an estimate for the (mean square) error that we make if we replace the noisy equation (1.3) by the equation where $k$ is replaced by its (constant) mean value.

Our main inspiration has come from the works of Ito (1951), Hida (1980) and Kuo (1983), but it has gradually become clear to us that also many other authors have discussed more or less related subjects, for example in connection with mathematical physics (renormalization etc.). We have included those that we know about in the reference list, and apologize to those that should have been mentioned that we are not yet aware of.

After this paper was written we have learned that ideas similar to ours based on the $\mathcal{S}$-transform (which is related to our Hermite transform, see §5) and the Wick product have already been adopted by Kuo and Potthoff (1990).

## §2. The white noise probability space

Since white noise is so fundamental for our construction, we recall some basic facts about this generalized (i.e. distribution valued) process:

For $n = 1, 2, \cdots$ let $\mathcal{S}(\mathbf{R}^n)$ be the Schwartz space of all rapidly decreasing smooth $(C^\infty)$ functions on $\mathbf{R}^n$. Then $\mathcal{S}(\mathbf{R}^n)$ is a Fréchet space under the family of seminorms

$$\|f\|_{N,\alpha} = \sup_{x \in \mathbf{R}^n} (1 + |x|^N)|\partial^\alpha f(x)|,$$

where $N \geq 0$ is an integer and $\alpha = (\alpha_1, \cdots, \alpha_k)$ is a multi-index of non-negative integers $\alpha_j$. The space of *tempered distributions* is the dual $S'(\mathbf{R}^n)$ of $S(\mathbf{R}^n)$, equipped with the weak star topology.

Now let $n = 1$ for the rest of this section and put $S = S(\mathbf{R}), S' = S'(\mathbf{R})$. By the Bochner-Minlos theorem (see e.g. Gelfand and Vilenkin (1964)) there exists a probability measure $\mu$ on $(S', B)$ (where $B = B(S')$ denotes the Borel subsets of $S'$) such that

$$(2.1) \qquad E^\mu[e^{i<\cdot,\phi>}] := \int_{S'} e^{i<\omega,\phi>} d\mu(\omega) = e^{-\frac{1}{2}\|\phi\|^2} \text{ for all } \phi \in S,$$

where $\|\phi\|^2 = \|\phi\|^2_{L^2(\mathbf{R})}$ and $< \omega, \phi > = \omega(\phi)$ for $\omega \in S'$. It follows from (2.1) that

$$(2.2) \qquad \int_{S'} f(<\omega, \phi >) d\mu(\omega) = (2\pi\|\phi\|^2)^{-\frac{1}{2}} \int_{\mathbf{R}} f(t) e^{-\frac{t^2}{2\|\phi\|^2}} dt; \phi \in S,$$

for all $f$ such that the integral on the right converges (It suffices to prove (2.2) for $f \in C_0^\infty(\mathbf{R})$, i.e. $f$ smooth with compact support. Such a function $f$ is the inverse Fourier transform of its Fourier transform $\hat{f}$ and we obtain (2.2) by (2.1) and the Fubini theorem). In particular, if we choose $f(t) = t^2$ we get from (2.2)

$$(2.3) \qquad E^\mu[< \omega, \phi >^2] = \|\phi\|^2; \phi \in S.$$

This allows us to extend the definition of $< \omega, \phi >$ from $\phi \in S$ to $\phi \in L^2(\mathbf{R})$ for a.a. $\omega \in S'$, as follows:

$$(2.4) \qquad < \omega, \phi >:= \lim_{k \to \infty} <\omega, \phi_k > \text{ for } \phi \in L^2(\mathbf{R}),$$

where $\phi_k$ is any sequence in $S$ such that $\phi_k \to \phi$ in $L^2(\mathbf{R})$ and the limit in (2.4) is in $L^2(S', \mu)$.

In particular, if we define

(2.5) $$\tilde{B}_t(\omega) := <\omega, \chi[0,t]>$$

then we see that $(\tilde{B}_t, \mathcal{S}', \mu)$ becomes a Gaussian process with mean 0 and covariance

$$E^\mu[\tilde{B}_t(\omega)\tilde{B}_s(\omega)] = \int_{\mathcal{S}'} <\omega, \chi_{[0,t]}> \cdot <\omega, \chi_{[0,s]}> d\mu(\omega)$$

$$= \int_{\mathbf{R}} \chi_{[0,t]}(x) \cdot \chi_{[0,s]}(x)dx = min(s,t), \text{ using (2.3)}.$$

Therefore $\tilde{B}_t$ is essentially a Brownian motion, in the sense that there exists a t-continuous version $B_t$ of $\tilde{B}_t$:

$$\mu(\{\omega; B_t(\omega) = \tilde{B}_t(\omega)\}) = 1 \quad \text{for all } t.$$

If $u \in L^2(\mathbf{R})$ we define, using (2.4)

(2.5) $$\int_{-\infty}^{\infty} \phi(t)dB_t(\omega) = <\omega, \phi>$$

which coincides with the classical Ito integral if $supp\phi \subset [0, \infty)$.

If we define *the white noise process* $W_\phi$ by

(2.6) $$W_\phi(\omega) = <\omega, \phi> \quad \text{for } \phi \in \mathcal{S}, \omega \in \mathcal{S}'$$

then $W_\phi$ may be regarded as the distributional derivative of $B_t$, in the sense that, if $\phi \in \mathcal{S}$

$$<\frac{d}{dt}B_t(\omega), \phi> = -\int_{-\infty}^{\infty} \phi'(t)B_t(\omega)dt = \int_{-\infty}^{\infty} \phi(t)dB_t(\omega)$$

$$= \lim_{\Delta t_j \to 0} \sum_j \phi(t_j)(B_{t_{j+1}} - B_{t_j}) = \lim_{\Delta t_j \to 0} \sum_j \phi(t_j) <\omega, \chi_{(t_j, t_{j+1}]}>$$

$$= \lim_{\Delta t_j \to 0} <\omega, \sum_j \phi(t_j)\chi_{(t_j, t_{j+1}]}> = <\omega, \phi> = W_\phi(\omega),$$

where the second identity is based on integration by parts for Ito integrals.

## §3. Generalized white noise functionals

In this section we summarize Hida's concept of generalized white noise functionals. They will be a natural starting point for our definition of functional processes in §4. For more details see Hida (1980) or Hida-Kuo-Potthoff-Streit (1991).

By the Wiener-Ito chaos theorem (Ito (1951)), we can write any function $f \in L^2(\mu) \, (= L^2(\mathcal{S}', \mu))$ on the form

$$(3.1) \qquad f = \sum_{n=0}^{\infty} \int f_n dB^{\otimes n},$$

where

$$(3.2) \qquad f_n \in \hat{L}^2(\mathbf{R}^n, dx),$$

i.e. $f_n \in L^2(\mathbf{R}^n, dx)$ and $f_n$ is symmetric (in the sense that $f_n(x_{\sigma_1}, x_{\sigma_2}, \cdots, x_{\sigma_n}) = f(x_1, \cdots, x_n)$ for all permutations $\sigma$ of $(1, 2, \cdots, n))$ and

$$
(3.3) \quad
\begin{aligned}
\int f_n dB^{\otimes n} &= \int_{\mathbf{R}^n} f_n(u) dB_u^{\otimes n} \\
&= n! \int_{-\infty}^{\infty} \left( \int_{-\infty}^{u_n} \cdots \int_{-\infty}^{u_3} \left( \int_{-\infty}^{u_2} f_n(u_1, \cdots, u_n) dB_{u_1} \right) dB_{u_2} \cdots dB_{u_{n-1}} \right) dB_{u_n}
\end{aligned}
$$

for $n \geq 1$, while the $n = 0$ term in (3.1) is just a constant $f_0$.

For a general (non-symmetric) $f \in L^2(\mathbf{R}^n)$ we define

$$(3.4) \qquad \int f dB^{\otimes n} := \int \hat{f} dB^{\otimes n}$$

where $\hat{f}$ is the symmetrization of $f$, defined by

$$(3.5) \qquad \hat{f}(u_1, \cdots, u_n) = \frac{1}{n!} \sum_{\sigma} f(u_{\sigma_1}, \cdots, u_{\sigma_n}),$$

the sum being taken over all permutations $\sigma$ of $(1, 2, \cdots, n)$.

With $f, f_n$ as in (3.1) we have

$$(3.6) \qquad \|f\|_{L^2(\mu)}^2 = \sum_{n=0}^{\infty} n! \|f_n\|_{L^2(\mathbf{R}^n)}^2$$

Note that (3.6) follows from (3.1) and (3.3) by the Ito isometry, since

$$E[(\int_{\mathbf{R}^n} f_n dB^{\otimes n})(\int_{\mathbf{R}^m} f_m dB^{\otimes m})] = 0 \quad \text{for } n \neq m$$

and

$$E[(\int_{\mathbf{R}^n} f_n dB^{\otimes n})^2] = (n!)^2 E[(\int_{-\infty}^{\infty} \cdots (\int_{-\infty}^{u_2} f_n(u_1, \cdots, u_n) dB_1) \cdots dB_{u_n})^2]$$

$$= (n!)^2 \cdot \int_{-\infty}^{\infty} \cdots (\int_{-\infty}^{u_2} f_n^2(u_1, \cdots, u_n) du_1) \cdots du_n = n! \int_{\mathbf{R}^n} f_n^2 dx$$

Here $B_u(\omega); u \geq 0, \omega \in \mathcal{S}'$ is the 1-dimensional Brownian motion associated with the white noise probability space $(\mathcal{S}', \mu)$ as explained in §2.

For $s \in \mathbf{R}$ we define the Sobolev space $H^s = H^s(\mathbf{R}^n)$ by

$$(3.7) \quad H^s = \{\psi \in \mathcal{S}'(\mathbf{R}^n); \|\psi\|^2_{H^s(\mathbf{R}^n)} := \int_{\mathbf{R}^n} |\hat{\psi}(y)|^2 (1 + |y|^2)^s dy < \infty\},$$

where $\hat{\psi}$ denotes the Fourier transform of $\psi$. Then the dual of $H^s$ is simply $H^{-s}$, for all $s \in \mathbf{R}$.

The *Hida test function space* $(L^2)^+ = (L^2)^+(\mu)$ is the subspace of $L^2(\mu)$ consisting of all functions $f \in L^2(\mu)$ of the form

$$(3.8) \quad f = \sum_{n=0}^{\infty} \int f_n dB^{\otimes n}$$

where $f_n \in \hat{H}^{\frac{n+1}{2}}(\mathbf{R}^n)$ and

$$(3.9) \quad \|f\|^2_{(L^2)+} := \sum_{n=0}^{\infty} n! \|f_n\|^2_{\hat{H}^{\frac{n+1}{2}}(\mathbf{R}^n)},$$

$\hat{H}^{\frac{n+1}{2}}(\mathbf{R}^n)$ being the space of symmetric functions in the Sobolev space $H^{\frac{n+1}{2}}(\mathbf{R}^n)$. Note that by the Sobolev imbedding theorem each $f_n$ has a continuous version, so we may - and will - assume from now on that each $f_n$ is continuous.

The *Hida space* $(L^2)^-$ *of generalized white noise functionals* (see Hida (1980), Kuo (1983)) is the dual of the space $(L^2)^+$. Formally we may represent an element $F$ of $(L^2)^-$ as follows

$$(3.10) \qquad F = \sum_{n=0}^{\infty} \int F_n dB^{\otimes n},$$

where

$$(3.11) \qquad F_n \in \hat{H}^{-\frac{n+1}{2}}(\mathbf{R}^n) \text{ for all } n,$$

and

$$(3.12) \qquad \|F\|_{(L^2)^-}^2 = \sum_{n=0}^{\infty} n! \|F_n\|_{\hat{H}^{-\frac{n+1}{2}}(\mathbf{R}^n)}^2,$$

the action $F(f) = <F, f>$ of $F$ on an element $f \in (L^2)^+$ being given by

$$(3.13) \qquad F(f) = \sum_{n=0}^{\infty} n! F_n(f_n)$$

Note that in the special case when $F \in L^2(\mu)$ then (3.13) coincides with the result of taking the inner product of $g$ and $f$ in $L^2(\mu)$, since, by (3.5) and (3.6),

$$(3.14)$$

$$E[Ff] = \sum_{n=0}^{\infty} E[(\int_{\mathbf{R}^n} F_n dB^{\otimes n})(\int_{\mathbf{R}^n} f_n dB^{\otimes n})] = \sum_{n=0}^{\infty} n! \int_{\mathbf{R}^n} F_n(u) f_n(u) du$$

$$= \sum_{n=0}^{\infty} n! F_n(f_n)$$

More generally one can consider the space $(\mathcal{S})^*$ of *generalized white noise functionals* consisting of elements of the form $\sum \int F_n dB^{\otimes n}$ where $F_n \in \mathcal{S}(\mathbf{R}^n)$, equipped with a certain topology. See Hida-Kuo-Potthoff-Streit (1991).

In addition to the natural vector space structure the space $(L^2)^-$ (and $(\mathcal{S})^*$) also has a multiplication called *Wick multiplication* $\diamond$, defined by

$$(3.15) \qquad (\int_{\mathbf{R}^n} F_n dB^{\otimes n}) \diamond (\int_{\mathbf{R}^m} G_m dB^{\otimes m}) = \int_{\mathbf{R}^{n+m}} F_n \hat{\otimes} G_m dB^{\otimes(n+m)},$$

where $F_n \hat{\otimes} G_m$ denotes the symmetrized tensor product of $F_n$ and $G_m$. (If $F_n$ and $G_m$ are functions, this means that we first form the usual tensor product

$$F_n \otimes G_m(z) = F_n \otimes G_m(x_1, \cdots, x_n, y_1, \cdots, y_m)$$
$$= F_n(x_1, \cdots, x_n) G_m(y_1, \cdots, y_m)$$

and then symmetrize the result, so that

$$F_n \hat{\otimes} G_m(z) = \frac{1}{(n+m)!} \sum_\sigma F_n \otimes G_m(z_{\sigma_1}, \cdots, z_{\sigma_{n+m}}),$$

the sum being taken over all permulations $\sigma$ of $(1, \cdots, n+m)$. This definition extends in the usual way to tempered distributions $F_n, G_m$).

Note that if $F_n \in \hat{H}^{-\frac{n+1}{2}}(\mathbf{R}^n), G_m \in \hat{H}^{-\frac{m+1}{2}}(\mathbf{R}^m)$ with $n \le m$ then

$$F_n \hat{\otimes} G_m \in \hat{H}^{-\frac{m+1}{2}}(\mathbf{R}^{n \times m})$$

and

(3.16)    $\|F_n \hat{\otimes} G_m\|_{\hat{H}^{-\frac{m+1}{2}}(\mathbf{R}^{n \times m})} \le \|F_n\|_{\hat{H}^{-\frac{n+1}{2}}(\mathbf{R}^n)} \cdot \|G_m\|_{\hat{H}^{-\frac{m+1}{2}}(\mathbf{R}^m)}$

Therefore we can extend in a natural way the product given in (3.13) to work for more general $F \in (L^2)^-, G \in (L^2)^-$ by defining

$$F \diamond G = \lim_{k \to \infty} \left( \sum_{n=0}^k \int F_n dB^{\otimes n} \right) \diamond \left( \sum_{m=0}^k \int G_m dB^{\otimes m} \right)$$

(3.17)

$$= \sum_{n,m=0}^\infty \int F_n \hat{\otimes} G_m dB^{\otimes(n+m)}.$$

In view of (3.16) we see that $F \diamond G \in (L^2)^-$ if (for example)

(3.18)    $\displaystyle\sum_{k=0}^\infty k \cdot k! \sum_{n+m=k} \|F_n\|^2_{\hat{H}^{-\frac{n+1}{2}}(\mathbf{R}^n)} \cdot \|G_m\|^2_{\hat{H}^{-\frac{m+1}{2}}(\mathbf{R}^m)} < \infty.$

## Remarks

1) In the literature the symbol $: FG :$ is often used for the Wick product $F \diamond G$. For more details on the connection between $::$ and $\diamond$ see the remark following Theorem 5.1.

2) Several motivations exist for this definition of multiplication of functional processes. We give some of them here:

a) First note that multiplication of a deterministic quantity with a random one reduces to the usual multiplication. Going one step further, one might say that the Wick product can be regarded as the product one would obtain if the two stochastic factors were independent. In general our quantities are of course dependent, but in a sense we ignore this when we form the Wick product. This point of view is related to the situation in thermodynamics for example, where technically the path of a single particle of course depends on the paths of the others, but the connection is so weak/chaotic that it is not unreasonable to assume that the effect of it averages to zero in time. The Wick product may be viewed as an axiomatic way of stating such a principle. Moreover, this special product has the technical advantage of preserving the martingale property. And - as shown in Lindstrøm, Øksendal and Ubøe (1991) - the product provides a new approach to the Ito calculus.

b) Wick multiplication can also be motivated from a renormalization point of view: Arguing heuristically we see that white noise $W_t$ may be represented as an element of $(L^2)^-$ as follows

$$W_t = \lim_{\Delta t \to 0} \frac{1}{\Delta t} \int_t^{t+\Delta t} dB_u = \int_{\mathbf{R}} \delta_t(u)dB_u,$$

where $\delta_t \in H^{-1}(\mathbf{R})$ is the Dirac measure at $t$.

Therefore one should have, by Ito's formula,

$$W_t^2 \approx (\frac{1}{\Delta t} \int_t^{t+\Delta t} dB_u)^2 = \frac{2}{(\Delta t)^2} \int_t^{t+\Delta t} (\int_t^v dB_u)dB_v) + \frac{1}{\Delta t}$$

The additive renormalization of this is

$$2 \int_t^{t+\Delta t} (\int_t^v \frac{dB_u}{\Delta t}) \frac{dB_v}{\Delta t} \to \int \int \delta_t(u)\delta_t(v)dB_u dB_v \text{ as } \Delta t \to 0$$

which motivates the definition

$$\int \delta_t dB \diamond \int \delta_t dB = \int \delta_t \otimes \delta_t dB^{\otimes 2},$$

in agreement with (3.15).

3) It should be pointed out that the argument above, as well as the notation

$$(3.19) \qquad \int_{\mathbf{R}^n} F_n dB^{\otimes n} \quad \text{for} \quad F_n \in \hat{H}^{-\frac{n+1}{2}}(\mathbf{R}^n)$$

from the general representation (3.10) of elements in $(L^2)^-$, is just formal: (3.19) is not defined as an Ito integral. Nevertheless, (3.19) - slightly modified and generalized - makes sense as a distribution valued process and this is the starting point for the type of *generalized white noise functional processes* (functional processes, for short) which we now construct.

## §4. Functional processes

Let $s \in \mathbf{R}$ and suppose $F \in H^{-s}(\mathbf{R}^n; L^2(\mathbf{R}^n))$ is an $L^2$-valued $H^{-s}$-distribution, i.e. $F$ is a linear map from $H^s(\mathbf{R}^n)$ into $L^2(\mathbf{R}^n)$ such that there exists $K < \infty$ with

$$(4.1) \qquad \| < F, \phi > \|_{L^2} \leq K \cdot \|\phi\|_{H^s} \text{ for all } \phi \in H^s$$

where $< F, \phi >:= F\phi := F(\phi)$.

Then

$$(4.2) \qquad Y\phi(\omega) = \int_{\mathbf{R}^n} F\phi(u)dB_u^{\otimes n}(\omega)$$

is defined for *a.a.$\omega$* as an $L^2(\mathcal{S}')$-limit in the usual way for each $\phi \in \mathcal{S} = \mathcal{S}(\mathbf{R}^n)$. Thus $Y$ is a *random linear functional* on $H^s$, in the sense that

$$Y(c_1\phi_1 + c_2\phi_2)(\omega) = c_1 Y\phi_1(\omega) + c_2 Y\phi_2(\omega) \text{ for a.a.}\omega,$$

for all $\phi_1, \phi_2 \in H^s$ and all $c_1, c_2 \in \mathbf{R}$.

It follows that $Y$ has a version with values in $H^{-r}$, if

$$r > s + \frac{n}{2}.$$

See Walsh (1984), theorem 4.1. In the following we will assume that this $H^{-r}$ version is chosen, so that we may regard

$$(4.3) \qquad Y(\cdot, \omega) = \int_{\mathbf{R}^n} F(\cdot) dB^{\otimes n}(\omega) \text{ with } F \in H^{-s}(\mathbf{R}^n; L^2(\mathbf{R}^n))$$

as an $H^{-r}(\mathbf{R}^n)$-valued stochastic process.

For notational simplicity we put

$$(4.4) \qquad\qquad H^{-\infty} = \bigcup_{k=1}^{\infty} H^{-k}$$

so that if $F \in H^{-\infty}$ then $F \in H^{-k}$ for some $k$ and then we write

$$\|F\|_{H^{-\infty}} = \|F\|_{H^{-k}}.$$

**DEFINITION 4.1.** A functional process $\{X(\cdot, \omega)\}_{\omega \in \mathcal{S}'}$ is a sum of distribution valued processes of the form

$$(4.5) \qquad X_\phi(\omega) = X(\phi, \omega) = \sum_{n=0}^{\infty} \int_{\mathbf{R}^n} F^{(n)}(\phi^{\otimes n}) dB^{\otimes n}(\omega); \phi \in \mathcal{S}, \omega \in \mathcal{S}'$$

where

$$F^{(n)}(\cdot) \in H^{-\infty}(\mathbf{R}^n; L^2(\mathbf{R}^n)) \text{ for all } n \geq 1$$

and

$$F^{(0)}(\cdot) \in H^{-\infty}(\mathbf{R}).$$

Moreover, we assume that

$$(4.6) \qquad E[|X(\phi, \omega)|^2] = \sum_{n=0}^{\infty} n! \int_{\mathbf{R}^n} < F^{(n)}, \phi^{\otimes n} >^2 (u) du < \infty$$

for all $\phi \in \mathcal{S}$ with $\|\phi\|_{L^2}$ sufficiently small.

To make the notation more suggestive we often write the functional process $X(\phi, \omega)$ on the form

$$(4.7) \qquad X_t(\omega) = \sum_{n=0}^{\infty} \int_{\mathbf{R}^n} F_{t,\cdots,t}^{(n)}(u) dB_u^{\otimes n}(\omega) = \sum_{n=0}^{\infty} \int_{\mathbf{R}^n} F_t^{(n)} dB^{\otimes n},$$

where each $F_t^{(n)}(u)$ is really an $L^2$-valued distribution in the $t$-variable, $t = (t_1, \cdots, t_n)$. The distributional derivative of $X_t$ with respect to $t$ is then defined by

$$(4.8) \qquad \frac{dX_t}{dt}(\omega) = \sum_{n=0}^{\infty} \int_{\mathbf{R}^n} \frac{d}{dt} F_{t,\cdots,t}^{(n)}(u) dB_u^{\otimes n}(\omega)$$

where

$$\frac{d}{dt} F_{t,t,\cdots,t}^{(n)} = \left( \sum_{j=1}^{n} \frac{\partial F^{(n)}}{\partial x_j} \right)_{x=(t,\cdots,t)},$$

$\frac{\partial}{\partial x_j}$ denoting the usual distributional derivative with respect to $x_j$, i.e.

$$< \frac{\partial F^{(n)}}{\partial x_j}, \psi >= - < F^{(n)}, \frac{\partial \psi}{\partial x_j} > \text{ for } \psi = \psi(x_1, \cdots, x_n) \in \mathcal{S}(\mathbf{R}^n).$$

**EXAMPLE 4.2.** The *white noise process* $W_t$ can be represented as a functional process as follows:

$$(4.9) \qquad W_t = \int_{-\infty}^{\infty} \delta_t(u) dB_u$$

where $\delta_t(u)$ is the usual Dirac measure, i.e.

$$< \delta_t(u), \phi(t) >= \phi(u)$$

To see this note that, according to the definition above, (4.9) means that

$$(4.10) \qquad W_\phi(\omega) = \int \phi(u) dB_u(\omega)$$

which is just a reformulation of (2.6).

**EXAMPLE 4.3.** By the previous example together with (3.15) we can represent *the square of white noise* as follows:

$$(4.11) \qquad W_t^{\diamond 2} = W_t \diamond W_t = \int_{\mathbf{R}^2} \delta_t(u)\delta_t(v)dB_u dB_v,$$

This means that as a distribution valued process $W_t^{\diamond 2}$ can be written

$$W^{\diamond 2}(\phi \otimes \psi, \omega) = \int_{\mathbf{R}^2} \phi(u)\psi(v)dB_u dB_v(\omega); \phi, \psi \in \mathcal{S}$$

In particular,

$$W^{\diamond 2}(\phi \otimes \phi, \omega) = \int_{\mathbf{R}^2} \phi(u)\phi(v)dB_u dB_v(\omega); \phi \in \mathcal{S}$$

so a more correct notation for $W_t^{\diamond 2}$ would be $W_{t,t}^{\diamond 2}$.

**EXAMPLE 4.4.** The functional process

$$X_t(\omega) = \int 1_t(u)1_t(v)dB_u dB_v = B_t^{\diamond 2}(\omega)$$

where

$$1_t(u) = \begin{cases} 0 & \text{if} \quad u > t \text{ or } u < 0 \\ 1 & \text{if} \quad 0 \le u \le t \end{cases}$$

has a derivative equal to

$$\frac{dX_t}{dt} = \int (1_t(u)\delta_t(v) + \delta_t(u)1_t(v))dB_u dB_v$$
$$= 2\int 1_t \hat{\otimes} \delta_t dB^{\otimes 2} = 2(\int 1_t dB) \diamond (\int \delta_t dB)$$
$$= 2B_t \diamond W_t,$$

which is in agreement with usual (not Ito) differentiation rules.

## §5. The Hermite transform $\mathcal{H}$

We now show that to a given functional process $X_\phi$ we can associate an (a.e. defined) function

$$\mathcal{H}(X_\phi) := \tilde{X}_\phi : \mathbf{C}_0^{\mathbf{N}} \to \mathbf{C}$$

where $C = \{x + iy; x, y \in \mathbf{R}\}$ denotes the set of complex numbers and $\mathbf{C}_0^{\mathbf{N}} = \{(z_1, z_2, \cdots, ); z_j \in \mathbf{C}, \exists N \text{ with } z_j = 0 \text{ for } \forall j > N\}$. This connection will be fundamental for the rest of this paper.

Fix an orthonormal family $\{\zeta_k\}_{k=1}^{\infty}$ in $L^2(\mathbf{R})$. Let

$$(5.1) \qquad\qquad X_\phi = \sum_{n=0}^{\infty} \int F_\phi^{(n)}(u)dB^{\otimes n}$$

be a functional process, and keep $\phi \in \mathcal{S}$ fixed. Using multi-index notation each function $F_\phi^{(n)}(u_1, \cdots, u_n)$ may be (uniquely) written

$$F_\phi^{(n)}(u_1, \cdots, u_n) = \sum_{|\alpha|=n} c_\alpha^{(n)} \zeta^{\otimes\alpha}(u_1, \cdots, u_n)$$

for suitable constants $c_\alpha^{(n)} = c_\alpha^{(n)}(\phi)$, where $\alpha = (\alpha_1, \cdots, \alpha_m), |\alpha| = \alpha_1 + \cdots + \alpha_m$ and

$$\zeta^{\otimes\alpha} = \zeta_1^{\otimes\alpha_1} \otimes \zeta_2^{\otimes\alpha_2} \otimes \cdots \otimes \zeta_m^{\otimes\alpha_m}$$

This gives the (unique) representation

$$(5.2) \qquad X_\phi = \sum_{n=0}^{\infty} \sum_{|\alpha|=n} c_\alpha^{(n)} \int \zeta^{\otimes\alpha}dB^{\otimes n} = \sum_\alpha c_\alpha \int \zeta^{\otimes\alpha}dB^{\otimes|\alpha|}$$

where $c_\alpha^{(n)}(\cdot) \in H^{-\infty}(\mathbf{R}^n)$ for $n \geq 1, c_\alpha^{(0)}(\cdot) \in H^{-\infty}(\mathbf{R})$.

By symmetrization of $\zeta^{\otimes\alpha}$ and use of the Ito isometry we see that with such a representation we have

$$(5.3) \qquad\qquad E[X_\phi^2] = \sum_\alpha \alpha! c_\alpha^2,$$

where $\alpha! = \alpha_1! \alpha_2! \cdots \alpha_m!$ if $\alpha = (\alpha_1, \cdots, \alpha_m)$.

The iterated Ito integrals on the right of (5.2) can be computed using the Hermite polynomials $h_n$ defined by

$$h_n(x) = (-1)^n e^{\frac{x^2}{2}} \cdot \frac{d^n}{dx^n}(e^{-\frac{x^2}{2}})$$

The result is the following:

**THEOREM 5.1** (Ito (1951), p. 162)

$$(5.4) \qquad \int \zeta_1^{\otimes \alpha_1} \otimes \zeta_2^{\otimes \alpha_2} \otimes \cdots \otimes \zeta_m^{\otimes \alpha_m} dB^{\otimes |\alpha|} = \prod_{j=1}^{m} h_{\alpha_j}(x_j) := h_\alpha(x)$$

where $x_j = \int \zeta_j(u) dB_u; j = 1, 2, \cdots, x = (x_1, x_2, \cdots)$.

**Remark.** Using Theorem 5.1 we can now explain more precisely the relation between the multiplication $\diamond$ defined by (3.15) and the classical Wick product:

If $X$ is a real Gaussian random variable with mean 0 and variance $E[X^2] = \sigma^2$ then the k'th Wick power of $X$ is defined by

$$: X^k := \sigma^k h_k(\frac{X}{\sigma})$$

(See e.g. Simon (1974) p.11).

For example, if $X = \int_{\mathbf{R}} f(t) dB_t$ with $f \in L^2(\mathbf{R})$ then

$$: X^k := \|f\|^k h_k(\frac{X}{\|f\|}) \text{ where } \|f\| = \|f\|_{L^2}.$$

On the other hand, applying Theorem 5.1 to $\alpha = \alpha_1 = k, \zeta_1 = \|f\|^{-1} f$, we see that

$$X^{\diamond k} = (\|f\| \cdot \int \|f\|^{-1} f dB)^{\diamond k} = \|f\|^k h_k(\frac{X}{\|f\|})$$

Thus

$$: X^k := X^{\diamond k} \quad \text{for such } X.$$

By polarization we obtain, if $Y = \int g(t) dB$ with $g \in L^2(\mathbf{R})$,
$$2 : XY :=: (X + Y)^2 : - : X^2 : - : Y^2 := 2X \diamond Y.$$

However, if we more generally consider two random variables $U, V$ of the form

$$U = \int_{\mathbf{R}^n} f_n dB^{\otimes n}, V = \int_{\mathbf{R}^m} g_m dB^{\otimes m} \text{ with } f_n \in L^2(\mathbf{R}^n), g_m \in L^2(\mathbf{R}^m),$$

then by the definition on p.12 in Simon (1974) we have

$$: UV := UV - E[UV]$$

This is not necessarily the same as $U \diamond V$. For example, if

$$U = \int_{\mathbf{R}^2} f^{\otimes 2} dB^{\otimes 2}, V = \int_{\mathbf{R}} f dB \text{ with } f = \chi_{[0,t]},$$

then $U = V^{\diamond 2}$ so by the above

$$U \diamond V = V^{\diamond 3} = \|f\|^3 h_3(\frac{V}{\|f\|}) = V^3 - 3\|f\|^2 V = B_t^3 - 3tB_t,$$

while

$$: UV := (B_t^2 - t)B_t - E[(B_t^2 - t)B_t] = B_t^3 - tB_t.$$

Nevertheless, we have found the relation between $\diamond$ and $: :$ so close that we feel that it is natural to use the name Wick product for $\diamond$.

The next representation of Hermite polynomials is well known:

(5.5) $$h_k(x) = \frac{1}{\sqrt{2\pi}} \int\limits_{-\infty}^{\infty} (x + iy)^k e^{-\frac{1}{2}y^2} dy \quad k = 1, 2, \cdots$$

Define a measure $\lambda$ on the product $\sigma$-algebra on $\mathbf{R}^{\mathbf{N}}$ by
(5.6)
$$\int f(y) d\lambda(y) = \int\limits_{-\infty}^{\infty} \cdots (\int\limits_{-\infty}^{\infty} (\int\limits_{-\infty}^{\infty} f(y_1, \cdots, y_n) e^{-\frac{1}{2}y_1^2} \frac{dy_1}{\sqrt{2\pi}}) e^{-\frac{1}{2}y_2^2} \frac{dy_2}{\sqrt{2\pi}})$$
$$\cdots e^{-\frac{1}{2}y_n^2} \frac{dy_n}{\sqrt{2\pi}}$$

if $f : \mathbf{R}^{\mathbf{N}} \to \mathbf{R}$ is a bounded function depending only on finitely many variables $y_1, \cdots, y_n$. (The formula (5.6) defines $\lambda$ as a premeasure on the algebra generated by finite products of sets in $\mathbf{R}$ and so $\lambda$ has a unique extension to the product $\sigma$-algebra (Folland (1984)).

Combining (5.3)-(5.6) we obtain

(5.7)
$$X_\phi = \sum_{n=0}^{\infty} (\int \sum_{|\alpha|=n} c_\alpha \zeta^{\otimes \alpha}) dB^{\otimes n} = \sum_{n=0}^{\infty} \sum_{|\alpha|=n} c_\alpha \int \zeta^{\otimes \alpha} dB^{\otimes n}$$
$$= \sum_{n=0}^{\infty} \sum_{|\alpha|=n} c_\alpha [\int (x + iy)^\alpha d\lambda(y)]_{x=\int \zeta dB}$$

where $(x + iy)^\alpha = (x_1 + iy_1)^{\alpha_1}(x_2 + iy_2)^{\alpha_2} \cdots (x_m + iy_m)^{\alpha_m}$ if
$\alpha = (\alpha_1, \cdots, \alpha_m)$ and we evaluate the right hand integrals at $x_j = \int \zeta_j dB$,
$j = 1, 2, \cdots$.

**DEFINITION 5.2** Let $X_\phi = \sum_{n=0}^{\infty} \sum_{|\alpha|=n} c_\alpha \int \zeta^{\otimes\alpha} dB^{\otimes n}$ be a functional process represented as in (5.2). Then the *Hermite transform* (or $\mathcal{H}$-*transform*) of $X_\phi$ is the formal power series in infinitely many complex variables $z_1, z_2, \cdots$ given by

$$(5.8) \qquad \mathcal{H}(X_\phi)(z) = \tilde{X}_\phi(z) = \sum_{n=0}^{\infty} \sum_{|\alpha|=n} c_\alpha z^\alpha = \sum_\alpha c_\alpha z^\alpha$$

where $z = (z_1, z_2, \cdots)$, $z^\alpha = z_1^{\alpha_1} \cdots z_m^{\alpha_m}$ if $\alpha = (\alpha_1, \cdots, \alpha_m)$ and $c_\alpha$ is defined by (5.3).

*Remark.* It turns out that there is a close relationship between the Hermite transform and the $\mathcal{S}$-transform as defined, e.g., in Hida-Kuo-Potthoff-Streit (1991); see Theorem 5.7 below.

We claim that the sum in (5.8) converges for all $z \in \mathbf{C}_0^N$, where

$$\mathbf{C}_0^N = \{(z_1, z_2, \cdots); \text{ only finitely many of the } z_j's \text{ are nonzero}\}$$

More generally, $\tilde{X}_\phi(z)$ is defined for all $z = (z_1, z_2, \cdots) \in \mathbf{C}^N$ such that

$$(5.9) \qquad \sum_\alpha \frac{1}{\alpha!}|z^\alpha|^2 < \infty$$

In fact, for such $z$ the series in (5.8) is absolutely convergent, since

$$\sum_{n=0}^{\infty} (\sum_{|\alpha|=n} |c_\alpha| \cdot |z^\alpha|) \leq \sum_{n=0}^{\infty} (\sum_{|\alpha|=n} \alpha! c_\alpha^2)^{\frac{1}{2}} (\sum_{|\alpha|=n} \frac{1}{\alpha!}|z^\alpha|^2)^{\frac{1}{2}}$$

$$\leq (\sum_{n=0}^{\infty} \sum_{|\alpha|=n} \alpha! c_\alpha^2)^{\frac{1}{2}} \cdot (\sum_{n=0}^{\infty} \sum_{|\alpha|=n} \frac{1}{\alpha!}|z^\alpha|^2)^{\frac{1}{2}} < \infty$$

$$= E[X_\phi^2] \cdot (\sum_\alpha \frac{1}{\alpha!}|z^\alpha|^2)^{\frac{1}{2}} < \infty.$$

In particular, if $z = (z_1, \cdots, z_N, 0, 0, \cdots)$ and $|z_k| \leq M$ for all $k$ then in order to check (5.9) it is enough to sum over those $\alpha$ such that $\alpha_j = 0$ for

$j > N$ (otherwise $z_{\alpha_j} = 0$). For such $\alpha$ with $|\alpha| = n$ there must exist an $\alpha_j \geq \frac{n}{N}$ and hence

$$\frac{1}{\alpha!} \leq \frac{1}{[\frac{n}{N}]!},$$

where $[\cdot]$ denotes integer value. This gives

$$\sum_{n=0}^{\infty} \sum_{|\alpha|=n} \frac{1}{\alpha!} |z^\alpha|^2 \leq \sum_{n=0}^{\infty} \frac{1}{[\frac{n}{N}]!} M^{2n} \binom{N+n-1}{n}$$

$$\leq \sum_{n=0}^{\infty} \frac{1}{[\frac{n}{N}]!} M^{2n} (N+n)^N < \infty.$$

We conclude that (5.8) converges absolutely for such $z$. Therefore, if we for all $N$ put

$$z^{(N)} := (z_1, \cdots, z_N, 0, 0, \cdots)) \text{ if } z = (z_1, \cdots, z_N, z_{N+1}, \cdots) \in \mathbf{C}^{\mathbf{N}}$$

and define

(5.10) $$\tilde{X}_\phi^{(N)}(z) = \tilde{X}_\phi(z^{(N)})$$

then the power series for $\tilde{X}_\phi^{(N)}$ converges uniformly on compacts in the variables $z_1, \cdots, z_N$. Hence we have proved:

**LEMMA 5.3** $\tilde{X}_\phi^{(N)}$ is an analytic function in $(z_1, \cdots, z_N) \in \mathbf{C}^N$ for all $N$ and all $\phi$.

From (5.7) we get the following connection between $X_\phi$ and its $\mathcal{H}$-transform:

(5.11) $$X_\phi(\omega) = [\int \tilde{X}_\phi(x+iy)d\lambda(y)]_{x=\int \zeta dB}$$

Since $\tilde{X}_\phi(z)$ is not defined for all $z$, the integral above requires some explanation. It is defined by

$$\int \tilde{X}_\phi(x+iy)d\lambda(y) = \lim_{N\to\infty} \int \tilde{X}_\phi^{(N)}(x+iy)d\lambda(y)$$

It follows from (5.7) that this limit exists (in $L^2(\mu)$) if $x_j = \int \zeta_j dB$.

**EXAMPLE 5.4** Let $X_\phi = W_\phi = \int \phi dB$ be the white noise process. If $\phi$ is fixed, $\rho = \|\phi\|_{L^2} \neq 0$, we may regard $\rho^{-1}\phi$ as the first function $\zeta_1$ in our orthonormal basis $\{\zeta_k\}$. Then

$$W_\phi = \int \rho \zeta_1 dB \quad \text{so} \quad \tilde{W}_\phi(z) = \rho z_1$$

and (5.11) says that

$$\int \phi dB = [\int\limits_{-\infty}^{\infty} \rho \cdot (x+iy)e^{-\frac{1}{2}y^2}\frac{dy}{\sqrt{2\pi}}]_{x=\int \zeta_1 dB}$$

If we don't wish to connect $\phi$ to $\{\zeta_k\}$ we may write

$$\phi = \sum_{k=1}^{\infty}(\phi,\zeta_k)\zeta_k \quad \text{where} \quad (\phi,\zeta_k) = \int\limits_{\mathbf{R}} \phi(t)\zeta_k(t)dt$$

This gives

$$W_\phi = \sum_k (\phi,\zeta_k)\int \zeta_k(u)dB_u$$

and therefore

(5.12) $$\tilde{W}_\phi(z) = \sum_{k=1}^{\infty}(\phi,\zeta_k)z_k$$

Hence

$$\int \tilde{W}_\phi(x+iy)d\lambda(y) = \sum_k (\phi,\zeta_k)\int\limits_{-\infty}^{\infty}(x_k+iy_k)e^{-\frac{1}{2}y_k^2}\frac{dy_k}{\sqrt{2\pi}} = \sum_k (\phi,\zeta_k)x_k,$$

which, after the substitution $x_k = \int \zeta_k dB$ gives

$$\sum_k (\phi,\zeta_k)\int \zeta_k dB = \int (\sum_k(\phi,\zeta_k)\zeta_k(u))dB_u = \int \phi(u)dB_u.$$

The representation (5.11) is useful for explicit computations, but the main reason for the importance of the $\mathcal{H}$-transform is the following property:

**THEOREM 5.5** Let $X_\phi, Y_\phi$ be functional processes such that $X_\phi \diamond Y_\phi$ is a functional process. Then

(5.13) $$\mathcal{H}(X_\phi \diamond Y_\phi) = \mathcal{H}(X_\phi) \cdot \mathcal{H}(Y_\phi)$$

*Remark.* The product on the right hand side of (5.13) is the usual complex product in the complex variables $z_j$ but a *tensor product* in the coefficients, i.e.

$$c_\alpha(\phi)z^\alpha \cdot e_\beta(\psi)z^\beta = (c_\alpha \otimes e_\beta)(\phi \otimes \psi)z^{\alpha+\beta} = c_\alpha(\phi)e_\beta(\psi)z^{\alpha+\beta}$$

*Proof.* It suffices to prove this when $\alpha = (\alpha_1, \cdots, \alpha_m), \beta = (\beta_1, \cdots, \beta_k)$ and

$$X_\phi = \int \zeta^{\otimes\alpha} dB^{\otimes|\alpha|}, \quad Y_\phi = \int \zeta^{\otimes\beta} dB^{\otimes|\beta|}.$$

Then

$$X_\phi \diamond Y_\phi = \int \zeta^{\otimes\alpha} \otimes \zeta^{\otimes\beta} dB^{\otimes|\alpha+\beta|} = \int \zeta^{\otimes(\alpha+\beta)} dB^{\otimes|\alpha+\beta|}$$

and therefore

$$(X_\phi \diamond Y_\phi)^\sim(z) = z^{\alpha+\beta} = z^\alpha \cdot z^\beta = \tilde{X}_\phi(t) \cdot \tilde{Y}_\phi(t), ; \text{as claimed.}$$

**COROLLARY 5.6** Suppose $X_\phi^{\diamond k}$ is a functional process for all $k \leq n$. Then for all $a_k \in \mathbf{R}, k = 1, \cdots, n$, we have

$$\sum_{k=0}^n a_k X_\phi^{\diamond k} = [\int \sum_{k=0}^n a_k \tilde{X}_\phi^k(x + iy)d\lambda(y)]_{x=\int \zeta dB}$$

As promised above we shall end this section by briefly explaining the relationship between the Hermite transform and the $\mathcal{S}$-transform as defined in Hida-Kuo-Potthoff-Streit (1991); i.e.,

$$(5.14) \qquad \mathcal{S}f(\psi) = \int_{\mathcal{S}'} f(\omega + \psi)d\mu(\omega),$$

where $f \in L^2(\mu)$ and $\psi \in \mathcal{S}$. If $f$ has Wiener-Itô decomposition $f = \sum f_n dB^{\otimes n}$, it turns out that

$$(5.15) \qquad \mathcal{S}f(\psi) = \sum_n \int_{\mathbf{R}^n} f_n(u_1, \cdots, u_n)\psi^{\otimes n}(u_1, \cdots, u_n)du,$$

and using this formula, we easily establish the following relationship between the two transforms.

**THEOREM 5.7** If $\mathcal{H}$ denotes the Hermite transform with respect to the orthonormal basis $\zeta_1, \zeta_2, \cdots$, then for any $f \in L^2(\mu)$ and any sequence $z = (z_1, z_2, \cdots) \in \mathbf{C}_0^{\mathbf{N}}$, we have

$$(5.16) \qquad \mathcal{H}(f)(z) = \mathcal{S}f(z_1\zeta_1 + z_2\zeta_2 + \cdots).$$

*Proof:* If we write $f = \sum_n \sum_{|\alpha|=n} c_\alpha \int \zeta^{\hat{\otimes}\alpha} dB^{\otimes n}$, then according to (5.15)

$$\mathcal{S}f(\sum z_i\zeta_i) = \sum_n \sum_{|\alpha|=n} c_\alpha \int_{\mathbf{R}^n} \zeta^{\hat{\otimes}\alpha}(u_1,\cdots,u_n)(\sum z_i\zeta_i)^{\otimes n}(u_1,\cdots,u_n)du,$$

and hence all we have to prove is that

$$(5.17) \qquad \int_{\mathbf{R}^n} \zeta^{\hat{\otimes}\alpha}(u_1,\cdots,u_n)(\sum z_i\zeta_i)^{\otimes n}(u_1,\cdots,u_n)du = z^\alpha$$

Let $i_1, i_2, \cdots, i_k$ be the indices where $\alpha_{i_2}, \alpha_{i_2}, \cdots, \alpha_{i_k}$ are different from zero, and let $\sigma$ run over all permutations of the set $\{1, 2, \cdots, n\}$. The integral in (5.17) can then be written as

$$\frac{1}{n!}\sum_\sigma \sum_{j_1,\cdots,j_n} \int_{\mathbf{R}^n} \zeta_{i_1}(u_{\sigma(1)})\cdots\zeta_{i_k}(u_{\sigma(n)})(z_{j_1}\zeta_{j_1}(u_1)\cdot \ldots \cdot z_j\zeta_{j_n}(u_n))du.$$

Since the sequence $\zeta_1, \zeta_2, \cdots$ is orthonormal, there will for each permutation $\sigma$ be exactly one sequence $j_1, \cdots, j_n$ such that the integral is nonzero; we have to choose $j_1 = \sigma^{-1}(1), \cdots, j_n = \sigma^{-1}(n)$. With this choice of $j'$s, the integral takes the value $z^\alpha$, and (5.17) follows.

*Remark:* In the theory of the $\mathcal{S}$-transform, the ray analytic functions $z \to \mathcal{S}f(z\psi)$ play a fundamental role. The theorem above tells us that we may think of the Hermite transform as an extension of these ideas to several variables (in fact, to infinitely many).

## §6. A functional calculus on functional processes

We now use the Hermite transform to show that for a large class of functions $f$ and functional processes $X_\phi$ we can define a new functional process $f \circ X_\phi$.

The key to such a functional calculus is Corollary 5.6, which can be stated as follows (under the given conditions):

$$(6.1) \qquad \breve{p}(X_\phi) = [\int p(\tilde{X}_\phi)(x+iy)d\lambda(y)]_{x=\int \zeta dB}$$

for every (complex) polynomial $p(z) = \sum_{k=0}^{n} a_k z^k$ with *real* coefficients $a_k$, where $\breve{p}$ indicates that Wick powers of $X_\phi$ are used.

The idea is to extend (6.1) to a larger family of functions than the polynomials. The following result will be useful:

**LEMMA 6.1** Let $f(x_1, x_2, \cdots) \in L^1(\lambda)$. Then

$$E[f(\int \zeta_1 dB, \int \zeta_2 dB, \cdots)] = \int f(x_1, x_2, \cdots)d\lambda(x)$$

*Proof.* It suffices to prove this in the case when $f$ only depends on finitely many variables, and so we may assume that $f \in C_0^\infty(\mathbf{R}^n)$ for some $n$. Then $f$ is the inverse Fourier transform of its Fourier transform $\hat{f}$:

$$f(x_1, \cdots, x_n) = (2\pi)^{-\frac{n}{2}} \cdot \int_{\mathbf{R}^n} \hat{f}(y_1, \cdots, y_n) \cdot e^{i(x,y)} dy$$

Hence, by (2.1)

$$E[f(\int \zeta_1 dB, \cdots, \int \zeta_n dB)] = (2\pi)^{-\frac{n}{2}} \cdot \int_{\mathbf{R}^n} \hat{f}(y_1, \cdots, y_n) E[e^{i<\omega, \sum y_j \zeta_j>}] dy$$

$$= (2\pi)^{-\frac{n}{2}} \int_{\mathbf{R}^n} \hat{f}(y_1, \cdots, y_n) e^{-\frac{1}{2} \sum y_j^2} dy = (2\pi)^{-n} \int_{\mathbf{R}^n} (\int_{\mathbf{R}^n} f(x_1, \cdots, x_n)$$

$$e^{-i(x,y)} dx) e^{-\frac{1}{2}|y|^2} dy = (2\pi)^{-n} \int_{\mathbf{R}^n} f(x_1, \cdots, x_n) \cdot (\int_{\mathbf{R}^n} e^{-i(x,y)-\frac{1}{2}|y|^2} dy) dx$$

$$= (2\pi)^{-n} \int_{\mathbf{R}^n} f(x_1, \cdots, x_n) \cdot (2\pi)^{\frac{n}{2}} \cdot e^{-\frac{1}{2}|x|^2} dx = \int f(x)d\lambda(x), \text{ as claimed.}$$

Here we have used the well-known formula

$$(6.2) \qquad \int_{-\infty}^{\infty} e^{i\alpha t - \beta t^2} dt = (\frac{\pi}{\beta})^{\frac{1}{2}} \cdot e^{-\frac{\alpha^2}{4\beta}}$$

**COROLLARY 6.2** Let $g : \mathbf{R} \to \mathbf{R}$ be measurable and such that $g(X_\phi) \in L^1(\mu)$. Then

$$E[g(X_\phi)] = \int g(\int \tilde{X}_\phi(x + iy)d\lambda(y))d\lambda(x).$$

In particular,

$$E[|X_\phi|^2] \le \int \int |\tilde{X}_\phi|^2 (x + iy)d\lambda(x)d\lambda(y)$$

**DEFINITION 6.3** a) We let $A^2 = A^2(\lambda \times \lambda)$ be the set of all formal power series

$$f(z_1, z_2, \cdots) = \sum_{n=0}^{\infty} (\sum_{|\alpha|=n} c_\alpha z^\alpha)$$

with real coefficients $c_\alpha$ for all $\alpha = (\alpha_1, \cdots, \alpha_m)$, such that

$$f^{(N)}(z) := f(z^{(N)}) \text{ is analytic for each } N$$

(using the notation of (5.10)) and $f \in L^2(\lambda \times \lambda)$, i.e.

$$(6.3) \qquad \begin{aligned} \|f\|^2_{L^2(\lambda \times \lambda)} &:= \int \int |f(x + iy)|^2 d\lambda(x)d\lambda(y) \\ &:= \lim_{N \to \infty} \int \int |f^{(N)}(x + iy)|^2 d\lambda(x)d\lambda(y) < \infty \end{aligned}$$

b) Define $\mathcal{G}$ to be the set of all functional processes $X_\phi$ such that $\tilde{X}_\phi^k \in A^2(\lambda \times \lambda)$ for all $k = 1, 2, 3, \cdots$

Remark:

$\mathcal{G}$ is an algebra (under Wick multiplication).

To prove this it suffices to show that if $X_\phi \in \mathcal{G}$, then $X_\phi^{\diamond 2} \in \mathcal{G}$ also. And this follows from Theorem 5.4.

Our first main result in this section is the following:

**THEOREM 6.4** (Functional calculus)
Let $X_\phi \in \mathcal{G}$ and $f : \mathbf{C} \to \mathbf{C}$ be given and assume that there exist polynomials $p_k : \mathbf{C} \to \mathbf{C}$ with real coefficients such that

$$(6.4) \qquad p_k(\tilde{X}_\phi) \to f(\tilde{X}_\phi) \text{ in } L^2(\lambda \times \lambda) \text{ as } k \to \infty.$$

Then

$$(6.5) \qquad \check{f}(X_\phi) := [\int f(\tilde{X}_\phi)(x + iy)d\lambda(y)]_{x=\int \zeta dB}$$

defines a functional process.

*Proof.* For each $k$ we have that

$$\check{p}_k(X_\phi) = [\int p_k(\tilde{X}_\phi)(x + iy)d\lambda(y)]_{x=\int \zeta dB}$$

is a functional process and, by Lemma 6.1,

$$E[|\check{p}_k(X_\phi) - \check{p}_l(X_\phi)|^2] = E[|\int (p_k(\tilde{X}_\phi) - p_l(\tilde{X}_\phi)) \circ (\int \zeta dB + iy)d\lambda(y)|^2]$$

$$\leq E[\int |p_k(\tilde{X}_\phi) - p_l(\tilde{X}_\phi)|^2 \circ (\int \zeta dB + iy)d\lambda(y)]$$

$$= \|p_k(\tilde{X}_\phi) - p_l(\tilde{X}_\phi)\|_{L^2(\lambda \times \lambda)} \to 0 \text{ as } k, l \to \infty.$$

We conclude that $\{\check{p}_k(X_\phi)\}$ constitute a Cauchy sequence in $L^2(\mu)$. If we write

$$\check{p}_k(X_\phi) = \sum_n \int F_\phi^{(n,k)} dB^{\otimes n} \quad \text{then}$$

$$E[|\check{p}_k(X_\phi) - \check{p}_l(X_\phi)|^2] = \sum_n n! \|F_\phi^{(n,k)} - F_\phi^{(n,l)}\|_{L^2(\mathbf{R}^n)}^2,$$

so for all fixed $n$ we see that $\{F_\phi^{(n,k)}\}_{k=1}^\infty$ converges to a limit $G_\phi^{(n)}$, say, in $L^2(\mathbf{R})$. Hence

$$\lim \check{p}_k(X_\phi) = \sum_n \int G_\phi^{(n)} dB^{\otimes n},$$

which is a functional process.

**EXAMPLE 6.5** *The exponential of white noise* is defined by

$$(6.6) \qquad \mathrm{Exp}(W_t) := \sum_{n=0}^{\infty} \frac{1}{n!} W_t^{\diamond n} := [\int \exp(\tilde{W}_t)(x + iy)d\lambda(y)]_{x = \int \zeta dB}$$

With $\phi \in S$ given we can choose $\zeta_1 = \rho^{-1}\phi$ where $\rho = \|\phi\|_{L^2}$ and this gives (see Ex. 5.3)

$$\tilde{W}_t(z) = \rho z_1 \in L^p(\lambda \times \lambda) \quad \text{for all } p < \infty$$

Hence $W_t \in \mathcal{G}$ and clearly

$$\exp(\tilde{W}_t)(z) = \exp(\rho z_1) = \lim_{k \to \infty} \sum_{n=0}^{k} \frac{1}{n!}(\rho z_1)^n = \lim_{k \to \infty} \sum_{n=0}^{k} \frac{1}{n!}(\tilde{W}_t)^n(z),$$

the limit being taken in $L^2(\lambda \times \lambda)$. So Theorem 6.4 applies to define (6.6) as a functional process.

**EXAMPLE 6.6** The square of white noise is given by

$$Y_\phi = W_\phi^{\diamond 2} = \int_{\mathbf{R}^2} \phi \otimes \phi dB^{\otimes 2},$$

so, with $\phi = \rho\zeta_1$, $\tilde{Y}_\phi(z) = \rho^2 z_1^2$ and hence $Y_\phi \in \mathcal{G}$. However,

$$\exp(\tilde{Y}_\phi) = e^{\rho^2 z_1^2} \text{ which belongs to } L^2(\lambda \times \lambda) \text{ only for } \rho < \frac{1}{2},$$

but for such $\phi$ we see that $\mathrm{Exp}(W_\phi^{\diamond 2})$ is well-defined.

*The inverse $\mathcal{H}$-transform*

The $\mathcal{H}$-transform assigns to each functional process $X_\phi$ a function $\tilde{X}_\phi :$ $\mathbf{C}_0^N \to \mathbf{C}$. Conversely, starting with a function $g : \mathbf{C}_0^N \to \mathbf{C}$ it is useful to be able to find a functional process $X$ whose $\mathcal{H}$-transform is $g$. Clearly the candidate for $X$ is

$$(6.7) \qquad X = [\int g(z)d\lambda(y)]_{x = \int \zeta dB}$$

but the question is for what $g$ this makes sense and defines a functional process.

Suppose that we are given distributions $c_\alpha(\cdot)$ such that
(6.8)  $c_0(\cdot) \in H^{-\infty}(\mathbf{R})$ and
(6.9)  $c_\alpha(\cdot) \in H^{-\infty}(\mathbf{R}^n)$ for each multiindex $\alpha$ with $|\alpha| = n \geq 1$, such that
(6.10)  $\sum_\alpha \alpha! c_\alpha^2(\phi) < \infty$ for all $\phi \in \mathcal{S}$.

Then, as shown by the proof of Lemma 5.3, the function

$$(6.11) \qquad g_\phi(z) := \sum_{n=0}^\infty \sum_{|\alpha|=n} c_\alpha(\phi) z^\alpha$$

is defined and analytic on $\mathbf{C}_0^N$, for each $\phi \in \mathcal{S}$.

**DEFINITION 6.7**
The set of all such functions $g_\phi$ is denoted by $\mathcal{P}$.

**THEOREM 6.8** (Inverse $\mathcal{H}$-transform)
Let $g_\phi \in \mathcal{P}$. Then

$$X_\phi := [\int g_\phi(z) d\lambda(y)]_{z=\int \zeta dB}$$

defines a functional process whose $\mathcal{H}$-transform is $g_\phi$.

We will use $\mathcal{H}^{-1}(g_\phi)$ or $\breve{g}_\phi$ as notation for the inverse Hermite transform. (Strictly speaking this notation is slightly in conflict with the notation of (6.5), but this should not cause any difficulties).

*Proof of Theorem 6.8.* If $g_\phi(z) = \sum_n \sum_{|\alpha|=n} c_\alpha(\phi) z^\alpha$ then by (5.5)

$$\int g_\phi(z) d\lambda(y) = \sum_n \sum_{|\alpha|=n} c_\alpha(\phi) h_\alpha(x),$$

where

$$h_\alpha(x) = h_{\alpha_1}(x_1) h_{\alpha_2}(x_2) \cdots h_{\alpha_m}(x_m) \text{ if } \alpha = (\alpha_1, \cdots, \alpha_m)$$

So by (5.4) we get

$$X_\phi := [\int g_\phi(z) d\lambda(y)]_{z=\int \zeta dB} = \sum_n \sum_{|\alpha|=n} c_\alpha(\phi) \int \zeta^{\otimes\alpha} dB^{\otimes n}$$

Since $E[X_\phi^2] = \sum_\alpha \alpha! c_\alpha^2(\phi) < \infty$, we conclude that $X_\phi$ is a functional process.

## §7. Positive noise

We now return to the concept of *positive noise*, mentioned in the introduction:

**DEFINITION 7.1.** A functional process

$$X_t(\omega) = \sum_{n=0}^\infty \int F_t^{(n)}(u) dB_u^{\otimes n}(\omega)$$

is called a *positive noise* if

(7.1)      $X_\phi(\omega) \geq 0$ a.s. for all $\phi \in \mathcal{S}$ with $\|\phi\|_{L^2}$ sufficiently small.

**EXAMPLE 7.2.**

$$W_t^{\diamond 2}(\omega) = \int \delta_t^{\otimes 2} dB^{\otimes 2}$$

is *not* a positive noise, since

$$W_\psi^{\diamond 2}(\omega) = 2 \int\limits_{-\infty}^{\infty} \int\limits_{-\infty}^{v} \psi(u, v) dB_u(\omega) dB_v(\omega) \text{ for all } \psi \in L^2(\mathbf{R}^2),$$

so choosing e.g. $\psi(u, v) = \chi_{[0,1] \times [0,1]}(u, v)$ we get

$$W_\psi^{\diamond 2}(\omega) = 2 \int_0^1 (\int_0^v dB_u(\omega)) dB_u(\omega) = 2 \int_0^1 B_v(\omega) dB_v(\omega) = B_1^2(\omega) - 1,$$

which is not positive a.s.

However, we have the following:

**EXAMPLE 7.3** The exponential of white noise,

$$N_t := \mathrm{Exp}(W_t) = \sum_{n=0}^\infty \frac{1}{n!} (\int \delta_t^{\otimes n} dB^{\otimes n})$$

is a positive noise.

To see this we use the computation from Example 6.5:

$$N_\phi(\omega) = [\int_{-\infty}^{\infty} \exp(\rho x_1 + i\rho y_1) \cdot e^{-\frac{1}{2}y_1^2} \frac{dy_1}{\sqrt{2\pi}}]_{x_1 = \int \zeta_1 dB}$$

$$= \exp(\int \phi dB) \cdot \frac{1}{\sqrt{2\pi}} \int_{-\infty}^{\infty} e^{i\rho y - \frac{1}{2}y^2} dy,$$

which combined with (6.2) gives

$$(7.2) \qquad \mathrm{Exp}(W_\phi) = \exp(\int \phi dB - \frac{1}{2}\|\phi\|_{L^2}^2); \quad \phi \in \mathcal{S}.$$

In particular, $\mathrm{Exp}(W_\phi) \geq 0$ a.s.

Computer simulations of $\mathrm{Exp}[W_\phi]$ for 3 choices of "averages" $\phi_1, \phi_2, \phi_3$ with supports $[t, t+\frac{1}{10}], [t, t+\frac{1}{25}]$ and $[t, t+\frac{1}{100}]$, respectively are shown on Figure 1 (In all cases $\|\phi_j\|_{L^1} = 1$).

Returning to the examples (1.2) and (1.3) in the introduction, we see that a crucial question for the usefulness of the concept of positive functional processes is the following:

If $X_\phi, Y_\phi$ are two positive functional processes and $X_\phi \diamond Y_\phi$ is defined, is $X_\phi \diamond Y_\phi$ also a positive functional process?

The main result in this section is an affirmative answer to this question, together with a characterization of positiveness for functional processes in terms of positive definiteness of its Hermite transform. This characterization is analogous to Theorem 4.1 in Potthoff (1987), but there the setting, the transform and the methods are different.

**THEOREM 7.4** Let $X$ be a functional process and fix $\phi \in \mathcal{S}$. Then $X_\phi(\omega) \geq 0$ a.s. if and only if

$$(7.3) \qquad g_n(y) = \tilde{X}_\phi^{(n)}(iy)e^{-\frac{1}{2}y^2}; y \in \mathbf{R}_0^N$$

is *positive definite* for all $n$, where $\tilde{X}_\phi^{(n)}(z)$ is defined by (5.10).

### Figure 1

A plot of $Exp[W_\phi]$ showing all values. $Supp[\phi] = [t, t + 1/10]$.

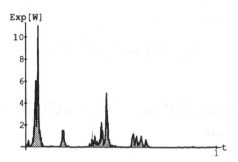

A plot of $Exp[W_\phi]$ showing "fine" structure. $Supp[\phi] = [t, t + 1/25]$.

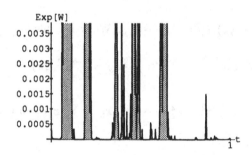

A plot of $Exp[W_\phi]$ showing "micro" structure. $Supp[\phi] = [t, t + 1/100]$. Now the graph is torn completely apart.

Remark: Condition (7.3) means the following:

For all positive integers $m$ and all $y^{(1)}, \cdots, y^{(m)} \in \mathbf{R}_0^N, a = (a_j) \in \mathbf{C}_0^N$ we have

$$(7.4) \qquad \sum_{j,k}^{m} a_j \bar{a}_k g_n(y^{(j)} - y^{(k)}) \geq 0.$$

Proof. Recall that $F(z) := F_n(z_1, \cdots, z_n) := \tilde{X}_\phi^{(n)}(z)$ is analytic in each of the variables $z_1, \cdots, z_n$.

Let

$$(7.5) \qquad \begin{aligned} H_n(x) = H_n(x_1, \cdots, x_n) &= \int \tilde{X}_\phi^{(n)}(x + iy) d\lambda(y) \\ &= \int F(x + iy) e^{-\frac{1}{2}y^2} (2\pi)^{-\frac{n}{2}} dy \end{aligned}$$

where $y = (y_1, \cdots, y_n), dy = dy_1 \cdots dy_n$. We write this as

$$(7.6) \quad e^{-\frac{1}{2}x^2} \cdot \int F(z) e^{\frac{1}{2}z^2} \cdot e^{-ixy} (2\pi)^{-\frac{n}{2}} dy = e^{-\frac{1}{2}x^2} (2\pi)^{-\frac{n}{2}} \int G(z) e^{-ixy} dy,$$

where $z = (z_1, \cdots, z_n), z^2 = z_1^2 + \cdots + z_n^2, xy = \sum_{j=1}^{n} x_j y_j$ and $G(z) = G_n(z) = F(z) e^{\frac{1}{2}z^2}$ is analytic.

Consider the function

$$(7.7) \qquad f(x; \xi) = \int G(x + iy) e^{-i\xi y} dy; x, \xi \in \mathbf{R}^n$$

By the Cauchy-Riemann equations we have

$$\frac{\partial f}{\partial x_1} = \int \frac{\partial G}{\partial x_1} \cdot e^{-i\xi y} dy = \int (-i) \frac{\partial G}{\partial y_1} e^{-i\xi y} dy$$

Now

$$\int_{-\infty}^{\infty} (-i) \frac{\partial G}{\partial y_1} \cdot e^{-i\xi_1 y_1} dy_1 = i \int_{-\infty}^{\infty} G(z) e^{i\xi_1 y_1} (-i\xi_1) dy_1$$

and this gives

$$\frac{\partial f}{\partial x_1} = \xi_1 f(x; \xi)$$

Hence

$$f(x_1, x_2, \cdots, x_n; \zeta) = f(0, x_2, \cdots, x_n; \zeta)e^{\zeta_1 x_1}$$

and similarly for $x_2, \cdots, x_n$. Therefore

(7.8)
$$f(x; \zeta) = f(0; \zeta)e^{\zeta x} = e^{\zeta x} \int G(iy)e^{-i\zeta y}dy$$

We conclude from (7.5)-(7.8) that

(7.9)
$$H_n(x) = \int \tilde{X}_\phi^{(n)}(x + iy)d\lambda(y) = e^{\frac{1}{2}x^2}(2\pi)^{-\frac{n}{2}} \int \tilde{X}_\phi^{(n)}(iy)e^{-\frac{1}{2}y^2}e^{-ixy}dy$$
$$= e^{\frac{1}{2}x^2}\hat{g}_n(x),$$

where $\hat{g}_n(\xi) = (2\pi)^{-n/2} \int g_n(y)e^{-i\xi y}dy$ is the Fourier transform of $g_n$.

Note that $g_n \in S$ and hence $\hat{g}_n \in S$. Therefore we can apply the Fourier inversion to obtain

(7.10)
$$g_n(\xi) = (2\pi)^{-\frac{n}{2}} \int \hat{g}_n(-x)e^{ix\xi}dx$$
$$= \int e^{i\xi x}H_n(x)d\lambda(x), \xi \in \mathbf{R}^n.$$

Hence, if $\xi^{(1)}, \xi^{(2)} \cdots \in \mathbf{R}_0^N$ and $a = (a_j) \in \mathbf{C}_0^N$ we have

$$\sum_{j,k}^{M} a_j\bar{a}_k g_n(\xi^{(j)} - \xi^{(k)}) = \int |\gamma(x)|^2 H_n(x)d\lambda(x),$$

where $\gamma(x)$ is the vector $[a_j e^{i\xi^{(j)}x}]_{1 \leq j \leq m}$.

We conclude that

(7.11)
$$\sum_{j,k}^{m} a_j\bar{a}_k g_n(\xi^{(j)} - \xi^{(k)}) = E[|\gamma(\int \zeta dB)|^2(\int \tilde{X}_\phi^{(n)}(\int \zeta dB + iy)d\lambda(y))]$$
$$\to E[|\gamma(\int \zeta dB)|^2 X_\phi(\omega)] \text{ as } n \to \infty.$$

So if $X_\phi(\omega) \geq 0$ for $a.a.\omega$ we deduce that

(7.12)
$$\lim_{n \to \infty} \sum_{j,k}^{m} a_j\bar{a}_k g_n(\xi^{(j)} - \xi^{(k)}) \geq 0$$

But with $\xi^{(1)}, \cdots \xi^{(m)} \in \mathbf{R}_0^N$ fixed $g_n(\xi^{(j)} - \xi^{(k)})$ becomes eventually constant as $n \to \infty$ so (7.12) is equivalent to (7.4).

Conversely, if $g_n$ is positive definite then $H_n(x) \geq 0$ for $\eta - a.a.x$ and if this holds for all $n$ we have $X_\phi(\omega) \geq 0$ a.s. by (7.5).

**COROLLARY 7.5** Let $X, Y$ be two functional processes such that $X \diamond Y$ is a functional process. If $X \geq 0$ and $Y \geq 0$ then $X \diamond Y \geq 0$.

**Proof.** For $\phi \in S$ consider $\tilde{X}_\phi^{(n)}(iy)e^{-\frac{1}{2}y^2}$ as before and similarly $\tilde{Y}_\phi^{(n)}(iy)e^{-\frac{1}{2}y^2}$. Replacing $\phi$ by $\rho\phi$ where $\rho > 0$ we obtain from Theorem 7.4 that

$$g_n^{(\rho)}(y) := \tilde{X}_\phi(i\rho y)e^{-\frac{1}{2}y^2} \text{ is positive definite,}$$

hence

(7.13) $$\sigma_n(y) := \tilde{X}_\phi^{(n)}(iy)e^{-\frac{1}{2}(\frac{y}{\rho})^2} \text{ is positive definite,}$$

and similarly

(7.14) $$\gamma_n(y) := \tilde{Y}_\phi^{(n)}(iy)e^{-\frac{1}{2}(\frac{y}{\rho})^2} \text{ is positive definite.}$$

Therefore the product $\sigma_n \gamma_n(y) = (\tilde{X}_\phi^{(n)} \cdot \tilde{Y}_\phi^{(n)})(iy)e^{-(\frac{y}{\rho})^2}$ is positive definite.

Choosing $\rho = \sqrt{2}$ this gives that

$$(X \diamond Y)_\phi^{\sim(n)}(iy)e^{-\frac{1}{2}y^2} \text{ is positive definite,}$$

so from Theorem 7.4 we have $X \diamond Y \geq 0$.

**EXAMPLE 7.6** To illustrate Theorem 7.4 let us check that the exponential of white noise, $X_t = \mathrm{Exp}(W_t)$, is positive: Choosing $\phi = \rho\zeta$, we have

$$\tilde{X}_t(z) = \exp(\rho z_1) \text{ so that}$$
$$g(y) = \exp(i\rho y_1 - \frac{1}{2}y^2).$$

Now

$$\sum a_j \bar{a}_k g(y^{(j)} - y^{(k)}) = \sum a_j \bar{a}_k e^{i\rho y_1^{(j)}} \cdot e^{-i\rho y_1^{(k)}} \cdot e^{-\frac{1}{2}(y^{(j)} - y^{(k)})^2}$$
$$= \sum (a_j e^{i\rho y_1^{(j)}}) \overline{(a_k e^{i\rho y_1^{(k)}})} e^{-\frac{1}{2}(y^{(j)} - y^{(k)})^2} \geq 0,$$

since $f(y) = e^{-\frac{1}{2}y^2}$ is positive definite. Hence $g$ is positive definite and therefore $X_t \geq 0$.

**EXAMPLE 7.7** Returning to equation (1.2) in the introduction, we see that Corollary 7.5 is necessary for our concept of a positive noise to make sense in such a model: If for example $r = 0, \alpha = 1$ so that

$$\frac{dX_t}{dt} = N_t \diamond X_t, X_0 = 1$$

with $N_t$ a positive noise, then Corollary 7.5 gives that $\frac{dX_t}{dt}$ is indeed positive if $X_t$ is.

## §8. The solution of stochastic differential equations involving functionals of white noise.

As pointed out in the introduction it is important that our generalized noise concept leads to stochastic differential equations which can be handled mathematically. In this section we show that this is indeed the case. The main idea is to transform the original noise equation into a (deterministic) differential equation for the Hermite transform. By applying the inverse Hermite transform to the solution of this deterministic equation we will - under some conditions - obtain a functional process which solves the original equation.

To illustrate this method we look at the 1-dimensional version of equation (1.3), i.e.

(8.1)
$$(k(t)X_t')' = 0$$

where

$$k(t) = \text{Exp}(\varepsilon W_t), \varepsilon > 0 \text{ constant},$$

is our model for the positive noise $k(t)$, and the product is interpreted as a Wick product $\diamond$. Taking $\mathcal{H}$-transform we get the equation

$$\exp(\varepsilon \tilde{W}_t) \cdot \tilde{X}_t' = C_1 \ (C_1 \text{ constant})$$

From (5.12) we have

$$\tilde{W}_\phi(z) = \sum_j (\phi, \zeta_j) z_j; \ \phi \in \mathcal{S}$$

So

$$\tilde{X}'_\phi(z) = C_1 \exp(-\varepsilon \sum_j (\phi, \zeta_j) z_j)$$

which gives

$$X'_\phi(z) = C_1 \cdot \lim_{k \to \infty} [\int \exp(-\varepsilon \sum_{j=1}^k (\phi, \zeta_j) z_j) d\lambda(y)]_{x = \int \zeta dB}$$

$$(\text{by formula 6.2}) = C_1 \lim_{k \to \infty} [\exp(-\varepsilon \sum_{j=1}^k ((\phi, \zeta_j) x_j + \frac{\varepsilon^2}{2} (\phi, \zeta_j)^2))]_{x = \int \zeta dB}$$

$$= C_1 \cdot \exp(-\varepsilon \int \phi dB - \frac{\varepsilon^2}{2} \|\phi\|_{L^2}^2)$$

In other words,

$$(8.2) \qquad X'_\phi = C_1 \text{Exp}(-\varepsilon W_\phi) = C_1 \sum_{n=0}^\infty \frac{(-1)^n \varepsilon^n}{n!} \int \phi^{\otimes n} dB^{\otimes n},$$

as we would have obtained by direct formal computation from (8.1).

Note that if we define $G : \mathcal{S}(\mathbf{R}^2) \to L^2(\mathbf{R}^2)$ by

$$(8.3) \ < G, \phi \otimes \psi > = \frac{1}{2} \int_0^\infty \{\phi(u+s)\psi(v+s) + \phi(v+s)\psi(u+s)\} ds \cdot \chi_{\{u>0, v>0\}}$$

then if supp $\phi \subset \mathbf{R}^+$, supp $\psi \subset \mathbf{R}^+$ we have

$$< \frac{\partial G}{\partial x_1}, \phi \otimes \psi > = - < G, \phi' \otimes \psi >$$

$$= -\frac{1}{2} \int_0^\infty \{\phi'(u+s)\psi(v+s) + \phi'(v+s)\psi(u+s)\} ds,$$

so

$$< \frac{\partial G}{\partial x_1}, \phi \otimes \phi > = -\frac{1}{2} \int_0^\infty \frac{d}{ds}(\phi(u+s)\phi(v+s)) ds = \frac{1}{2}\phi(u)\phi(v)$$

Hence

(8.4) $$\frac{d}{dt}G_t = (\frac{\partial G}{\partial x_1} + \frac{\partial G}{\partial x_2})_{x=(t,t)} = \delta_t \otimes \delta_t \text{ for } t > 0.$$

Similarly we see that if we more generally define

(8.5)

$$G_\phi^{(n)} :=< G^{(n)}, \phi^{\otimes n} >:= \int_0^\infty \phi(u_1 + s)\phi(u_2 + s)\cdots\phi(u_n + s)ds \cdot \chi_{\{u_i>0,\forall i\}}$$

then

$$\frac{d}{d\phi}G_\phi^{(n)} = \phi^{\otimes n} \text{ if supp } \phi \subset \mathbf{R}^+$$

In other words

(8.6) $$\frac{d}{dt}G_t^{(n)} = \delta_t^{\otimes n} \text{ for } t > 0.$$

Using this in (8.2) we conclude that

(8.7)

$$X_t = X_t^{(\epsilon)} = C_2 + C_1 t + C_1 \cdot \sum_{n=1}^\infty \frac{(-1)^n \epsilon^n}{n!} \int G_t^{(n)}(u)dB_u^{\otimes n} \ (C_2 \text{ constant})$$

is the solution of (8.1).

We want to write this on a more transparent form. A test function $\phi$ should be substituted for $t$, and we then generate a sample path inserting $\phi_t(u) = \phi(u - t)$. This gives

$$X_{\phi_t}^{(\epsilon)} = C_2 + C_1 t + C_1 \sum_{n=1}^\infty \frac{(-1)^n \epsilon^n}{n!} \int G_{\phi_t}^{(u)}(u)dB_u^{\otimes n}$$

Using the definition (8.5) of $G_{\phi_t}^{(n)}$, we get

$$X_{\phi_t}^{(\epsilon)} = C_2 + C_1 t + C_1 \sum_{n=1}^\infty \frac{(-1)^n \epsilon^n}{n!} \times$$

$$\int \int_0^\infty \chi_{[0,\infty)}(u_1)\phi(u_1 + s - t)\cdots\chi_{[0,\infty)}(u_n)\phi(u_n + s - t)dsdB_u^{\otimes n}$$

$$= C_2 + C_1 t + C_1 \int_0^\infty \sum_{n=1}^\infty \frac{(-1)^n \epsilon^n}{n!} \left[\int \chi_{[0,\infty)}(u)\phi_{t-s}(u)dB(u)\right]^{\circ n} ds$$

$$= C_2 + C_1 t + C_1 \int_0^\infty \{\text{Exp}[-\epsilon W_{\chi_{[0,\infty)}\phi_{t-s}}] - 1\} ds$$

If we assume that $\phi$ has support in the interval $[0, h]$, the integral is zero unless $s \leq t + h$. After a change of variables $r = t - s$, we therefore get

$$X_{\phi_t}^{(\varepsilon)} = C_2 + C_1 t + C_1 \int_{-h}^{t} \{\mathrm{Exp}[-\varepsilon W_{\chi_{[0,\infty)}\phi_r}] - 1\} dr$$

If $r \geq 0$, then $\chi_{[0,\infty)}\phi_r = \phi_r$. Hence

$$X_{\phi_t}^{(\varepsilon)} = C_2 + C_1 \int_{-h}^{0} \{\mathrm{Exp}[-\varepsilon W_{\chi_{[0,\infty)}\phi_r}] - 1\} dr + C_1 \int_{0}^{t} \mathrm{Exp}[-\varepsilon W_{\phi_r}] dr$$

The first two terms are just constants. So we get

$$(8.8) \qquad X_{\phi_t}^{(\varepsilon)} = C_1 \int_{0}^{t} \mathrm{Exp}[-\varepsilon W_{\phi_r}] dr + C_2(\varepsilon, \phi, \omega)$$

which is nearly what we would get from a formal solution of the equation (8.1).

Computer simulations of this solution with $C_1 = 1$ and $C_2 = 0$ and various choices of $\varepsilon$ are shown on Figure 2. There we have chosen $\phi(u) = \phi_t(u) = \frac{1}{h} \cdot \chi_{[t,t+h]}(u)$ with $h = 0.1$, so $X_t$ really means $X_{\phi_t}$. The following question is of interest:

How does the stochastic solution $X_t^{(\varepsilon)}$ differ from the no-noise solution $X_t^{(0)} = C_2 + C_1 t$? How big error do we make if we replace the stochastic permeability $k(t) = \mathrm{Exp}(\varepsilon W_t)$ by its average $k_0(t) = 1$? The answer depends heavily on the ratio between $\varepsilon^2$ and the support $h$ of the averaging function $\phi(u) = \frac{1}{h} \cdot \chi_{[t,t+h]}(u)$:

**THEOREM 8.1** For $\varepsilon > 0$ let
(8.9)

$$X_t^{(\varepsilon)} = t + \sum_{n=1}^{\infty} \frac{(-1)^n \varepsilon^n}{n!} \int G_t^{(n)}(u) dB_u^{\otimes n} = \int_{0}^{t} \mathrm{Exp}\,[-\varepsilon W_{\phi_r}] dr + C(\varepsilon, \phi, \omega)$$

## Figure 2

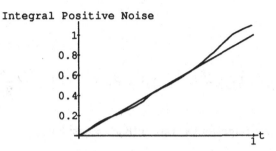

Integral Positive Noise

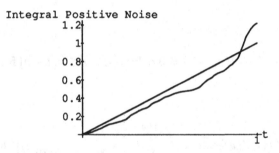

Integral Positive Noise

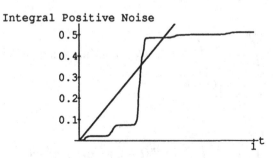

Integral Positive Noise

be the solution of (8.1) with $C_1 = 1, C_2 = 0$. Then

(8.10) $$E[X_t^{(\varepsilon)}] = t$$

and

(8.11) $$\max\{\varepsilon^2(t+h/3), \frac{th}{2}(e^{\frac{\varepsilon^2}{2h}} - 1)\} \leq E[(X_t^\varepsilon - t)^2] \leq \frac{(t+h)h}{2}(e^{\frac{2\varepsilon^2}{h}} - 1)$$

*Proof.* (8.10) is straightforward from (8.9). Consider

(8.12)
$$E[(X_t^{(\varepsilon)} - t)^2] = E[(\sum_{n=1}^{\infty} \frac{(-1)^n \varepsilon^n}{n!} \int G_t^{(n)}(u) dB_u^{\otimes n})^2]$$
$$= \sum_{n=1}^{\infty} \frac{\varepsilon^{2n}}{n!} \int (G_t^{(n)}(u))^2 du$$

Now

$$\phi(u_1 + s) \cdots \phi(u_n + s) \neq 0 \text{ only if } u_i + s \in [t, t+h] \text{ for all } i.$$

Hence

$$\int (G_t^{(n)}(u))^2 du = \int_{\mathbf{R}_+^n} [\int_0^{t+h} h^{-n} \chi_{\{u_i + s \in [t, t+h], \forall i\}} ds]^2 du$$

$$= \int_{[0,t+h] \times \cdots \times [0,t+h]} [\int_0^{t+h} h^{-n} \chi_{\{t - \min_i(u_i) \leq s \leq t - \max_i(u_i) + h\}} ds]^2 du$$

(8.13)
$$= \int_{\substack{0 \\ \max(u_i) - \min(u_i) \leq h \\ \min(u_i) \leq t}}^{t+h} \cdots \int_0^{t+h} h^{-2n}(h - \max_i(u_i) + \min_i(u_i))^2 du$$

$$+ \int_{\substack{0 \\ \max(u_i) - \min(u_i) \leq h \\ \min(u_i) \geq t}}^{t+h} \cdots \int_0^{t+h} h^{-2n}(t - \max_i(u_i) + h)^2 du$$

If $n = 1$, we have by (8.13)

$$(8.14) \quad \int (G_t^{(1)}(u))^2 du = \int_0^t h^{-2} \cdot h^2 du + \int_t^{t+h} h^{-2}(t + h - u_1) du = t + \frac{h}{3}$$

For general $n$, we easily get

$$\underset{\substack{0 \\ \max_i(u_i)-\min_i(u_i) \le \frac{h}{2} \\ u_1 \le t}}{\int^{t+h}} \cdots \underset{0}{\int^{t+h}} h^{-2n} \cdot (\frac{h}{2})^2 du \le \int (G_t^{(n)}(u))^2 du \le$$

$$\underset{\substack{0 \\ \max_i(u_i)-\min_i(u_i) \le h}}{\int^{t+h}} \cdots \underset{0}{\int^{t+h}} h^{-2n} h^2 du$$

This again gives the estimates

$$(8.15) \quad \int (G_t^{(n)}(u))^2 du \ge \int_0^t \int_{u_1-\frac{h}{4}}^{u_1+\frac{h}{4}} \cdots \int_{u_1-\frac{h}{4}}^{u_1+\frac{h}{4}} h^{-2n} \cdot \frac{h^2}{4} du = \frac{th}{2} (\frac{1}{2h})^n$$

$$(8.16) \quad \int (G_t^{(n)}(u))^2 du \le \int_0^{t+h} \int_{u_1-h}^{u_1+h} \cdots \int_{u_1-h}^{u_1+h} h^{-2n} h^2 du = \frac{(t+h)h}{2} (\frac{2}{h})^n$$

If we substitute (8.14), (8.15) and (8.16) in (8.12), we get (8.11).

**Remark.**

It is interesting to note that the estimate (8.11) resembles some properties expected in a real world experiment. Using a very small averaging window (molecule-size?) the world is looking very discontinuous with rapid, chaotic variations. On the other hand, with a very large averaging window (much larger than the object we study), the image get blurred and matter again looks less structured. Somewhere in the middle there should be a scale where we see matter clearly and more or less independent of the size of the averaging window. Comparing with (8.11) we see that

$$\lim_{h \to 0} E[(X_t^{(\varepsilon)} - t)^2] \to \infty \text{ and } \lim_{h \to \infty} E[(X_t^{(\varepsilon)} - t)^2] \to \infty$$

corresponding to the first two properties.

On the other hand, if $\varepsilon^2 << h << t$, we have

$$E[[X_t^{(\varepsilon)} - t]^2] \approx \varepsilon^2 t$$

more or less independent of the size of $h$. If e.g. $\varepsilon = 0.1, t = 100$ then the variance is essentially equal to 1 when $h$ ranges from 0.1 to 10.

*Acknowledgements*

We are grateful to S. Albeverio, P. Malliavin and J. Potthoff for useful communications.

This work is supported by VISTA, a research cooperation between The Norwegian Academy of Science and Letters and Den Norske Stats Oljeselskap A.S. (Statoil).

**REFERENCES**

E. Dikow and U. Hornung (1987):   A random boundary value problem modelling spatial variability in porous media flow. IMA Preprint Series ♯ 309, April 1987, Minneapolis, Minnesota 55455, USA.

G.B. Folland (1984):   Real Analysis. Wiley.

I.M. Gelfand and N. Ya. Vilenkin (1964):   Generalized Functions, Vol. 4: Applications of Harmonic Analysis. Academic Press (English translation).

T. Hida (1980):   Brownian Motion. Springer-Verlag.

T. Hida, H.-H. Kuo, J. Potthoff and L. Streit (1991): White Noise Analysis (Forthcoming book).

K. Ito (1951):   Multiple Wiener integral. J. Math. Soc. Japan 3 (1951), 157-169.

H.-H. Kuo (1983):   Brownian functionals and applications.   Acta Appl. Math. 1, 175-188.

H.-H. Kuo and J. Potthoff (1990): Anticipating stochastic integrals and stochastic differential equations. In T. Hida, H.-H. Kuo, J. Potthoff and L. Streit (editors): White Noise Analysis-Mathematics and Applications. World Scientific, 256-273.

T. Lindstrøm, B. Øksendal and J. Ubøe (1991): Wick multiplication and Ito-Skorohod stochastic differential equations. To appear in S. Albeverio, J.E. Fenstad, H. Holden and T. Lindstrøm (editors): Ideas and Methods in Mathematics and Physics. Proceedings of a conference dedicated to the memory of Raphael Høegh-Krohn. Cambridge University Press.

D. Nualart and M. Zakai (1986): Generalized stochastic integrals and the Malliavin calculus. Prob. Th. Rel. Fields 73, 255-280.

D. Nualart and M. Zakai (1988): Generalized multiple stochastic integrals and the representation of Wiener functionals. Stochastics 23, 311-330.

J. Potthoff (1987): On positive generalized functionals. J. Funct. Anal. 74, 81-95.

B. Simon (1974): The $P(\phi)_2$ Euclidean (Quantum) Field Theory. Princeton Univ. Press.

J.B. Walsh (1986): An introduction to stochastic partial differential equations. In P.L. Hennequin (editor): École d'Été de Probabilités de Saint-Flour XIV-1984, Springer LNM 1180, 265-437.

B. Øksendal (1990): A stochastic approach to moving boundary problems. To appear in M. Pinsky (editor): Diffusion Processes and Related Problems in Analysis. Birkhäuser.

Dept. of Mathematics, University of Oslo

Box 1053, Blindern, N-0316 Oslo 3

NORWAY

# Feeling the Shape of a Manifold with Brownian Motion—The Last Word in 1990

## Mark A. Pinsky

Northwestern University

Dedicated to Henry McKean on the occasion of his sixtieth birthday

## 1. Introduction.

In 1984 [P1] we introduced the theme of inverse problems for Brownian motion on Riemannian manifolds, in terms of the mean exit time from small geodesic balls. Since that time a number of works have appeared on related stochastic problems as well as on some classical, non-stochastic quantities which may be treated by the same methods. Most recently H.R. Hughes [Hu] has shown that in six dimensions one *cannot* recover the Riemannian metric from the exit time distribution, thereby answering in a strong sense the main question posed in [P1].

The general area of "inverse spectral theory" was initiated by Mark Kac in his now famous paper [Ka] on the two-dimensional drumhead. In the intervening years a large literature has developed on inverse spectral problems in higher dimensional Euclidean space and differentiable manifolds; for recent surveys see ([Be],[Br], [Go]). In these approaches one is given the *entire* spectrum of eigenvalues, from which one asks various geometric questions. Our approach, by contrast, is able to obtain strong geometric information from the sole knowledge of the *principal eigenvalue* of a parametric family of geodesic balls (see section 6, below).

It will be our goal here to give an account of the above mentioned developments, focusing on the following topics:

    i) Exit time distribution of Brownian motion
    ii) Exit place distribution of Brownian motion
    iii) Joint distribution of exit time and exit place
    iv) Principal eigenvalue of the Laplacian.

For various reasons we are not pretending a survey of all the recent activity; for example we refer to the work of Stafford [St] which charts out some promising new areas for development.

## 2a. Notations and definitions.

The setting is as follows: $(M, g)$ is a $p$-dimensional Riemannian manifold, not necessarily complete. The Laplacian is defined locally by the coordinate formula

$$\Delta = (1/\sqrt{g})\, \partial/\partial x_i \left(\sqrt{g}\, g^{ij} \partial/\partial x_j\right)$$

with the usual notation for the metric tensor $g_{ij}$, its inverse $g^{ij}$ and its determinant $g$. The Brownian motion process $(X_t, P_x)$ is the (local) diffusion generated by $\frac{1}{2}\Delta$. The exit time from a ball of radius r about $m \in M$ is defined by

$$T_r = \inf\{t > 0 : dist(X_t, m) = r\}.$$

The homogeneous decomposition of the Laplacian in a normal coordinate chart $(x_1, \ldots, x_p)$ about $m \in M$ is written

$$\Delta = \Delta_{-2} + \sum_{k=0}^{\infty} \Delta_k$$

where $\Delta_{-2} = \sum_{i=1}^{p} \partial^2/\partial x_i^2$ is the standard Laplacian of $\mathbf{R}^p$ and $\{\Delta_k\}_{k \geq 0}$ are second-order differential operators whose coefficients are homogeneous polynomials in the normal coordinates $(x_1, \ldots, x_p)$ with the following additional properties:

2a i): $\Delta_k$ maps a homogeneous polynomial of degree $n$ to a homogeneous polynomial of degree $n + k$.

2a ii): The coefficients of $\Delta_k$ depend on the curvature tensor and its derivatives of order $\leq k$.

$\Delta_0$, $\Delta_1$ and $\Delta_2$ were computed in [GP] and $\Delta_3$ was computed in [KO].

The curvature tensor is computed from the Taylor expansion in a normal coordinate chart [G] by the formula

$$g_{ij}(x_1, \ldots, x_p) = \delta_{ij} - \frac{1}{3} \sum_{a,b=1}^{p} R_{iajb}(m) x_a x_b + O(|x|^3) \quad (|x| \downarrow 0).$$

The Ricci tensor is defined by the first contraction:

$$\rho_{ij}(m) = \sum_{a=1}^{p} R_{iaja}(m)$$

and the scalar curvature is defined by the second contraction: $\tau(m) = \sum_{i=1}^{p} \rho_{ii}(m)$. In this notation, we can display the operator $\Delta_0$ as

$$\Delta_0 = (1/3) \sum_{i,a,j,b=1}^{p} R_{iajb}(m) x_a x_b \partial^2 / \partial x_i \partial x_j - (2/3) \sum_{i,a=1}^{p} \rho_{ia}(m) x_a \partial / \partial x_i.$$

In the following sections we'll discuss results on the probability law of the random variables $T_r$ and $X_{T_r}$ for a general manifold. In order to put these in perspective, we first recall the familiar properties of Brownian motion of Euclidean space.

## 2b. Brownian motion of Euclidean space and related asymptotics.

In the Euclidean space $(\mathbf{R}^p, g_0)$ we have the following well-known properties for paths starting at the center $m$ of a ball of radius $r$.

2b i): The exit time $T_r$ has the *Brownian scaling property*: $T_r/r^2 \equiv T_1$ in the sense of the probability law $P_m$,

2b ii): The exit place $X_{T_r}$ is uniformly distributed on the sphere: $X_{T_r} \equiv Leb(\mathbf{S}^{p-1})$ in the sense of the probability law $P_m$,

2b iii): The exit time and exit place are $P_m$-independent random variables: $\forall \alpha > 0, \psi \in C(\mathbf{R}^p)$

$$E_m\left(e^{-\alpha T_r} \psi(X_{T_r})\right) = E_m\left(e^{-\alpha T_r}\right) E_m\left(\psi(X_{T_r})\right),$$

2b iv): The principal eigenvalue of $\Delta$ with zero boundary conditions on the ball of radius $r$ is inversely proportional to the radius:

$$\lambda_1(B_r) = z^2/r^2$$

where $z^2$ is the square of the first positive zero of the Bessel function $J_{(p-2)/2}$.

In the following sections we will describe results which show, for example that the property 2b i) is characteristic of $\mathbf{R}^p$ if $p < 6$, but fails to discriminate in case the dimension $p \geq 6$.

For any Riemannian manifold it is not difficult to show that the above properties take hold in the limit $r \downarrow 0$, in the following sense:

2b $i'$): The scaled exit time $T_r/r^2$ converges in law when $r \downarrow 0$ to a limit $\mathbf{T}$ which is the exit time of Euclidean Brownian motion from the unit ball in $\mathbf{R}^p$.

2b $ii'$): The exit place $X_{T_r}$ converges in law when $r \downarrow 0$ to the normalized Lebesgue measure $Leb$ of the unit sphere in $\mathbf{R}^p$.

2b $iii'$): The random variables $T_r$, $X_{T_r}$ are asymptotically independent:

$$\forall \alpha > 0, \psi \in C(M), \quad E_m\big(e^{-\alpha T_r}\psi(X_{T_r})\big) =$$
$$E_m(e^{-\alpha T_r})E_m\psi(X_{T_r})(1 + o(1)), \quad r \downarrow 0.$$

These probabilistic statements reflect the *locally Euclidean* nature of a Riemannian manifold; the corresponding properties of Euclidean space take over in the limit of small radius. The asymptotic behavior of the principal eigenvalue is the statement

2b $iv'$): $\lambda_1(B_r) \sim z^2/r^2$ $\quad (r \downarrow 0)$.

In the following sections we investigate to what extent these quantities may be used to detect the intrinsic geometry of the Riemannian manifold $(M, g)$. It is emphasized that in the formulations of the theorems, all considerations are *local*; there is no need to be concerned with the cut-locus or other global issues.

## 3. Exit time distribution of Brownian motion

The first general result on the exit time was obtained in [GP], from which it follows that if $p < 6$ the *mean exit time* alone is sufficient to detect the geometry of Euclidean space or other model spaces (constant curvature or more general rank-one symmetric spaces).

**Theorem 3.1.** For any $p$-dimensional Riemannian manifold $(M,g)$, the mean exit time has the asymptotic expansion

$$E_m(T_r) = r^2/p + c_1 r^4 \tau_m + c_2 r^6 [|R|_m^2 - |\rho|_m^2 + (5/p)\tau_m^2 + 6\Delta\tau_m] + O(r^8)(r \downarrow 0)$$

where $c_1, c_2$ are constants which depend on the dimension $p$. Here $|R|_m^2$ (resp. $|\rho|_m^2$) is the sum of squares of the components of the curvature tensor (resp. Ricci tensor).

**Corollary 3.2.** Suppose that $p = 2$ and $\forall m \in M$ the mean exit time satisfies $E_m(T_r) = r^2/2 + o(r^4), r \downarrow 0$. Then $(M,g)$ is locally isometric to the Euclidean plane.

Indeed, the scalar curvature (=Gaussian curvature) is the only local geometric invariant of a two-dimensional Riemannian manifold.

In higher dimensions, we may obtain converse results by writing the third term of the expansion in a more transparent fashion. This may be expressed in terms of the *Weyl conformal curvature tensor* $C_{ijkl}$ ([BGM], p. 83) through the identity

$$|R|^2 - |\rho|^2 = |C|^2 + \frac{6-p}{p-2}|\rho^2|$$

If this term is zero and $p < 6$ then both coefficients $|C|$ and $|\rho|$ must be zero and the curvature tensor is identically zero. These remarks serve to prove the following

**Corollary 3.3.** Suppose that $p < 6$ and that $\forall m \in M$ the mean exit time satisfies $E_m(T_r) = r^2/p + o(r^6)\ r \downarrow 0$. Then $(M,g)$ is locally isometric to the Euclidean space $\mathbf{R}^p$.

**Corollary 3.4.** *Suppose that $p < 6$ and that $\forall m \in M$ the exit time distribution has the Brownian scaling property: the $P_m$ law of $T_r/r^2$ doesn't depend on $r$ for $r < r(m)$. Then $(M, g)$ is locally isometric to the Euclidean space $\mathbf{R}^p$.*

One might infer from the above discussion that, in order to prove suitable generalizations to dimensions $p \geq 6$, it would suffice to take additional terms in the asymptotic expansion of the mean exit time. This is *false* as was shown by the following example of H.R. Hughes [Hu]. This is the product manifold $M = \mathbf{S}^3 \times \mathbf{H}^3$ with the product metric; here $\mathbf{S}^3$ is the three-dimensional sphere of constant sectional curvature $= +k^2$ and $\mathbf{H}^3$ is the three dimensional hyperbolic space of constant sectional curvature $= -k^2$ where $k$ is any positive number. The following result will be proved.

**Proposition 3.5.** [Hu]. *For any $r < \pi/k$ the probability law of the exit time $T_r$ of $\mathbf{S}^3 \times \mathbf{H}^3$ coincides with the probability law of the exit time $T_r$ of the six-dimensional Euclidean space $\mathbf{R}^6$.*

This example depends on the structure of the *bi-radial Laplacian* of $\mathbf{S}^3 \times \mathbf{H}^3$. This is the operator on $\mathbf{R}_+^2$ which results from the Laplacian restricted to functions which depend only on the combinations $r_1^2 = x_1^2 + x_2^2 + x_3^2$, and $r_2^2 = x_4^2 + x_5^2 + x_6^2$. It will be shown below that this operator is *conjugate* to the bi- radial Laplacian of $\mathbf{R}^3 \times \mathbf{R}^3$ from which many interesting conclusions follow.

Suitable counterexamples may be obtained in any dimension $p > 6$ by taking an additional flat factor in the form $M = \mathbf{S}^3 \times \mathbf{H}^3 \times \mathbf{R}^{p-6}$.

## 4. Exit Place Distribution of Brownian Motion.

Brownian motion of Euclidean space *cannot* be characterized by the law of its exit place, even in dimension two (to see this, think of a brownian particle on the surface of the earth starting at the North Pole and hitting the Arctic Circle). More generally, on any space of *constant sectional curvature*, the exit place distribution from a geodesic sphere is the uniform distribution. Therefore, in the general case, all we can hope for is to use the exit place distribution of Brownian motion to study the *variation* of curvature. This may be measured exactly by the covariant derivatives of the curvature tensor over the manifold, for example. To detect the variation of curvature in different directions, it is useful to examine the Ricci tensor as a first approximation, as we shall see below.

In order to study the distribution of the exit place from a ball of radius $r$, it is convenient to work on the unit sphere of Euclidean space by means of the exponential mapping $exp_m$. This is the differentiable map

$$exp_m : \mathbf{R}^p \to M$$

which sends $0 \in \mathbf{R}^p$ to $m \in M$ and maps straight lines to geodesics emanating from $m \in M$. If $\Psi$ is a continuous function on the sphere $\mathbf{S}^{p-1}$ we define the *harmonic measure* $\mu_m(r, d\theta)$ on the sphere by the operator

$$H_r\Psi(m) \equiv E_m\Psi(r^{-1}exp_m^{-1}X_{T_r}) = \int_{\mathbf{S}^{p-1}} \Psi(\theta)\mu_m(r,\, d\theta)$$

From property 2b $ii'$) these measures converge to the uniform measure $Leb(d\theta)$ when $r \downarrow 0$. A two-term correction formula was discovered by Ming Liao in 1988.

**Theorem 4.1.** [ML]. *For any p-dimensional Riemannian manifold, the harmonic measure operator has the asymptotic expansion*

$$H_r\Psi(m) = \int_{\mathbf{S}^{p-1}} [1 - (1/12)r^2\rho_m^\sharp(\theta) - (1/24)r^3\rho_m^{\sharp\sharp}(\theta)]\Psi(\theta)Leb(d\theta) + O(r^4)$$

*where $\rho^\sharp$ denotes the traceless Ricci tensor defined by $\rho_m^\sharp(\theta) = (\rho_{ij} - \delta_{ij}\tau_m/p)\theta_i\theta_j$ and $\rho_m^{\sharp\sharp}(\theta) = \partial\rho_{ij}/\partial x_k\theta_i\theta_j\theta_k - \theta_k[\partial\tau/\partial x_k]/(p+2)$ where repeated indices imply a summation.*

In the case of two dimensions the traceless Ricci tensor is zero and we can infer that the uniform distribution of exit place implies constant curvature, as follows.

**Corollary 4.2.** *Suppose that $p = 2$ and the exit place distribution satisfies $H_r\Psi(m) = \int_{\mathbf{S}^1} \Psi(\theta)Leb(d\theta) + o(r^3)$, $r \downarrow 0$ $\forall m \in M$. Then $g$ has constant curvature.*

In the case of higher dimensions we have the following general result.

**Corollary 4.3.** *Suppose that for some $m \in M$ the exit place distribution satisfies $H_r \Psi(m) = \int_{S^{p-1}} \Psi(\theta) Leb(d\theta) + o(r^2) \quad r \downarrow 0$. Then $\rho_m^\sharp = 0$. In particular if the condition holds $\forall m \in M$, then $\rho^\sharp \equiv 0$, i.e. $g$ is an Einstein metric.*

Combining this with the results of the previous section, we see that the combined hypotheses of these two sections suffice in *all* dimensions:

**Corollary 4.4.** *Suppose that $\forall m \in M$ the exit distribution satisfies $H_r \Psi(m) = \int_{S^{p-1}} \Psi(\theta) Leb(d\theta) + o(r^2)$ and that the mean exit time satisfies $E_m(T_r) = r^2/p + o(r^6), r \downarrow 0$. Then $(M, g)$ is locally isometric to the Euclidean space $\mathbf{R}^p$.*

To see this last fact, note that the second hypothesis implies that $\tau_m \equiv 0$ which, taken with the first hypothesis, shows that the entire Ricci tensor $\rho_{ij}(m) \equiv 0$. Using again the second hypothesis and the expansion in theorem 3.1 shows that the curvature tensor $R_{ijkl}(m) \equiv 0$, which was to be proved.

## 5. Joint distribution of exit time and exit place.

On any space of constant sectional curvature it may be shown that the exit time and exit place from a ball are independent random variables, for Brownian paths starting at the center. This is easily seen by the separation-of- variables solution associated with the Laplace operator of such a space. In general, one may inquire to what extent independence characterizes a general Riemannian manifold. For a recent discussion of independence in the Euclidean case, see [Pt].

The random variables $(T_r, X_{T_r})$ may be studied jointly through the Laplace transform which defines a family of measures $\mu_m^\alpha(r, \cdot)$ on $S^{p-1}$ by

$$E_m \left( e^{-\alpha T_r/r^2} \Psi(r^{-1} exp_m^{-1} X_{T_r}) \right) = \int_{S^{p-1}} \Psi(\theta) \mu_m^\alpha(r, d\theta)$$

When $r \downarrow 0$ property 2b $iii'$) indicates the factorization into the product of the uniform measure with the Laplace transform $E_0 \, e^{-\alpha T}$ in the Euclidean case. The following results have been proved:

**Proposition 5.1.** ([ML], [KO], [P2]) *Suppose that $\forall m \in M$ the random variables $T_r$, $X_{T_r}$ are independent $\forall r < r(m)$. Then the scalar curvature is constant.*

One might conjecture that independence implies stronger curvature conditions, such as the Einstein condition obtained in the previous section. The following example of H.R. Hughes shows that one may have independence without the Einstein condition.

**Proposition 5.2.** [Hu] *For any $r < \pi/k$ the exit time and exit place from the ball of radius r in the product manifold $\mathbf{S}^3 \times \mathbf{H}^3$ are independent random variables.*

## 6. Principal eigenvalue of the Laplacian.

Quite apart from considerations of Brownian motion, one may study this classical quantity, which may be defined as a Rayleigh quotient

$$\lambda_1(B_r) = \inf_{(f : f \not\equiv 0, f = 0 \text{ on } \partial B_r)} \frac{\int_{B_r} |df|^2}{\int_{B_r} |f|^2}.$$

The following three-term expansion was obtained in [KP].

**Theorem 6.1.** *For any p-dimensional Riemannian manifold $(M, g)$, the principal eigenvalue $\lambda_1$ has the asymptotic expansion*

$$\lambda_1(B_r) = z^2/r^2 - \tau_m/6 + const. \, r^2[|R|_m^2 - |\rho|_m^2 + 6\Delta\tau_m] + O(r^4), \quad r \downarrow 0$$

*where the constant only depends on the dimension p.*

From the principal eigenvalue one obtains *exactly* the same conclusions as for the mean exit time in section 3.

**Corollary 6.2.** *Suppose that $p = 2$ and $\forall m \in M$ the principal eigenvalue satisfies $\lambda_1(B_r) = const./r^2 + o(1), \quad r \downarrow 0$. Then $(M, g)$ is locally isometric to the Euclidean plane.*

**Corollary 6.3.** *Suppose that $p < 6$ and that $\forall m \in M$ the principal eigenvalue satisfies $\lambda_1(B_r) = const./r^2 + o(r^2)$, $r \downarrow 0$. Then $(M,g)$ is locally isometric to the Euclidean space $\mathbf{R}^p$.*

As with the previous studies, a counter-example is available in dimension six:

**Proposition 6.4.** *For any $r < \pi/k$ the principal eigenvalue of the ball of radius $r$ in $\mathbf{S}^3 \times \mathbf{H}^3$ satisfies $\lambda_1(B_r) \equiv const/r^2$, where the constant is the the square of the first positive zero of the Bessel function $J_2$.*

## 7. Computations on $M = \mathbf{S}^3 \times \mathbf{H}^3$.

In this section we give the proofs of the above statements concerning the six-dimensional counter-example. The metric is defined as the product of the corresponding metrics in the respective spaces, which are most conveniently written in terms of *geodesic polar coordinates* $(r_1, \theta_1)$, $(r_2, \theta_2)$. In $\mathbf{S}^3$ we have

$$ds_1^2 = dr_1^2 + \frac{\sin^2 kr_1}{k^2} d\theta_1^2,$$

the latter being the standard metric on the 2-sphere. In $\mathbf{H}^3$ we have

$$ds_2^2 = dr_2^2 + \frac{\sinh^2 kr_2}{k^2} d\theta_2^2.$$

One may note that in the product space the geodesic distance is correctly given by $r^2 = r_1^2 + r_2^2$, where we must respect the cut-locus by requiring that $r < \pi/k$ to obtain a ball which is diffeomorphic to a standard ball of $\mathbf{R}^6$.

The full Laplacian of the product space is written

$$\Delta_M f = \partial^2/\partial r_1^2 + (2k \cot kr_1)\partial/\partial r_1 + (k^2 \csc^2 kr_1)\Delta_{\mathbf{S}^2}^1$$
$$+ \, \partial^2/\partial r_2^2 + (2\,k \coth kr_2)\partial/\partial r_2 + (k^2 \operatorname{csch}^2 kr_2)\Delta_{\mathbf{S}^2}^2$$

where the operators $\Delta_{\mathbf{S}^2}^1$, resp. $\Delta_{\mathbf{S}^2}^2$ refer to the Laplace operators of the standard 2-spheres in the respective spaces. This formula follows from the radial decomposition of the Laplacian of a Riemannian manifold and the properties of product spaces(e.g. [He]). Let $q(r_1, r_2)$ be defined by

$$q(r_1, r_2) = \frac{\sin kr_1}{r_1} \frac{\sinh kr_2}{r_2}$$

The Brownian motion may be constructed in terms of independent radial processes $r_t^1$ $r_t^2$ and the associated angular processes $\theta_t^1$, $\theta_t^2$. The following statement gives the required between the relation between the generators.

**Proposition 7.1.** *For any twice-differentiable function $f(r_1, r_2)$ on $\mathbf{R}^2_+$ we have*

$$\Delta_M \left[ \frac{f(r_1, r_2)}{q(r_1, r_2)} \right] = \frac{1}{q(r_1, r_2)} \Delta_{\mathbf{R}^6} f(r_1, r_2)$$

*where the last term refers to the Euclidean Laplacian acting on "bi-radial functions":*

$$\Delta_{\mathbf{R}^6} f(r_1, r_2) = \partial^2 f / \partial r_1^2 + (2/r_1)\partial f / \partial r_1 + \partial^2 f / \partial r_2^2 + (2/r_2)\partial f / \partial r_2$$

In probabilistic terms we have two independent Bessel processes whose generator is *conjugate* to the generator of a pair of independent radial diffusions on the sphere and hyperbolic spaces. The proof by explicit computation is given at the end of the section. One may note that by taking $f = q$ the formula implies that $q$ is a harmonic function for the pair of Bessel processes. Probabilistically we have an *h-transform* by the function $q$. This formula can equivalently be paraphrased as the statement that $1/q(r_1, r_2)$ is $\Delta_M$-harmonic. This fact and Proposition 7.1 could be proved directly by Ito's formula applied to the pair of radial diffusions. At the end of this section we have included a classical, non- stochastic proof.

In order to use this to study the exit time, we may characterize the Laplace transform $u = E_\bullet^M(e^{-\alpha T_r})$ as the solution of of the equation $\Delta_M u = \alpha u$ which assumes the value $u = 1$ on the boundary where $r_1^2 + r_2^2 = r^2$ [D]. From the rotational invariance of the respective components it follows that u only depends on the radii $(r_1, r_2)$. Therefore we may use the above structure of the bi-radial Laplacian to reduce to a problem on $\mathbf{R}^6$. Defining a new function by $v = qu$ we have

$$\alpha \frac{v}{q} = \alpha u = \Delta_M u = \Delta_M \left( \frac{v}{q} \right) = \frac{1}{q} \Delta_{\mathbf{R}^6} v$$

Therefore $v$ satisfies the Euclidean equation $\Delta_{\mathbf{R}^6} v = \alpha v$ in the Euclidean ball with the boundary condition that $v = q$ on the sphere $r_1^2 + r_2^2 = r^2$ The probabilistic representation in terms of Euclidean Brownian motion is

$v(0,0) = E_{(0,0)}^{\mathbf{R}^6}[q(r_{T_r}^1, r_{T_r}^2)e^{-\alpha T_r}]$. Invoking the independence property 2b iii) this factors and we obtain

$$E_{0,0}^M e^{-\alpha T_r} = u(0,0) = [1/q(0,0)]E_{(0,0)}^{\mathbf{R}^6}[e^{-\alpha T_r}]E_{(0,0)}[q(r_{T_r}^1, r_{T_r}^2)]$$

Taking $\alpha = 0$ shows that the product of the q-terms equals 1 and we obtain the desired conclusion in the form

$$E_{(0,0)}^M e^{-\alpha T_r} = E_{(0,0)}^{\mathbf{R}^6} e^{-\alpha T_r}$$

which proves Proposition 3.5.  □

To study the independence, we examine the joint Laplace transform

$$u_\alpha = E_\bullet(e^{-\alpha T_r}\psi(X(T_r))$$

We must show that this factors into a function of $\alpha$ and an operator on $\psi$. To prove independence, it suffices to prove the factorization property of the Laplace transform for $\psi$ of the form $\psi = \psi_1(r_1)\psi_2(r_2)\phi_1(\theta_1)\phi_2(\theta_2)$, where $\theta_1, \theta_2$ are the angular components. We first do the case of $\phi_1 = 1, \phi_2 = 1$.

To do this we characterize $u_\alpha$ as the solution of the equation $\Delta_M u = \alpha u$ which assumes the value $\psi$ on the boundary of the ball, where $r_1^2 + r_2^2 = r^2$ [D]. For $\psi$ of the type $\psi = \psi_1(r_1)\psi_2(r_2)$ the solution is a bi-radial function.

Defining $v = qu$ as above, we see that $v$ satisfies the Euclidean equation $\Delta_{\mathbf{R}^6}v = \alpha v$ with $v = q\psi$ on the surface. To prove the required independence we write, as above

$$u(0,0) = E_{(0,0)}^M[\psi_1(r_{T_r}^1)\psi_2(r_{T_r}^2)e^{-\alpha T_r}]$$
$$= \frac{1}{q(0,0)}E_{(0,0)}^{\mathbf{R}^6}[q(r_{T_r}^1, r_{T_r}^2)\psi_1(r_{T_r}^1)\psi_2(r_{T_r}^2)e^{-\alpha T_r}]$$
$$= \frac{1}{q(0,0)}E_{(0,0)}^{\mathbf{R}^6}[q(r_{T_r}^1, r_{T_r}^2)\psi_1(r_{T_r}^1)\psi_2(r_{T_r}^2)]E_{(0,0)}^{\mathbf{R}^6}[e^{-\alpha T_r}]$$

where we have used the independence property 2b iii) of Brownian motion on Euclidean space. Setting $\alpha = 0$ shows that the product of the first two factors in the final expression reduces to the expectation on the manifold $E_{(0,0)}^M(\psi_1(r_{T_r}^1)\psi_2(r_{T_r}^2))$ which proves the independence of $T_r$ from the pair $(r_{T_r}^1, r_{T_r}^2)$.

To prove the independence in general, note that the angular components of the M-diffusion may be represented in terms of the respective radial processes $r^1, r^2$ and independent $S^2$- valued brownian motions. Given $r^1, r^2$ these are conditionally independent, by the usual skew-product decomposition [McK]. The joint Laplace transform is written

$$E_{0,0}^M \left[ \psi_1(r_{T_r}^1) \psi_2(r_{T_r}^2) \phi_1(\theta_{T_r}^1) \phi_2(\theta_{T_r}^2) e^{-\alpha T_r} \right]$$

$$= E_{(0,0)}^M \left[ E_{(0,0)}^M [\phi_1(\theta_{T_r}^1)|r_1, r_2] E_{(0,0)}^M [\phi_2(\theta_{T_r}^2)|r_1, r_2] \psi_1(r_{T_r}^1) \psi_2(r_{T_r}^2) e^{-\alpha T_r} \right]$$

Starting at the center $(0,0)$ the conditional hitting laws of $\theta_{T_r}^1, \theta_{T_r}^2$ are uniform on the respective spheres, so that the $\phi_1, \phi_2$ factors are constant, equal to the respective averages with respect to normalized Lebesgue measure.

Therefore the independence property property has been reduced to the independence for radial functions, which has been proved above. This proves Proposition 5.3. $\square$

To study the first eigenvalue, we write the numerator of the Rayleigh characterization in the form

$$\int_{B_r} |df|^2 = \int_{B_r} (\partial f/\partial r_1)^2 + (\partial f/\partial r_2)^2 \, dvol$$

$$+ \int_{B_r} (|d_{\theta_1} f|^2 + |d_{\theta_2} f|^2) \, dvol$$

which is greater than or equal to the term obtained by ignoring the angular derivatives. Therefore $\lambda_1(B_r)$ is greater than or equal to the quantity obtained when we minimize *only over bi-radial functions*. But for bi-radial functions we may use the conjugacy proved in Proposition 7.1 to determine the minimum in terms of the Euclidean Laplacian, which yields the minimum as $z^2/r^2$, where $z$ is the first zero of the Bessel function $J_2$, hence the inequality $\lambda_1(B_r) \geq z^2/r^2$. To show equality, it suffices to *transplate* the exact eigenfunction $v(r_1, r_2) = J_2(r)/r^2$ by letting $f = v/q$. Substituting this into the Rayleigh quotient, we obtain the inequality $\lambda_1(B_r) \leq z^2/r^2$. The proof is complete. $\square$

*Proof of Proposition 7.1.* In the sphere $S^3$ we have for any $F = F(r_1, r_2)$

$$\Delta^{S^3} F(r_1, r_2) = \partial^2 F/\partial r_1^2 + (2k \cot kr_1)\partial F/\partial r_1$$
$$= (1/\sin kr_1)\partial^2/\partial r_1^2(\sin kr_1 F) + k^2 F$$

In the hyperbolic space $\mathbf{H}^3$ we have

$$\Delta^{\mathbf{H}^3} F(r_1, r_2) = \partial^2 F/\partial r_2^2 + (2k \coth kr_2)\partial F/\partial r_2$$
$$= (1/\sinh kr_2)\partial^2/\partial r_2^2(\sinh kr_2 F) - k^2 F$$

Adding these cancels the zero-order term and we have

$$\Delta^M F(r_1, r_2)$$
$$= (1/\sin kr_1)\partial^2/\partial r_1^2(\sin kr_1 F) + (1/\sinh kr_2)\partial^2/\partial r_2^2(\sinh kr_2 F)$$
$$= (1/\sin kr_1 \sinh kr_2)[\partial^2/\partial r_1^2 + \partial^2/\partial r_2^2](\sin kr_1 \sinh kr_2 F).$$

Writing $F = f/q$ and noting the definition of $q$ yields the relation

$$\Delta^M[f/q] = [1/\sin kr_1 \sinh kr_2][\partial^2/\partial r_1^2 + \partial^2/\partial r_2^2](r_1 r_2 f)$$

On the other hand we may compute the six-dimensional Euclidean Laplacian on bi-radial functions as

$$\Delta^{\mathbf{R}^3 \times \mathbf{R}^3} f(r_1, r_2) = \partial^2 f/\partial r_1^2 + (2/r_1)\partial f/\partial r_1 + \partial^2 f/\partial r_2^2 + (2/r_2)\partial f/\partial r_2$$
$$= (1/r_1)[\partial^2 f/\partial r_1^2(r_1 f)] + (1/r_2)[\partial^2 f/\partial r_2^2(r_2 f)]$$
$$= (1/r_1 r_2)[\partial^2/\partial r_1^2 + \partial^2/\partial r_2^2](r_1 r_2 f).$$

Comparing the two forms obtained yields the desired result.   □

**Acknowledgement** We would like to thank Rick Durrett, Randy Hughes, Joseph B. Keller, Ming Liao and the referee for some helpful comments on a preliminary version of this paper.

# References

[Be] P. Bérard, Variétés Riemanniennes isospectrales non isometriques, Asterisque, 177-178 (1989), 127-154.

[BGM] M. Berger, P. Gauduchon and E. Mazet, Le spectre d'une variété Riemannienne, Lecture Notes in Mathematics, vol. 194, 1971.

[Br] R. Brooks, Constructing isospectral manifolds, American Mathematical Monthly, 95 (1988), 823-839.

[D] E. B. Dynkin, *Markov Processes*, 2 volumes, Springer Verlag, 1965.

[G] A. Gray, The volume of a small geodesic ball in a Riemannian manifold, Michigan Mathematics Journal, 20 (1973) 329-344.

[Go] C. S. Gordon, When you can't hear the shape of a manifold, Math. Intelligencer II, no. 3 (1989), 39-47.

[GP] A. Gray and M. Pinsky, Mean exit time from a small geodesic ball in a Riemannian manifold, Bulletin des Sciences Mathématiques, 107 (1983), 345- 370.

[He] S. Helgason, *Groups and Geometric Analysis*, Academic Press, 1984.

[Hu] H. R. Hughes, Brownian exit distributions from normal balls in $S^3 \times H^3$, Annals of Probabilty, to appear.

[Ka] M. Kac, Can one hear the shape of a drum, American Mathematical Monthly, 73 (1966), 1-23.

[KO] M. Kozaki and Y. Ogura, On the independence of exit time and exit position from small geodesic balls on Riemannian manifolds, Mathematische Zeitschrift 197 (1988), 561–581.

[KP] L. Karp and M. Pinsky, First eigenvalue of a small geodesic ball in a Riemannian manifold, Bulletin des Sciences Mathématiques 111 (1987), 222- 239.

[McK] H. P. McKean, *Stochastic Integrals*, Academic Press, 1969.

[ML] M. Liao, Hitting distributions of small geodesic spheres, Annals of Probability 16 (1988), 1029–1050.

[P1] M. Pinsky, Can you feel the shape of a manifold with Brownian motion, Expositiones Mathematicae 2 (1984), 273–281.

[P2] M. Pinsky, Local Stochastic Differential Geometry, Contemporary Mathematics, vol. 73 (1988), 263–272.

[Pt] L. Pitt, On a problem of H. P. McKean: Independence of Brownian hitting times and places, Annals of Probability 17 (1989), 1651–1658.

[St] S. Stafford, A stochastic criterion for Yang-Mills fields, in *Diffusion Proceeses and Related Problems in Analysis*, vol. 1, Birkhauser Boston Inc. 1990, 313-324 (ed. M. Pinsky).

Department of Mathematics
Northwestern University
Evanston Illinois 60208-2730

# DECOMPOSITION OF DIRICHLET PROCESSES ON HILBERT SPACE

M. Röckner
Institut für Angewandte Mathematik
Universität Bonn
Wegelerstrasse 6
D-5300 Bonn 1
Germany

Zhang T-Sheng [*]
Department of Mathematics
University of Edinburgh
The King's Buildings
Mayfield Road
Edinburgh EH 9 3JZ
Scotland

## 1. FRAMEWORK AND INTRODUCTION

The purpose of this note is to extend results in [3] on the decomposition of diffusion processes associated with Dirichlet forms on infinite dimensional state space to the case where the martingale part is not necessarily a Brownian motion. Our main techniques are those developed in [3] and the stochastic integral on Banach space defined in [6]. We also prove a corresponding decomposition in a forward and a backward martingale in the spirit of [7], [8]. To state our results precisely we need some preparations. We start with describing our framework.

Let $E$ be a separable real Banach space. Denote by $\mathcal{B}(E)$ the Borel $\sigma$-field of $E$. Let $\mu$ be a probability measure on $(E, \mathcal{B}(E))$. $L^2(E, \mu)$ is the corresponding real $L^2$-space with the usual inner product $(\ ,\ )_\mu$. The topological dual of $E$ is denoted by $E'$. Define the linear space

[*] The second-named author has been supported by the British SERC.

(1.1)   $\mathcal{F} C_b^\infty(E) = \Big\{ u : E \to R \mid u = f(\ell_1,...,\ell_m) \,,\, \text{for some } m \in \mathbb{N} \,,$
$$f \in C_b^\infty(R^m) \,,\, \ell_1,...,\ell_m \in E' \Big\} .$$

For $u \in \mathcal{F} C_b^\infty(E)$ , $k \in E$ we introduce

(1.2)          $\dfrac{\partial u}{\partial k} \triangleq \dfrac{d}{ds} u(z+sk)\big|_{s=0}$ ,   $z \in E$ .

We say that $k \in E$ is well–($\mu$–)admissible if there exists $\beta_k \in L^2(E,\mu)$ such that

(1.3)          $\int \dfrac{\partial u}{\partial k} \, d\mu \;=\; -\int u \, \beta_k \, d\mu$   for all $u \in \mathcal{F} C_b^\infty(E)$ ,

(cf. [1] for more details on this notion.) In addition, we assume that there exists a separable real Hilbert space $(H, < , >_H)$ satisfying

(1.4)                    $H \subset E$ densely and continuously.

Identifying $H$ with its dual we obtain $E' \subset H \subset E$ densely and continuously. Note that hence the dualization $_{E'}< , >_E$ between $E'$ and $E$ coincides with $< , >_H$ when restricted to $E' \times H$ . For convenience below we also sometimes write $l(z)$ for $l \in E'$ , $z \in E$ , instead of $_{E'}<l,z>_E$ . We denote for $z \in E$ and $u \in \mathcal{F} C_b^\infty(E)$ by $\nabla u(z)$ the element in $H$ representing the bounded linear functional $h \mapsto \dfrac{\partial u}{\partial h}(z)$ , $h \in H$ , and denote the Banach space of all bounded linear operators on $H$ by $L^\infty(H)$ equipped with the usual operator norm. Let $A : E \to L^\infty(H)$ be a continuous map such that

(1.5)   For each $z \in E$ , $A(z)$ is symmetric and $c\,\mathrm{Id}_H \leq A(z) \leq c^{-1}\,\mathrm{Id}_H$ (in quadratic form sense). Here $c$ is a positive constant (indep. of $z$).

From now on we also assume that any $k \in E'$ $(\subset H \subset E)$ is well–($\mu$–)admissible and $E' \subset L^2(E,\mu)$ . We introduce the following symmetric form $\mathcal{E}(u,v)$ , on $L^2(E,\mu)$ ,

(1.6)      $\mathcal{E}(u,v) = \dfrac{1}{2} \int\limits_E \big\langle A(z)\nabla u(z), \nabla v(z) \big\rangle_H \mu(dz)$   for $u,v \in \mathcal{F} C_b^\infty(E)$ .

Under our assumptions, it follows from [3, Theorem 3.1] (see also [4]) that the symmetric form $\mathcal{E}(u,v)$ is closable on $L^2(E,\mu)$ . If we denote the closure of $\mathcal{E}$ by $(\mathcal{E}_A, \mathcal{D}(\mathcal{E}_A))$ , then by [10, Proposition 3.2] and [2, Theorem 2.7] (see also [11]) we know that there exists a diffusion process $\mathcal{M} =$

$\left[\Omega, \mathcal{F}, (\mathcal{F}_t)_{t \geq 0}, (X_t)_{t \geq 0}, (P_z)_{z \in E}\right]$ associated with $(\mathcal{E}_A, \mathcal{D}(\mathcal{E}_A))$, that is for every bounded $\mathcal{B}(E)$–measurable function u on E and all $t > 0$

$$\int u(X_t) dP_z \;=\; e^{tL_A} u(z) \quad \mu\text{–a.e. } z \in E .$$

$L_A$ is the generator of $(\mathcal{E}_A, \mathcal{D}(\mathcal{E}_A))$ on $L^2(E, \mu)$, i.e. the unique negative definite self–adjoint operator on $L^2(E, \mu)$ such that

$$\mathcal{E}_A(u, v) \;=\; (-L_A u, v)_\mu \quad \text{for all } u \in \mathcal{D}(L_A), v \in \mathcal{D}(\mathcal{E}_A) .$$

With respect to $(\mathcal{E}_A, \mathcal{D}(\mathcal{E}_A))$ we have the usual notions such as (1–)capacity Cap , q.e. etc. The reader can find the corresponding details in [3] (and of course [5]). Let $P_\mu$ be the probability measure on $(\Omega, \mathcal{F})$ defined by

(1.7) $$P_\mu(\cdot) \;=\; \int_E P_z(\cdot) \mu(dz) .$$

Since $E' \subset L^2(E, \mu)$, it is easily shown as in [3, Lemma 5.1] that $E' \subset \mathcal{D}(\mathcal{E}_A)$. Hence by [3, Theorem 4.3] the following decomposition holds:

(1.8) $$k(X_t) - k(X_0) \;=\; M_t^k + N_t^k , \quad t \geq 0 , P_z\text{–a.s., q.e. } z \in E .$$

Here $(M_t^k)_{t \geq 0}$ is a square integrable continuous martingale additive functional of $\mathcal{M}$ of finite energy, $(N_t^k)_{t \geq 0}$ is a continuous additive functional of $\mathcal{M}$ of zero energy (cf. [3], [5]). When $A(z) = \text{Id}_H$ ("collecting" the components of the decomposition above) it was proved in [3] that there is an E–valued $(\mathcal{F}_t)_{t \geq 0}$–Brownian motion $(W)_{t \geq 0}$ with covariance given by $< , >_H$ and an E–valued continuous process $(N_t)_{t \geq 0}$ both starting at $0 (\in E)$ such that for q.e. $z \in E$

(1.9) $$X_t = z + W_t + N_t \quad \text{and} \quad {}_{E'}{<}k, W_t{>}_E = M_t^k , \quad t \geq 0 , P_z\text{–a.e. .}$$

Let $\sigma(z)$ be the positive square root of $A(z)$. Note that $z \mapsto \sigma(z)$ is again continuous from E to $L^\infty(H)$ (cf. e.g. [9]). In this paper, using the stochastic integral in [6] we prove an extension of (1.9) in case E is a Hilbert space.

Assume from now on that

(1.10)    E is a Hilbert space with inner product $< , >_E$ and $H \subset E$ is Hilbert–Schmidt .

The stochastic integrals below are in the sense of [6] (see (2) in Section 2 below).

**Theorem 1.** There exist continuous E–valued $(\mathcal{F}_t)_{t\geq 0}$–adapted processes $(W_t)_{t\geq 0}$ and $(N_t)_{t\geq 0}$ starting at $0(\in E)$ such that

(i)     for q.e. $z \in E$ under $P_z$, $(W_t)_{t\geq 0}$ is an $(\mathcal{F}_t)_{t\geq 0}$–Brownian motion on $E$ with covariance $<\,,\,>_H$.

(ii)    $X_t = z + \int\limits_0^t \sigma(X_s)dW_s + N_t$, $t \geq 0$, $P_z$–a.s., q.e. $z \in E$

(iii)   If $M_t := \int\limits_0^t \sigma(X_s)dW_s$, $t \geq 0$, then for all $k \in E'$,

$_{E'}<k,M_t>_E = M_t^k$ and $_{E'}<k,N_t>_E = N_t^k$, $t \geq 0$, $P_z$–a.s.
for all $z \in E$ outside some capacity zero set (possibly depending on $k$) where $M_t^k$, $N_t^k$ are as in (1.8).

**Theorem 2.** Fix $T > 0$. Let $\mathcal{F}_t = \sigma(X_{T-s}, s\leq t)$, $t \in [0,T]$, then under $P_\mu$ there exist an E–valued $(\mathcal{F}_t)_{t\in[0,T]}$–Brownian motion $(W_t)_{t\in[0,T]}$ and an E–valued $(\mathcal{F}_t)_{t\in[0,T]}$–Brownian motion $(\overline{W}_t)_{t\in[0,T]}$ with covariance $<\,,\,>_H$ starting at $0(\in E)$ such that

$$X_t - X_0 = \frac{1}{2}\int\limits_0^t \sigma(X_s)dW_s - \frac{1}{2}\int\limits_{T-t}^T \sigma(X_{T-s})d\overline{W}_s \quad \text{for } 0 \leq t \leq T, P_\mu\text{–a.s.}.$$

The proofs will be given in Section 2 below.

## 2. PROOFS

### Proof of Theorem 1.
(1) <u>Construction of the Brownian motion</u> $(W_t)_{t\geq 0}$.
Since $H \subset E$ densely by a Hilbert–Schmidt map we can find an orthonormal

basis $\{e_n \mid n \in \mathbb{N}\}$ of $(H, <, >_H)$ and $\lambda_n > 0$, with $\displaystyle\sum_{n=1}^{\infty} \lambda_n^2 < +\infty$ such

that $\{e_n/\lambda_n \mid n \in \mathbb{N}\}$ is an orthonormal basis of $E$, and

(2.1) $\quad <e_n, h>_E = \lambda_n^2 <e_n, h>_H$ for all $h \in H \, (\subset E)$, $n \in \mathbb{N}$.

Define $\bar{e}_n \in E'$ by $\bar{e}_n(z) = <e_n/\lambda_n^2, z>_E$, $z \in E$. Observe that by (2.1), $\bar{e}_n(h) = <e_n, h>_H$ if $h \in H$, i.e. $\bar{e}_n$ corresponds to $e_n \in H$ under the embedding $E' \subset H$.

We can find $S \subset E$ with $\mathrm{Cap}(S) = 0$ such that (1.8) with $k = \bar{e}_n$ holds for all $z \in E \backslash S$ and all $n \in \mathbb{N}$. By [3, Proposition 4.5] we also know that for all $i, j \in \mathbb{N}$, $z \in E \backslash S$,

(2.2) $\quad \left\langle M^{\bar{e}_i}, M^{\bar{e}_j} \right\rangle_t = \int_0^t \left\langle A(X_s)e_i, e_j \right\rangle_H ds$, $\quad t \geq 0$, $P_z$–a.s.

Now fix $z \in E \backslash S$, and let $\sigma^{-1}(z)$ denote the inverse of $\sigma(z)$ on $H$. Fix also $k \in E'$, and for $n \in \mathbb{N}$ define a square integrable martingale $(M_t^n)_{t \geq 0}$ by the following stochastic integral

(2.3) $\quad W_t^{k,n} = \displaystyle\sum_{i=1}^n \int_0^t \left\langle \sigma^{-1}(X_s)k, e_i \right\rangle_H dM_s^{\bar{e}_i}$, $\quad t \geq 0$.

Under $P_z$ for $n \geq m$, $t \geq 0$

$<W^{k,n} - W^{k,m}, W^{k,n} - W^{k,m}>_t$

$= \left\langle \displaystyle\sum_{i=m+1}^n \int_0^t \left\langle \sigma^{-1}(X_s)k, e_i \right\rangle_H dM_s^{\bar{e}_i}, \displaystyle\sum_{i=m+1}^n \int_0^t \left\langle \sigma^{-1}(X_s)k, e_i \right\rangle_H dM_s^{\bar{e}_i} \right\rangle_t$

$= \displaystyle\sum_{i=m+1}^n \sum_{j=m+1}^n \int_0^t \left\langle \sigma^{-1}(X_s)k, e_i \right\rangle_H \left\langle \sigma^{-1}(X_s)k, e_j \right\rangle_H \left\langle A(X_s)e_i, e_j \right\rangle_H ds$

$= \displaystyle\sum_{i=m+1}^n \sum_{j=m+1}^n \int_0^t \left\langle \sigma^{-1}(X_s)k, e_i \right\rangle_H \left\langle \sigma^{-1}(X_s)k, e_j \right\rangle_H \left\langle \sigma(X_s)e_i, \sigma(X_s)e_j \right\rangle_H ds$

$= \displaystyle\int_0^t \left\langle \sigma(X_s)\left[ \sum_{i=m+1}^n \left\langle \sigma^{-1}(X_s)k, e_i \right\rangle_H e_i \right], \sigma(X_s)\left[ \sum_{i=m+1}^n \left\langle \sigma^{-1}(X_s)k, e_i \right\rangle_H e_i \right] \right\rangle_H ds$

$$\leq c^{-1} \int_0^t \sum_{i=m+1}^n \left\langle \sigma^{-1}(X_s)k, e_i \right\rangle_H^2 ds \ .$$

Since $\displaystyle\sum_{i=m+1}^n \left\langle \sigma^{-1}(X_s)k, e_i \right\rangle_H^2 \to 0$     $P_z$–a.e.     as     $m, n \to \infty$     and

$\displaystyle\sum_{i=1}^\infty \left\langle \sigma^{-1}(X_s)k, e_i \right\rangle_H^2 \leq c^{-1} \|k\|_H^2$ , by the dominated convergence theorem we

get that for all $t \geq 0$,

$$E_z < W^{k,n} - W^{k,m} \ , \ W^{k,n} - W^{k,m} >_t \ \to 0 \ \text{ as } m \to \infty \, , \, n \to \infty \ .$$

Hence by Doob's inequality, a square integrable continuous process $(W^k_t)_{t \geq 0}$ can be defined such that for all $z \in E\backslash S$ , it is a square integrable $(\mathcal{F}_t)_{t \geq 0}$–martingale under $P_z$ and for all $t \geq 0$,

$$(2.4) \qquad W^k_t = \lim_{n \to \infty} W^{k,n}_t \ \text{ in } L^2(\Omega, P_z) \ .$$

Now we compute the bracket of $W^k$ . From the definition, for $t \geq 0$ , we have that in $L^2(\Omega, P_z)$

$$< W^k, W^k >_t$$

$$= \lim_{n \to \infty} < W^{k,n}, W^{k,n} >_t$$

$$= \lim_{n \to \infty} \sum_{i=1}^n \sum_{j=1}^n \int_0^t \left\langle \sigma^{-1}(X_s)k, e_i \right\rangle_H \left\langle \sigma^{-1}(X_s)k, e_j \right\rangle_H \left\langle A(X_s)e_i, e_j \right\rangle_H ds$$

$$= \lim_{n \to \infty} \int_0^t \left\langle A(X_s)\Big[\sum_{i=1}^n \left\langle \sigma^{-1}(X_s)k, e_i \right\rangle_H e_i\Big] \, , \, \sum_{i=1}^n \left\langle \sigma^{-1}(X_s)k, e_i \right\rangle_H e_i \right\rangle_H ds \ .$$

Since $A(X_s), \sigma^{-1}(X_s) \in L^\infty(H)$ , by (1.5), and the dominated convergence theorem we conclude that

$$(2.5) \qquad < W^k, W^k >_t = \int_0^t \left\langle A(X_s)\sigma^{-1}(X_s)k, \sigma^{-1}(X_s)k \right\rangle_H ds$$

$$= t\|k\|_H^2 \ , \quad t \geq 0 \ .$$

From the construction of $W^k$ , it is easy to see that for fixed $t \geq 0$, $k \mapsto W^k_t$ is $P_z$–a.s. linear on $E'$ . In addition, (2.5) implies that

$$(2.6) \qquad E_z\!\left[e^{iW_t^k}\right] \;=\; \exp\!\left[-\tfrac{1}{2}\,t\|k\|_H^2\right], \quad t \geq 0,\, k \in E'\,.$$

Consequently, (the proof of) Theorem 6.2 in [3] applies to $(W_t^k)_{t\geq 0}$ and we conclude that there exists an E–valued $(\mathcal{F}_t)_{t\geq 0}$–adapted continuous process $(W_t)_{t\geq 0}$ on $(\Omega,\mathcal{F})$ such that, for all $z \in E\backslash S$, under $P_z$, $(W_t)_{t\geq 0}$ is an $(\mathcal{F}_t)_{t\geq 0}$–Brownian motion starting at $0 (\in E)$ with covariance $<\,,\,>_H$. Furthermore,

$$(2.7) \qquad {}_{E'}\!<k,W_t>_E = W_t^k \;,\quad t \geq 0,\, P_z\text{–a.e. for all } k \in E'\,, \text{ and all } z \in E$$
outside some capacity zero set (possibly depending on $k$).

## (2) <u>Stochastic integral with respect to</u> $(W)_{t\geq 0}$.

Below for a linear operator $T$ on $E$ we denote its restriction to $H$ by $\tilde{T}$. Define finite dimensional orthogonal projections $Q_n$, $n \in \mathbb{N}$, on $E$ as follows.

$$(2.8)\; Q_n z \;=\; \sum_{i=1}^{n} {}_{E'}\!<\bar{e}_i,z>_E\, e_i \quad \left[\;=\; \sum_{i=1}^{n} <e_i/\lambda_i,z>_E\, e_i/\lambda_i\right], z \in E\,.$$

Then by (2.1), $\tilde{Q}_n$, $n \in \mathbb{N}$, are also orthogonal projections on $H$ and both $(Q_n)_{n\in\mathbb{N}}$, and $(\tilde{Q}_n)_{n\in\mathbb{N}}$ converge strongly to $\mathrm{Id}_E$ on $E$, $\mathrm{Id}_H$ on $H$ respectively. Denote the set of all Hilbert–Schmidt operators from $H$ to $E$ by $L_{(2)}(H,E)$; the usual norm in $L_{(2)}(H,E)$ is denoted by $\|\cdot\|_2$. Below again we fix $z \in E\backslash S$. Let $\mathcal{L}$ be the space of all $(\mathcal{F}_t)_{t\geq 0}$–adapted processes $(\zeta(t))_{t\geq 0}$ with state space $L_{(2)}(H,E)$ such that $E_z \int_0^t \|\zeta(s)\|_2^2\, ds < +\infty$ for each $t \geq 0$. Using $(Q_n)_{n\in\mathbb{N}}$, for any $\zeta \in \mathcal{L}$ the stochastic integral of $\zeta$ w.r.t. $(W_t)_{t\geq 0}$ denoted by $J_\zeta(t) = \int_0^t \zeta(s)dW_s$, $t \geq 0$, can be defined as in [6]. $J_\zeta$ is a continuous E–valued $(\mathcal{F}_t)_{t\geq 0}$–martingale (i.e. all components ${}_{E'}\!\left\langle k, J_\zeta(t)\right\rangle_E$, $k \in E'$, are martingales) with the property

$$(2.9) \qquad E_z\, J_\zeta(t) \;=\; 0 \;,\quad E_z\|J_\zeta(t)\|_E^2 \;=\; E_z \int_0^t \|\zeta(s)\|_2^2\, ds\,.$$

Clearly, the embedding $H \subset E$ has $\|\cdot\|_2$–norm equal to $\left[\sum\limits_{n=1}^{\infty} \lambda_n^2\right]^{1/2}$ and

for each $T \in L^{\infty}(H)$ we have that

$$(2.10) \qquad T \in L_{(2)}(H,E) \text{ with } \|T\|_2 \leq \|T\|_{L^{\infty}(H)} \left[\sum\limits_{n=1}^{\infty} \lambda_n^2\right]^{1/2}.$$

In particular, it follows that $(\sigma(X_t))_{t \geq 0} \in \mathcal{L}$ and by the above discussion, we get a continuous E–valued, $(\mathcal{F}_t)_{t \geq 0}$–martingale $(M_t)_{t \geq 0}$ defined by

$$(2.11) \qquad M_t := \int\limits_0^t \sigma(X_s)dW_s \ , \ t \geq 0.$$

(3) Identification.

In this step, we will show that

$$_{E'}\langle k, M_t \rangle_E = M_t^k \ , \ t \geq 0 \ , \ P_z\text{–a.e. for q.e. } z \in E \backslash S$$

(where $M_t^k$ is as in (1.8)). We need the following simple Lemma.

**Lemma.** Let $T_n , T \in L_{(2)}(H,E)$ , $n \in \mathbb{N}$ , such that $\|T-T_n\|_2 \xrightarrow[n \to \infty]{} 0$ , then $\|T-T_n \tilde{Q}_n\|_2 \xrightarrow[n \to \infty]{} 0$ .

**Proof.** For $n \in \mathbb{N}$ ,

$$\sum\limits_{j=1}^{\infty} \left\|(T-T_n \tilde{Q}_n)e_j\right\|_E^2 = \sum\limits_{j=1}^{n} \left\|(T-T_n)e_j\right\|_E^2 + \sum\limits_{j=n+1}^{\infty} \|T e_j\|_E^2 . \quad \blacksquare$$

Take a sequence $\tau^n = \{0 = t_0^n < t_1^n < t_2^n \ ...\}$ of partitions of $[0,\infty[$ such that $\delta(\tau^n) = \max\limits_i \left[t_{i+1}^n - t_i^n\right] \to 0$ as $n \to \infty$ . For each $n \geq 1$ we define a process $\zeta_n \in \mathcal{L}$ by

$$(2.12) \qquad \zeta_n(t) = \sigma\left[X_{t_i^n}\right] \text{ if } t_i^n \leq t < t_{i+1}^n .$$

Fix $z \in E \backslash S$ , then, for $t \geq 0$ by the Lemma, (2.10) and the dominated convergence theorem

$$E_z \int_0^t \left\| \zeta_n(s)\tilde{Q}_n - \sigma(X_s) \right\|_2^2 ds \;\longrightarrow\; 0 \quad \text{as } n \to \infty.$$

Thus, $J_{\zeta_n}(t) \to M_t$ in $L^2(\Omega \to E, P_z)$ as $n \to \infty$, consequently

$$_{E'}\!\left\langle \bar{e}_j, J_{\zeta_n}(t) \right\rangle_E \;\longrightarrow\; _{E'}\!<\bar{e}_j, M_t>_E \quad \text{in } L^2(\Omega, P_z) \text{ as } n \to \infty \text{ for all } j \in \mathbb{N}.$$

Fix $j, n \in \mathbb{N}$. Now we calculate $_{E'}\!\left\langle \bar{e}_j, J_{\zeta_n}(t) \right\rangle_E$. For notational convenience the index $n$ is omitted in the following. By definition of $J_\zeta$ if $t_{N-1} \leq t \leq t_N$ we have that $P_z$-a.s.

$$_{E'}\!\left\langle \bar{e}_j, J_\zeta(t) \right\rangle_E = \sum_{i=1}^{N-1} {}_{E'}\!\left\langle \bar{e}_j, \zeta(t_i)Q\left[W(t_{i+1}) - W(t_i)\right] \right\rangle_E$$
$$+ {}_{E'}\!\left\langle \bar{e}_j, \zeta(t_N)\left[W(t) - W(t_N)\right] \right\rangle_E,$$

and

$$_{E'}\!\left\langle \bar{e}_j, \zeta(t_i)Q\left[W(t_{i+1}) - W(t_i)\right] \right\rangle_E$$

$$= \sum_{k=1}^{n} {}_{E'}\!\left\langle \bar{e}_k, W(t_{i+1}) - W(t_i) \right\rangle_E \; {}_{E'}\!\left\langle \bar{e}_j, \zeta(t_i)e_k \right\rangle_E$$

$$= \sum_{k=1}^{n} \left[ W^{\bar{e}_k}(t_{i+1}) - W^{\bar{e}_k}(t_i) \right] \left\langle e_j, \zeta(t_i)e_k \right\rangle_H$$

$$= \sum_{k=1}^{n} \sum_{\ell=1}^{\infty} \int_{t_i}^{t_{i+1}} \left\langle \sigma^{-1}(X_s)e_k, e_\ell \right\rangle_H \left\langle \zeta(t_i)e_k, e_j \right\rangle_H \, dM_s^{\bar{e}_\ell}$$

$$= \sum_{\ell=1}^{\infty} \int_{t_i}^{t_{i+1}} \left\langle \sigma^{-1}(X_s)e_\ell, \tilde{Q}\,\zeta(t_i)e_j \right\rangle_H \, dM_s^{\bar{e}_\ell}$$

where we used (2.4) and the symmetry of $\sigma^{-1}(X_s)$ and $\zeta(t_i)$. Hence

$$_{E'}\!\left\langle \bar{e}_j, J_{\zeta_n}(t) \right\rangle_E = \sum_{\ell=1}^{\infty} \int_0^t \left\langle \sigma^{-1}(X_s)e_\ell, \tilde{Q}_n\, \zeta_n(s)e_j \right\rangle_H \, dM_s^{\bar{e}_\ell}$$

where the sum converges in $L^2(\Omega, P_z)$. Define $T_n(s) := \zeta_n(s)\tilde{Q}_n\,\sigma^{-1}(X_s)$

$- \mathrm{Id}_H$ , $s \geq t$ , then since $M_t^{\bar{e}_j} = \sum_{\ell=1}^{\infty} \int_0^t <e_\ell, e_j>_H \, dM_s^{\bar{e}_\ell}$ , we obtain that

$$E_z \left[ \left\langle \bar{e}_j, J_{\zeta_n}(t) \right\rangle_E - M_t^{\bar{e}_j} \right]^2$$

$$= E_z \left[ \sum_{\ell=1}^{\infty} \int_0^t \left\langle T_n(s) e_\ell, e_j \right\rangle_H dM_s^{\bar{e}_\ell} \right]^2$$

$$= E_z \left[ \sum_{\ell=1}^{\infty} \sum_{m=1}^{\infty} \int_0^t \left\langle T_n(s) e_\ell, e_j \right\rangle_H \left\langle T_n(s) e_m, e_j \right\rangle_H \left\langle A(X_s) e_\ell, e_m \right\rangle_H ds \right]$$

$$= E_z \int_0^t \left\langle A(X_s) T_n(s) e_j, T_n(s) e_j \right\rangle_H ds$$

$$\longrightarrow 0 \quad \text{as } n \longrightarrow \infty,$$

since $T_n(s) \xrightarrow[n\to\infty]{} 0$ strongly on $H$ and by the dominated convergence theorem. Hence

$$_{E'}<\bar{e}_j, M_t>_E = \lim_{n\to\infty} {}_{E'}\left\langle \bar{e}_j, J_{\zeta_n}(t) \right\rangle_E = M_t^{\bar{e}_j} \, , \quad t \geq 0 \, , \, P_z\text{–a.e.} \, ,$$

and consequently, if $k \in E'$ is a finite linear combination of elements in $\{\bar{e}_j | j \in \mathbb{N}\}$ ,

(2.13) $$_{E'}<k, M_t>_E = M_t^k \, , \quad t \geq 0 \, , \, P_z\text{–a.e.}$$

For general $k \in E'$ , by approximation we also get (2.13) $P_z$–a.e. for q.e. $z \in E \backslash S$ (cf. [F80, Corollary 1(ii), p.139]). Combining (2), (1), (3), and letting $N_t = X_t - X_0 - M_t$ , $t \geq 0$ , we finish the proof of Theorem 1.

#### Proof of Theorem 2.

Since $E' \subset \mathcal{D}(\mathcal{E}_A)$ we know by [7], [8] that for all $k \in E'$

(2.14) $\quad k(X_t) - k(X_0) = \frac{1}{2} M_t^k - \frac{1}{2} \left[ \overline{M}_T^k - \overline{M}_{T-t}^k \right] \quad P_\mu$–a.e. for $0 \leq t \leq T$ .

Here $M^k, \overline{M}^k$ is a square integrable $\mathcal{F}_t$–martingale, $\overline{\mathcal{F}}_t$–martingale, respectively. Moreover,

$$\left\langle M^{k_1}, M^{k_2} \right\rangle = \int_0^t \left\langle A(X_s)k_1, k_2 \right\rangle_H ds \ , \ \left\langle \overline{M}^{k_1}, \overline{M}^{k_2} \right\rangle = \int_0^t \left\langle A(X_{T-s})k_1, k_2 \right\rangle_H ds$$

$$\text{for } 0 \leq t \leq T .$$

As in (2.4), we can define

$$W_t^k = \sum_{i=1}^\infty \int_0^t \left\langle \sigma^{-1}(X_s)k, e_i \right\rangle_H dM_s^{\overline{e}_i}$$

and

$$\overline{W}_t^k = \sum_{i=1}^\infty \int_0^t \left\langle \sigma^{-1}(X_{T-s})k, e_i \right\rangle_H d\overline{M}_s^{\overline{e}_i} .$$

By the same procedure as that in (1), there exists an E–valued $\mathcal{F}_t$–Brownian

motion $W_t$ , and an E–valued $\overline{\mathcal{F}}_t$–Brownian motion $\overline{W}_t$ such that for all $k \in E'$

$$_{E'}\!<k, W_t>_E \ = \ W_t^k \ , \quad _{E'}\!<k, \overline{W}_t>_E \ = \ \overline{W}_t^k \ P_\mu\text{–a.e.}$$

and the stochastic integrals $M_t := \int_0^t \sigma(X_s)dW_s$ ; $\overline{M}_t := \int_0^t \sigma(X_{T-s})d\overline{W}_s$

are well defined. Furthermore, for all $k \in E'$

$$_{E'}\!<k, M_t>_E \ = \ M_t^k \ , \quad _{E'}\!<k, \overline{M}_t>_E \ = \ \overline{M}_t^k \ \text{for } 0 \leq t \leq T , \ P_\mu\text{–a.s.}$$

Set $Z_t = \frac{1}{2}\int_0^t \sigma(X_s)dW_s - \frac{1}{2}\int_{T-t}^T \sigma(X_{T-s})d\overline{W}_s$ ; then Z is a continuous

E–valued process such that

$$k(Z_t) \ = \ k(X_t - X_0) \ P_\mu\text{–a.e. for all } k \in E' .$$

By the continuity of $X_t - X_0$ , we finally have that

$$X_t - X_0 \ = \ \frac{1}{2}\int_0^t \sigma(X_s)dW_s - \frac{1}{2}\int_{T-t}^T \sigma(X_{T-s})d\overline{W}_s \ , \quad 0 \leq t \leq T , \ P_\mu\text{–a.e.}.$$

REFERENCES


[1]  Albeverio,S., Kusuoka,S., Röckner,M.: Partial integration on infinite
     dimensional space and application to Dirichlet forms. J. London Math.
     Soc. 42, 122–136 (1990).

[2]  Albeverio,S., Röckner,M.: Classical Dirichlet forms on topological vec-
     tor space – construction of an associated diffusion process. Probab. Th.
     Rel. Fields 83, 405–434 (1989).

[3]  Albeverio,S., Röckner,M.: Stochastic differential equations in infinite
     dimensions: solutions via Dirichlet forms. Preprint Edinburgh 1989. To
     appear in Probab. Th. Rel. Fields.

[4]  Albeverio,S., Röckner,M.: Classical Dirichlet forms on topological vec-
     tor spaces – closability and a Cameron–Martin formula. J. Funct.
     Anal. 88, 395–436 (1990).

[5]  Fukushima,M.: Dirichlet forms and Markov processes. Amsterdam–
     Oxford–New York: North–Holland 1980.

[6]  Hui–Hsiung Kuo,H.H.: Gaussian Measures in Banach Spaces. Sprin-
     ger–Verlag, New York 1975.

[7]  Lyons,T.J., Zhang Tu–sheng: Decomposition of Dirchlet processes and
     its application. Preprint Edinburgh 1990. Publication in preparation.

[8]  Lyons,T.J., Zheng, Weian: A crossing estimate for the canonical pro-
     cess on a Dirichlet space and a tightness result. Colloque Paul Lévy sur
     les processus stochastique, Asterisque 157–158, 249–272 (1988).

[9]  Reed,M., Simon,B.: Methods of modern mathematical pyhsics. Vol. I.
     Functional Analysis. New York, Academic Press 1975.

[10] Röckner,M., Schmuland,B.: Tightness of $C_{1,p}$–capacities on Banach
     space. Preprint (1991), publication in preparation.

[11] Schmuland,B.: An alternative compactification for classical Dirichlet
     forms. Stochastics 33, 75–90 (1990).

# A Supersymmetric Feynman-Kac Formula

Alice Rogers

Department of Mathematics, King's College, Strand, London WC2R 2LS

## 1. INTRODUCTION

Over the years a number of non-commutative generalisations of Brownian motion have been developed. One such generalisation is the concept of fermionic Brownian motion, where paths in a space parametrised by anticommuting variables are introduced, following the original idea of Martin (1959), in order to study the evolution of quantum-mechanical systems which have fermionic degrees of freedom. As with conventional Brownian motion, the original motivation was physical, but the approach is also valuable in handling analytic questions on manifolds, extending standard probabilistic methods to spin bundles and differential forms. The aim of this article is to introduce probabilists to this extended family of physical and geometrical ideas, and particularly to the rôle of supersymmetry.

Recently physicists have given much attention to so-called supersymmetric systems. There are two (equivalent) ways of characterising such a system. One is to describe a system as supersymmetric if its quantum Hamiltonian operator $\hat{H}$ is the square of a Dirac-like operator $\hat{Q}$. (More details of both Hamiltonians and Dirac operators are given below.) The other characterisation is that there should exist an invariance of the classical action of the theory under transformations which intertwine bose and fermi degrees of freedom— it is this intertwining which makes the symmetry super in the usual physicist's terminology. Standard techniques in classical mechanics show these two characterisations to be equivalent, but for our purposes it will be most appropriate to concentrate on the first idea, and consider supersymmetric quantum mechanics.

The study of supersymmetry has lead to a variety of mathematical constructions which are, loosely, $Z_2$-graded extensions of conventional constructions. Such objects are referred to by the prefix 'super', which in this context has no connotations of superiority. For instance, superspace is a space parametrised by both commuting and anticommuting variables, while a supermanifold is a manifold which locally resembles superspace.

Before introducing the non-commuting, fermionic aspects of supersymmetric quantum mechanics, a brief summary of simple quantum mechanics and its relation to conventional Brownian motion will be given. This material is standard, but is presented so that the rôle of analogous constructions using anticommuting variables should be clear. Thus section 2 of this article describes conventional Brownian motion and bosonic quantum mechanics, while section 3 introduces the analysis of functions of anticommuting variables. At this stage, only finite-dimensional spaces will be considered. Section 4 describes how fermionic quantum mechanics may be formulated in terms of operators on spaces of functions of anticommuting variables, and in section 5 fermionic Brownian motion is introduced and its use in fermionic quantum mechanics established. In the following section fermionic Brownian motion is combined with conventional Brownian motion and thus paths in superspace are considered. The appropriate stochastic calculus for these paths is described. Also superpaths, which are parametrised by both a commuting time $t$ and an anticommuting time $\tau$, are introduced, leading to a supersymmetric Itô calculus. In section 7 a supersymmetric Feynman-Kac formula for the evolution operator $\exp(-\hat{H}t - \hat{Q}\tau)$ of a supersymmetric quantum-mechanical system in flat space is first described. Solutions to stochastic differential equations in superspace are used to construct more general superpaths, which are then used to extend the supersymmetric Feynman-Kac formula to a waider class of operators, particularly those which relate to curved superspace.

## 2. Brownian motion and quantum mechanics

In quantum mechanics the observables of the system are represented by self-adjoint operators on a Hilbert space, the space of states of the system. The particular choice of space and operators determines the physical content of the system, and is determined from the classical physics of the system by a process called quantisation. The quantisation process involves replacing classical observables such as position and momentum by quantum operators in such a way that the Poisson bracket algebra of the classical system is preserved as the commutator algebra of the observables. For a single particle with no intrinsic spin and moving in one dimension two fundamental observables are position $\hat{x}$ and momentum $\hat{p}$ corresponding to the classical canonical coordinates $x$ and $p$. Since $x$ and $p$ satisfy the canonical Poisson bracket relation

$$\{p, x\} = -1, \tag{2.1}$$

the quantum operators must obey the canonical commutation relation

$$[\hat{p}, \hat{x}] = -i\hbar. \tag{2.2}$$

In the Schrödinger reresentation of these operators the states (or wavefunctions, as they are often called in this context) belong to the Hilbert space $L^2(\mathbf{R})$ of complex-valued square-integrable functions of the real line, and the operators $\hat{p}$ and $\hat{x}$ are defined by

$$\begin{aligned} \hat{p}f(x) &= -i\hbar\frac{\partial f}{\partial x}(x), \\ \hat{x}f(x) &= x\,f(x). \end{aligned} \tag{2.3}$$

The dynamics of the system is defined by the Hamiltonian, which is the energy operator, and usually denoted $\hat{H}$. Classically, the Hamiltonian, which is a function of the canonical phase-space variables $x$ and $p$, generates infinitesimal time translations. According to the standard quantisation prescription the quantum Hamiltonian is obtained from the classical Hamiltonian by replacing the phase space variables $x$ and $p$ by the quantum operators $\hat{x}$ and $\hat{p}$. (Since classically $px = xp$ while in quantum mechanics $\hat{p}\hat{x} \neq \hat{x}\hat{p}$ this prescription may not be unambiguous.) In the Schrödinger picture the Hamiltonian determines the evolution of the system via the Schrödinger equation

$$i\hbar\frac{\partial f}{\partial t} = \hat{H}f, \tag{2.4}$$

(where units have been chosen such that the constant $\hbar = 1$). This equation may be solved if one can construct the evolution operator $\exp i\hat{H}t$. Feynman proposed a path integral approach to the construction of this operator for certain Hamiltonians; this is non-rigorous, involving ill-defined oscillatory integrals, but if one replaces $t$ by $it$, using imaginary time, the heat equation is obtained, and for many useful Hamiltonians Brownian motion and Wiener integrals provide representations of the solutions in the form of Feynman-Kac formulae. For instance, if $\hat{H} = \frac{1}{2}\hat{p}^2 + V(\hat{x})$, the Feynman-Kac formula is

$$(\exp -\hat{H}t)f(x) = \int d\mu \exp\left(-\int_0^t V(x + b_s)ds\right)f(x + b_t) \qquad (2.5)$$

where $d\mu$ denotes Wiener measure, $b_t$ is a Brownian path and units have been chosen so that Planck's constant $\hbar = 1$. While this is of course standard material for a probabilist, it has been spelt out in detail here to emphasise a point which is crucial in the generalisation to fermionic Brownian paths. This is the dissection of the Hamiltonian into two parts, the part $\frac{1}{2}\hat{p}^2$ catered for by the measure and the part $V(\hat{x})$ which appears explicitly in the integrand. Physically, $\frac{1}{2}\hat{p}^2$ is the Hamiltonian of a free particle (in other words, simply its kinetic energy) while $V(\hat{x})$ is the potential energy. Even in more complicated situations, such as that of a particle moving in curved space, an analogous dissection exists, and the fermionic Wiener measure constructed below incorporates the free Hamiltonian in a similar manner.

## 3. CALCULUS IN SUPERSPACE

In this section the analysis of functions of anticommuting variables is briefly described. Such functions carry a representation of the operators used in fermionic quantum mechanics which is an analogue of the Schrödinger representation for bosonic particles; this allows fermionic and bosonic variables to be treated in on an equal footing, which is clearly desirable in any supersymmetric theory. From a more mathematical point of view, representations of Clifford algebras can be built from differential operators on these spaces, which makes them the natural arena for various geometric constructions, such as Hodge de Rham theory on a Riemannian manifold.

Let $\theta^1, \ldots, \theta^n$ be $n$ anticommuting variables. There are various interpretations of this statement (Berezin and Leites (1975), Kostant (1977), Batchelor (1980), Rogers (1980)): one can take a concrete approach, considering anticommuting variables to be functions with values in the odd part of a Grassmann algebra; alternatively one may think of the $\theta^i$ as generators of an abstract algebra over the complex numbers, with relations

$$\theta^i \theta^j = -\theta^j \theta^i, \qquad i,j = 1,\ldots,n. \tag{3.1}$$

In either approach, one finds that the most general polynomial function, or element of the algebra, is of the form

$$f(\theta) = \sum_{\mu \in M_n} f_\mu \theta^\mu \tag{3.2}$$

where $\mu$ is a multi-index $\mu = \mu_1 \ldots \mu_k$ with $1 \leq \mu_1 < \ldots < \mu_k \leq n$, $M_n$ denotes the set of all such multi-indices (including the empty multi-index), each $f_\mu$ is a complex number and $\theta^\mu = \theta^{\mu_1} \ldots \theta^{\mu_k}$. For instance, if $n = 2$, the most general polynomial function has the form

$$f(\theta) = f_\emptyset + f_1 \theta^1 + f_2 \theta^2 + f_{12} \theta^1 \theta^2. \tag{3.3}$$

Differentiation with respect to $\theta^i$ is defined by

$$\frac{\partial}{\partial \theta^i} \theta^\mu = \begin{cases} (-1)^{r+1} \theta^{\mu_1} \ldots \theta^{\mu_{r-1}} \theta^{\mu_{r+1}} \ldots \theta^{\mu_k} & \text{if } \mu_r = i \\ = 0 & \text{otherwise.} \end{cases} \tag{3.4}$$

Integration is carried out according to the Berezin prescription where, if $f(\theta) = f_{1\ldots n} \theta^1 \ldots \theta^n +$ lower order terms, then

$$\int d^n\theta \, f(\theta) = f_{1\ldots n}. \tag{3.5}$$

This allows any linear operator $M$ on the space of polynomial functions to be represented by an integral kernel $M(\theta, \phi)$ with

$$Mf(\theta) = \int d^n\phi M(\theta,\phi) \, f(\phi). \tag{3.6}$$

(Here both $\theta = (\theta^1,\ldots,\theta^n)$ and $\phi = (\phi^1,\ldots,\phi^n)$ consist of variables all of which anticommute with one another. Indeed, throughout this paper, any anticommuting variable is defined to anticommute with any other anti-commuting variable.) One particular example is the $\delta$-function, which as usual is the kernel of the identity operator $I$:

$$
\begin{aligned}
\delta(\theta, \phi) &= I(\theta, \phi) \\
&= (\theta^1 - \phi^1) \ldots (\theta^n - \phi^n) \\
&= \int d^n \kappa \exp\left[-i\kappa^i(\theta^i - \phi^i)\right].
\end{aligned}
\tag{3.7}
$$

The Berezin integral is defined formally, and does not have any measure-theoretic interpretation as, for instance, the limit of a sum.

The analysis of functions of purely anticommuting variables is simply the analysis of a certain family of finite-dimensional vector spaces. It becomes useful when combined with functions of ordinary variables. (The notational convention will be that commuting variables are represented by lower case Roman letters, anticommuting variables by lower case Greek letters, while Roman capitals are used for quantities which may have either parity, or, in the case of multicomponent variables, both parities. Also the standard Einstein summation convention, where repeated indices are summed over their range, will be used.)

Given any class of functions on $\mathbf{R}^m$ (or on some subset of $\mathbf{R}^m$) there is a natural way to extend this to $(m, n)$-dimensional superspace. For instance $L^2(\mathbf{R}^m)$ may be extended to the space $L^{2\prime}(\mathbf{R}^{m,n})$ by defining this space to consist of functions of the form

$$
f(x^1, \ldots, x^m, \theta^1, \ldots \theta^n) = \sum_{\mu \in M_n} f_\mu(x)\theta^\mu,
\tag{3.8}
$$

where each of the coefficient functions $f_\mu$ is in $L^2(\mathbf{R}^m)$. Other spaces, such as $C^{\infty\prime}(\mathbf{R}^{m,n})$ can be defined in a similar way. Differentiation and integration may be defined by combining traditional operators with those described above for anticommuting variables. Each of these extended function spaces has a natural $Z_2$-grading, with an even function consisting entirely of terms containing the product of an even number of $\theta$'s and an odd function having terms which contain products of an odd number of $\theta$'s. There is a corresponding grading of differential operators on these spaces, with an operator said to be odd or even according as to whether it does or does not change the parity of a function. An important example of an odd operator is the supercharge $\hat{Q}$ of a supersymmetric system, as will appear below.

The particular case of calculus on a (1,1)-dimensional superspace (parametrised by a real variable $t$ and an anticommuting variable $\tau$), will now be described. Functions on this space will be termed superpaths, and play a considerable part in the supersymmetric stochastic calculus which follows. First, the superderivative $D_T$ is defined to be the operator

$$D_T = \frac{\partial}{\partial \tau} + \tau \frac{\partial}{\partial t}. \tag{3.9}$$

This operator may act on a function of the form $F(t, \tau) = A(t) + \tau B(t)$ where $A(t)$ has a time derivative. (As the notation implies, $A$ and $B$ may have either Grassmann parity.) One then has

$$D_T F(t, \tau) = B(t) + \tau \frac{\partial A}{\partial t}(t). \tag{3.10}$$

Where applicable straightforward calculation then shows that $D_T^2 = \frac{\partial}{\partial t}$. As will be seen below, in supersymmetric quantum mechanics there is a square root of the Schrödinger equation where the time derivative is replaced by $D_T$ and the Hamiltonian by its square root, the supercharge. Integration in (1,1)-dimensional superspace is exceptional in that one may define an integral which includes both commuting and anticommuting limits, in the following way (Rogers (1987a)):

$$\int_0^\tau d\sigma \int_0^t ds\, F(s, \sigma) =_{def} \int d\sigma \int_0^{t+\sigma\tau} ds\, F(s, \sigma). \tag{3.11}$$

The even integral on the right hand side of this equation is evaluated by regarding $\int_0^u ds\, F(s, \sigma)$ as a function of $u$, and evaluating this function when $u = t + \sigma\tau$ by Taylor expansion about $u = t$. (Here both $\sigma$ and $\tau$ are anticommuting variables, that is, they anticommute with one another and with all other anticommuting variables.) Integration with respect to the anticommuting variable $\sigma$ is then carried out according to the usual Berezin prescription. Thus if $F(t, \tau) = A(t) + \tau B(t)$ one has

$$\int_0^\tau d\sigma \int_0^t ds\, F(s, \sigma) = \tau A(t) + \int_0^t B(s)ds. \tag{3.12}$$

This definition of integration leads to a supersymmetric version of the fundamental theorem of calculus in the form

$$\int_0^\tau d\sigma \int_0^t ds \, D_S F(s,\sigma) = F(t,\tau) - F(0,0), \qquad (3.13)$$

and hence the natural rule for the superderivative of such an integral with respect to its upper limits is obtained as

$$D_T \int_0^\tau d\sigma \int_0^t ds \, G(s,\sigma) = G(t,\tau). \qquad (3.14)$$

The stochastic version of these integration results gives a supersymmetric Itô formula which in turn leads to the supersymmetric Feynman-Kac formula of section 8.

## 4. FERMIONIC QUANTUM MECHANICS

Intrinsic spin is a property of a particle in quantum mechanics which has no classical analogue. It may take integer or half-integer values, particles with integer spin being known as bosons and those with half-integer spin as fermions. To describe $n$ fermionic degrees of freedom one requires operators $\hat{\psi}^i$, $i = 1, \ldots, n$ satisfying canonical anti-commutation relations

$$\hat{\psi}^i \hat{\psi}^j + \hat{\psi}^j \hat{\psi}^i = 2. \qquad (4.1)$$

An analogue of the Schrödinger representation can then be constructed by letting the Hilbert space of states be the space of polynomial functions of $n$ anti-commuting variables $\theta^1, \ldots, \theta^n$ and setting

$$\hat{\psi}^i = \theta^i + \frac{\partial}{\partial \theta^i}, \qquad i = 1, \ldots, n. \qquad (4.2)$$

The dynamics is then specified by setting $\hat{H} = V(\hat{\psi})$. For a free particle the fermionic operators do not contribute to the Hamiltonian, and so in building the appropriate Wiener measure one splits the Hamiltonian $\hat{H} = V(\hat{\psi})$ into $\hat{H} = 0 + V(\hat{\psi})$ and forms the measure from the kernel of the identity operator. Despite its apparently vacuous nature, this measure, which is described in detail in the following section, has useful properties and can be used to build a Feynman-Kac formula.

This purely fermionic quantum mechanics only describes the spin degrees of freedom of a particle, and a full treatment must include position and momentum, so that the appropriate Hilbert space is a space of functions of both commuting and anticommuting variables, with the Hamiltonian a function of $\hat{x}$, $\hat{p}$ and $\hat{\psi}$. In particular, supersymmetric quantum mechanics in flat space is based on the free supercharge $\hat{Q}_0 = \frac{i}{\sqrt{2}} \hat{\psi}^i \frac{\partial}{\partial x^i}$ which satisfies

$$\hat{Q}_0^2 = \hat{H}_0, \tag{4.3}$$

where $\hat{H}_0$ is the Hamiltonian of a free particle,

$$\hat{H}_0 = -\tfrac{1}{2} \sum_{i=1}^{n} \frac{\partial}{\partial x^i} \frac{\partial}{\partial x^i}. \tag{4.4}$$

$\hat{Q}_0$ is the simplest example of a Dirac operator; a more general Dirac operator is described in section 7.

## 5. FERMIONIC BROWNIAN MOTION

Let $I$ be the real interval $[0, t_f)$ (where $t_f$ may possibly be infinite). Then, for any positive even integer $n$, fermionic Wiener measure is defined on the infinte-dimensional superspace $(\mathbf{R}^{0,2n})^I$ by specifying its finite distributions. As with conventional Wiener measure, these are built from the kernel of the evolution operator $\exp -\hat{H}_0 t$ of the free Hamiltonian. Thus, if $J = \{t_1, \ldots, t_N\} \subset I$ with $0 < t_1 < \ldots < t_N$, the distribution function $f_J$ on $\mathbf{R}^{0,2nN}$ is defined by

$$\begin{aligned} f_J({}^1\underline{\theta}, \ldots, {}^N\underline{\theta}, {}^1\underline{\rho}, \ldots, {}^N\underline{\rho}) \\ = \pi(0, {}^1\underline{\theta}, {}^1\underline{\rho})\pi({}^1\underline{\theta}, {}^2\underline{\theta}, {}^2\underline{\rho}) \ldots \pi({}^{N-1}\underline{\theta}, {}^N\underline{\theta}, {}^N\underline{\rho}), \end{aligned} \tag{5.1}$$

where ${}^r\underline{\theta} = (\theta^{1r}, \ldots, \theta^{nr})$ for $r = 1, \ldots, N$ and

$$\pi(\underline{\theta}, \underline{\phi}, \underline{\kappa}) = \exp\left[-i\kappa^i(\theta^i - \phi^i)\right]. \tag{5.2}$$

The space $(\mathbf{R}^{0,2n})^I$ equipped with this measure will be called fermionic Wiener space. Fermionic Brownian motion is then defined to be the $(0, 2n)$-dimensional process $\theta_t^j, \rho_t^j$, $j = 1, \ldots, n$, with $\theta_0$ and $\rho_0$ defined to be zero. For the purpose of representing the operators $\theta^i + \partial/\partial\theta^i$ the process

$$\xi_t = \theta_t + i\rho_t \tag{5.3}$$

is also useful. The principle property of this measure is that if $V(\hat{\psi}) = \hat{\psi}^{\mu_1} \ldots \hat{\psi}^{\mu_k}$ with $1 \le \mu_1 < \ldots < \mu_k \le n$, and $f$ is a function of $n$ anti-commuting variables $\theta^1, \ldots, \theta^n$,

$$(V(\hat{\psi}))f(\theta) = \int d\mu_f (\theta^{\mu_1} + \xi_s^{\mu_1}) \ldots (\theta^{\mu_k} + \xi_s^{\mu_k}) f(\theta + \theta_s). \qquad (5.4)$$

This result allows one to establish a Feynman-Kac formula in the expected form. That is, if $f$ is a polynomial function of $n$ anti-commuting variables and $\hat{H} = V(\hat{\psi})$ where $V(\hat{\psi}) = \sum_{\mu \in M_n} V_\mu \hat{\psi}^\mu$, the $V_\mu$ being complex numbers,

$$(\exp -\hat{H}t)f(\theta) = \int d\mu_f \exp\left(-\int_0^t V(\theta + \xi_s)ds\right) f(\theta + \theta_t). \qquad (5.5)$$

Details of the limiting process needed to define the integrand in this expression, together with other aspects of fermionic Brownian motion, may be found in Rogers (1987b).

When considering bosons and fermions together, one may use the super Wiener space of paths $(b_t, \theta_t, \rho_t)$ in $(m, 2n)$-dimensional superspace, with super Wiener measure $d\mu_s$ being the product of bosonic Wiener measure $d\mu_b$ and fermionic Wiener measure $d\mu_f$. This leads to a Feynman-Kac formula for Hamiltonians of the form $\hat{H} = \frac{1}{2}\hat{p}^2 + V(\hat{x}, \hat{\psi})$.

## 6. STOCHASTIC CALCULUS IN SUPERSPACE

As in the purely classical, bosonic case a much wider class of differential operators can be included if one considers solutions of stochastic differential equations, but first the appropriate stochastic calculus must be developed by incorporating fermionic paths into the Itô calculus. This stochastic calculus is outlined in this section; a more detailed account may be found in Rogers (1990).

Classically Wiener measure, Brownian motion and stochastic calculus can be formulated in a number of different ways. Only one of these approaches seems compatible with the formalism of fermionic Brownian motion, that of defining the measure from its finite-dimensional marginals, using the Kolmogorov extension theorem. A further restriction is that random variables must be defined as the limit of a set of functions $f_N$ defined on $(\mathbf{R}^{m,2n})^{J_N}$ where the $J_N$ are finite subsets of $I$, and the expectations of the $f_N$ tend to a limit. (The details may be found in Rogers (1987b).) A process $A_t$ corresponding to sequences of functions $A_{t(N)}$ on $(\mathbf{R}^{m,2n})^{J_{t(N)}}$ is said to be adapted if, for each $t \in I$ and for each positive integer $N$, $J_{t(N)} \subset (0,t]$.

It is not necessary to define an integral along fermionic Brownian paths because the fermionic measure already allows one to handle differential operators of any order; this contrasts with the classical, commuting case, where Itô integrals are needed to handle operators which contain first order derivatives, while stochastic differential equations are required for general elliptic second order differential operators. The case of fermionic Brownian paths is different because the phase space or fourier transform variable is retained in the paths. In fact it does not seem possible to define integrals along fermionic Brownian paths, because they are extremely irregular. This irregularity is most easily seen if one considers the Fourier mode expansion of $\xi_t$ given in Rogers (1987b). However, it is still necessary to give a meaning to $\int_0^t A_s db_s$ where $A_s$ is an adapted process on super Wiener space. Provided that it possesses suitable regularity properties, such a process can be integrated along a Brownian path according to the following definition.

## Definition 6.1

For $a = 1, \ldots, m$ suppose that $F_t^a$ is an adapted stochastic process on the super Wiener space $(\mathbf{R}_L^{m,2n})^I$. For each $N = 1, 2, \ldots$ let $J_N = \{t_1, \ldots, t_{2^N-1}\}$ with $t_r = \frac{rt}{2^N}$ for $r = 1, \ldots, 2^N - 1$. Then the process $F_t^a$ is defined by sequences of functions ${}^N F_t^a$ on $(\mathbf{R}_L^{m,2n})^{J_N}$. Let $G_N \in L^{2\prime}\left((\mathbf{R}_L^{m,2n})^{J_N}, \mathbf{C}\right)$ with

$$
\begin{aligned}
& G_N({}^1\underline{x}, \ldots, {}^{2^N-1}\underline{x}, {}^1\underline{\theta}, \ldots, {}^{2^N-1}\underline{\theta}, {}^1\underline{\rho}, \ldots, {}^{2^N-1}\underline{\rho}) \\
& = \sum_{a=1}^{m} \sum_{r=1}^{2^N-1} {}^N F^a({}^1\underline{x}, \ldots, {}^r\underline{x}, {}^1\underline{\theta}, \ldots, {}^r\underline{\theta}, {}^1\underline{\rho}, \ldots, {}^r\underline{\rho})({}^{r+1}x^a - {}^r x^a).
\end{aligned}
\tag{6.1}
$$

Then the sequence $(J_N, G_N)$ defines a random variable which is denoted

$$
\int_0^t \sum_{a=1}^{m} F_s^a \, db_s^a.
$$

It is also useful to define a stochastic integral to be a process $Z_t$ such that if $0 \leq t_1 \leq t_2 \leq t_f$,

$$Z_{t_1} - Z_{t_2} = \int_{t_1}^{t_2} A_s ds + \int_{t_1}^{t_2} \sum_{a=1}^{m} C_s^a db_s^a$$

where $A_s$ and $C_s^a$ are adapted processes on the super Wiener space, again obeying suitable regularity properties. Full details of these definitions may be found in Rogers (1990). A key theorem, underpinning the development of supersymmetric stochastic calculus and the supersymmetric Feynman-Kac formula, is the following generalised Itô theorem.

**Theorem 6.2**

Suppose that $Z_s^j$ $(j = 1, \ldots, p+q)$ are stochastic integrals on the super Wiener space $\left(\mathbf{R}_L^{m,2n}\right)^{(0,t)}$ with

$$dZ_s^j = A_s^j dt + \sum_{a=1}^{m} C_s^{aj} db^a(s) \tag{6.2}$$

and that $Z^j$ is even for $j = 1, \ldots, p$ and odd for $j = p+1, \ldots, p+q$. Also suppose that $\mathbf{E}(|A_s^j|)^2$ is bounded for each positive integer $r$.

Let $F \in C^{5\prime}(\mathbf{R}^{(p,q)}, \mathbf{C})$. Then, with ${}^N Z_s^j$ denoting the $N^{th}$ term of the sequence defining the random variable $Z_s^j$, the sequence $F({}^N Z_s)$ defines a stochastic process denoted $F_s$, and $F_s$ is also a stochastic integral with

$$dF_s = \sum_{j=1}^{p+q} \sum_{a=1}^{m} \partial_j F(Z_s) C_s^{aj} db_s^a$$

$$+ \left( \sum_{j=1}^{p+q} \partial_j F(Z_s) A_s^j + \frac{1}{2} \sum_{a=1}^{m} \sum_{j=1}^{p+q} \sum_{k=1}^{p+q} C_s^{ak} C_s^{aj} \partial_j \partial_k F(Z_s) \right) ds. \tag{6.3}$$

This theorem, which is proved in Rogers (1990), shows that the fermionic Brownian paths do not affect the Itô correction term. In particular, for the simple case where $C_t^{aj} = f(b_t, \theta_t, \rho_t)$ and $A_s^j = 0$, the correction term takes the familiar form $\frac{1}{2} \sum_{i=1}^{m} \frac{\partial}{\partial x^i} \frac{\partial}{\partial x^i} f(b_t, \theta_t, \rho_t)$.

The next step is to develop a stochastic version of the $(1,1)$-dimensional integrals $\int_0^\tau d\sigma \int_0^t ds$, and so obtain a supersymmetric Itô theorem with a correction term of the form $(\hat{Q}_0 - 2\tau \hat{H}_0) f(b(t,\tau), \xi(t,\tau))$. Here $b(t,\tau)$ and $\xi(t,\tau)$ are superpaths, which will now be defined.

### Definition 6.3

For $0 \le t \le t_f$ and $\tau$ an anticommuting variable let $Z_T$, $\zeta_T$ denote the processes

$$
\begin{aligned}
Z_T^a &= b_t + \tfrac{i}{\sqrt{2}} \tau \xi_t^a \\
\zeta_T^a &= \xi_t^a + \sqrt{2} i \sigma \dot{b}_t^a.
\end{aligned}
\tag{6.4}
$$

(Here the notation $\dot{b}_t$ is formal; it is to be interpreted in combination with $dt$ as $\dot{b}_t \, dt = db_t$. The placing of $\zeta_T^a$ in formulae will be such that only this combination will occur.) Before giving the supersymmetric Itô formula, a time integral for superpaths must be defined.

### Definition 6.4

Suppose that $G_s$ and $K_s$ are suitably regular adapted stochastic processes on super Wiener space. Then, if $F_S = G_s + \sigma K_s$,

$$
\int_0^\tau d\sigma \int_0^t ds \, F_S =_{def} \int_0^t ds \, K_s + \tau G_t.
\tag{6.5}
$$

The following supersymmetric Itô formula may then be proved by expanding each side as a series in $\tau$ and applying the classical Itô theorem together with (6.3).

### Theorem 6.5

Suppose that $f \in C^{5\prime}(\mathbf{R}^{p,m})$. Then

$$
f(z_T, \theta_t) - f(0,0) = \int_0^\tau d\sigma \int_0^t ds \, [Q_0 f(z_S, \theta_s)
$$
$$
- 2\sigma H_0 f(z_S, \theta_s)] + \int_0^\tau d\sigma \int_0^t db_s^a \, \sigma \partial_a f(z_S, \theta_s).
\tag{6.6}
$$

This Itô formula will be used in the following section to prove a supersymmetric Feynman-Kac formula. A further result in this extended stochastic calculus is the following theorem which establishes the existence of solutions of a useful class of stochastic differential equations. The full details of the regularity conditions on the functions $A_a^j$ and $B^j$ require an excursion into the theory of functions of anticommuting variables which, together with a proof of the theorem, may be found in Rogers (1990).

**Theorem 6.6**

Let $p$, $m$ and $n$ be positive integers. For $j = 1, \ldots, p$ and for $a = 1, \ldots, m$ let $A_a^j$ and $B^j$ be suitably regular functions on $\mathbf{R}^{p,2n} \times \mathbf{R}^+$. Also let $A$ be a fixed element of $\mathbf{C}^p$. Then, for $j = 1, \ldots, p$, there exists a stochastic process $Z_s^j$ such that

$$dZ_t^j = A_a^j(Z_s, \theta_s, \rho_s, s)db_s^a + B^j(Z_s, \theta_s, \rho_s, s)ds \quad (a)$$
$$Z_0 = A \quad (b)$$

$$(6.7)$$

(where $(b_s, \theta_s, \rho_s)$ are Brownian paths in $(m, n)$-dimensional super Wiener space). These solutions are effectively unique, in a sense which is defined in Rogers (1990).

By using solutions to carefully chosen stochastic differential equations a more general supersymmetric Itô formula may be proved, leading to a supersymmetric Feynman-Kac formula for a wider class of supersymmetric Hamiltonians.

## 7. EVOLUTION IN SUPERSYMMETRIC QUANTUM MECHANICS

In supersymmetric quantum mechanics the Hamiltonian $\hat{H}$ is the square of the supercharge $\hat{Q}$. The supercharge is always an odd operator, which means that the Schrödinger equation

$$\frac{\partial f}{\partial t} = -\hat{H}f \tag{7.1}$$

has a square root (Friedan and Windey (1984))

$$D_T f = \hat{Q}f. \tag{7.2}$$

Here $f$ is a function of $m+1$ commuting variables $x^1, \ldots, x^m, t$ and $n+1$ anticommuting variables $\theta^1, \ldots, \theta^n, \tau$. Since, by definition, $\theta\tau = -\tau\theta$, one finds that $\hat{Q}$ anti-commutes with $D_T$, and hence, recalling that $D_T^2 = \frac{\partial}{\partial t}$ one finds that (7.2) implies (7.1). One may also check by direct calculation that, provided $\hat{H}$ is suitably regular,

$$
\begin{aligned}
D_T\big(\exp(-\hat{H}t - \hat{Q}\tau)\big) &= \hat{Q}\exp(-\hat{H}t - \hat{Q}\tau) \\
&= \big(\exp(-\hat{H}t - \hat{Q}\tau)\big)\,(\hat{Q} - 2\tau\hat{H}).
\end{aligned}
\tag{7.3}
$$

Thus $\exp(-\hat{H}t - \hat{Q}\tau)$ is the evolution operator for the square root of the Schrödinger equation.

For a supersymmetric particle in flat, Euclidean space, wave-functions are defined on an $(m, m)$-dimensional superspace and $\hat{Q}$ has the form

$$
\hat{Q} = \tfrac{i}{\sqrt{2}}\hat{\psi}^a \frac{\partial}{\partial \hat{x}^a} + \phi(x, \hat{\psi})
\tag{7.4}
$$

where $\phi$ is an odd function. By applying the supersymmetric Itô formula (6.6) to a carefully chosen function one obtains the following supersymmetric Feynman-Kac formula.

**Theorem 7.1**

Let $F \in C^{\infty\prime}(\mathbf{R}^{m,m})$. Then

$$
\begin{aligned}
&\exp(-\hat{H}t - \hat{Q}\tau)F(x, \theta) \\
&= \int d\mu_s \exp\left\{ \int_0^\tau d\sigma \int_0^t ds\, \phi(x + \tfrac{i}{\sqrt{2}}\tau\theta + z_S, \theta + \zeta_S) \right\} \\
&\quad \times F(x + \tfrac{i}{\sqrt{2}}\tau\theta + z_T, \theta + \theta_t).
\end{aligned}
\tag{7.5}
$$

*Proof*     Let $\bar{U}_{t,\tau}$ be the operator on $L^{2\prime}(\mathbf{R}^{m,m})$ defined by the completion of $U_{t,\tau}: C_0^{\infty\prime}(\mathbf{R}^{m,m}) \to L^{2\prime}(\mathbf{R}^{m,m})$ with

$$
\begin{aligned}
&U_{t,\tau}\, G(x, \theta) \\
&= \mathbf{E}\Bigg[ \exp\left\{ \int_0^\tau d\sigma \int_0^t ds\, \phi(x + \tfrac{i}{\sqrt{2}}\sigma\theta + z_S, \theta + \zeta_S) \right\} \\
&\quad \times G(x + \tfrac{i}{\sqrt{2}}\tau\theta + Z_T, \theta + \theta_t) \Bigg]
\end{aligned}
\tag{7.6}
$$

for $G \in C_0^{\infty'}(\mathbf{R}_L^{m,m})$. Then, applying the supersymmetric Itô formulae, one finds

$$U_{t,\tau} G(x,\theta) - G(x,\theta)$$
$$= \int_0^\tau d\sigma \int_0^t ds \, U_{s,\sigma} \hat{Q} G(x,\theta) - 2\sigma U_{s,\sigma} \hat{H} G(x,\theta). \tag{7.7}$$

Thus

$$D_T U_{t,\tau} G(x,\theta) = U_{t,\tau}(\hat{Q} - 2\tau\hat{H})G(x,\theta), \tag{7.8}$$

and hence, using equation (7.3) and appealing to the uniquness of solutions to differential equations such as (7.8), one may deduce that

$$U_{t,\tau} G(x,\theta) = \exp(-\hat{H}t - \hat{Q}\tau)G(x,\theta) \tag{7.9}$$

as required.

These methods can be extended to more general theories, where the supercharge has the form

$$\hat{Q} = \hat{\psi}^a e_a^i(x)\frac{\partial}{\partial x_i} + \phi(x,\hat{\psi}), \tag{7.10}$$

(with the function $\phi$ linear in $\hat{\psi}$) by introducing modified superpaths. The Hamiltonians of such theories takes the form

$$\hat{H} = -\tfrac{1}{2}g^{ij}(x)\frac{\partial}{\partial x_i}\frac{\partial}{\partial x_j} + \text{ lower order terms} \tag{7.11}$$

where $g^{ij}(x) = \sum_{a=1}^m e_a^i(x)e_a^j(x)$. As in the case without spin operators, paths which satisfy stochastic differential equations are required.

**Definition 7.2**

For $a, i = 1,\ldots,m$ let

$$z_S^i = x_s^i + \tfrac{i}{\sqrt{2}}\sigma\xi_s^a e_a^i(x_s)$$
$$\zeta_S^a = \xi_s^a + \sqrt{2}i\sigma b_s^a \tag{7.12}$$
$$\text{and} \quad \eta_s^i = \theta^a e_a^i(x_s).$$

where $x_s^i$ satisfies

$$dx_s^i = e_a^i(x_s)db_s^a + \tfrac{1}{2}(\delta^{ab} + \xi_s^a\xi_s^b)e_a^j(x_s)\partial_j e_b^i(x_s)ds$$

$$x_0 = 0. \tag{7.13}$$

In terms of the superpaths $z_s^i, \zeta_s^i$, one has the following supersymmetric Feynman-Kac formula for the supercharge $\hat{Q}$ of (7.10):

**Theorem 7.3**

If $F \in C^{5\prime}(\mathbf{R}_L^{m,m})$ and $(x, \theta) \in \mathbf{R}_L^{m,n}$,

$$\exp\left(-\hat{H}t - \hat{Q}\tau\right) F(x, \theta)$$

$$= \mathbf{E}\left[\exp\left\{\int_0^\tau d\sigma \int_0^t ds\, \phi(x + z_s + \tfrac{i}{\sqrt{2}}\sigma\eta_s, \theta + \zeta_s)\right\}\right. \tag{7.14}$$

$$\left. F(x + z_T + \tfrac{i}{\sqrt{2}}\tau\eta_t, \theta + \theta_t)\right].$$

This theorem is proved in Rogers (1990), using a generalisation of the supersymmetric Itô formula (6.6).

For geometrical applications, such as the investigation of the Dirac operator of a spin bundle or the Hodge de Rham operator on a Riemannian manifold, it is necessary to formulate this approach in a globally valid manner on a supermanifold constructed over a conventional manifold. As in the classical treatment of path integration on manifolds described in Elworthy (1982) and Ikeda and Watanabe (1981), the approach depends on stochastic differential equations which transform covariantly under change of coordinate, as will be described in a forthcoming paper (Rogers, in preparation).

## 8. Conclusion

This article describes some new stochastic techniques for handling Dirac-like operators. It is primarily addressed to probabilists, and it seems appropriate to conclude with a brief indication of the motivation for this work. Physically, symmetry in a model is desirable both as an aesthetic criterion for selecting models, and for the practical reason that models tend to become more tractable as they become more symmetric. One great hope of supersymmetry is that cancellations between the fermi and bose sectors will remove the infinities which plague most models in quantum field theories. So far this hope has not been realised, but neither has it been abandoned—currently superstring models are receiving most attention as possible candidates for a unified theory. Supersymmetry is essential to these models, but the technical demands are enormous, and it will be some time before a full assessment of them can be made. However it is clear that rigorous, rather than purely formal, manoevres with anticommuting variables should play a part in handling the fermionic aspects of these theories, and this is one motivation for the work described in this article.

Mathematically, supersymmetry has had some interesting consequences. Use has been made in various contexts of the positivity of $\hat{H}$ implied by the formula $\hat{H} = \hat{Q}^2$, but of particular relevance to the material of this article are the supersymmetric proofs of the index theorem due to Alvarez-Gaumé (1983)and Friedan and Windey (1984). These proofs use fermionic path-integral manipulations which, while standard for physicists, are not entirely rigorous. The techniques descrbed in this article allow one to make these arguments rigorous, as will be shown in a forthcoming paper (Rogers, in preparation).

## References

Alvarez-Gaumé, L. (1983) Supersymmetry and the Atiyah-Singer index theorem. Comm. Math. Phys. **90** 161-173

Batchelor, M. (1980) Two approaches to supermanifolds. Trans. Amer. Math. Soc. **258** 257-270

Berezin, F.A. and Leites, D. (1975) Supervarieties. Sov. Math. Dokl. **16** 1218-1222

Elworthy, K.D. (1982) Stochastic differential equations on manifolds. London Mathematical Society Lecture notes in Mathematics, Cambridge.

Friedan, D. and Windey, P. (1984) Supersymmetric derivation of the Atiyah-Singer index theorem and the chiral anomaly. Nuc. Phys. B **235** 395-416

Ikeda, N. and Watanabe, S. (1981) Stochastic differential equations and diffusion processes. North-Holland, Amsterdam

Kostant, B. (1977) Graded manifolds, graded Lie theory and prequantization. Differential geometric methods in mathematical physics, Lecture notes in mathematics **570** 177-306, Springer

Martin, J. (1959) The Feynman principle for a Fermi system. Proc. Roy. Soc. A **251** 543-549

Rogers, A. (1980) A global theory of supermanifolds. Journ. Math. Phys. **21** 1352-1365

Rogers, A. (1985) Integration and global aspects of supermanifolds, in Topological properties and global structure of spacetime, eds Bergmann, P. and De Sabbata, V, Plenum, New York

Rogers, A. (1986) Graded manifolds, supermanifolds and infinite- dimensional Grassmann algebras. Comm.Math. Phys. **105** 375-384

Rogers, A. (1987a) Supersymmetric path integration. Phys. Lett. B **193** 48-54

Rogers, A. (1987b) Fermionic path integration and Grassmann Brownian motion. Comm. Math. Phys. **113** 353-368

Rogers, A. (1990) Stochastic calculus in superspace I: supersymmetric Hamiltonians, King's College London preprint

Rogers, A. (in preparation) Stochastic Calculus in Superspace II: spin bundles, supermanifolds and the index theorem

# ON LONG EXCURSIONS OF BROWNIAN MOTION AMONG POISSONIAN OBSTACLES

Alain-Sol Sznitman
Courant Institute of Mathematical Sciences
New York University
251 Mercer St., New York, NY 10012

## 0. Introduction

Consider random obstacles on $\mathbb{R}^d$, $d \geq 1$, given by means of closed balls of fixed radius $a > 0$, centered at the points of a Poisson cloud, with constant intensity $\nu > 0$, and law $\mathbb{P}(d\omega)$. Let $Z.$ be an independent Brownian motion starting from the origin, $T$ stand for its entrance time in the obstacles, and $P_0$ for the standard Wiener measure, Donsker-Varadhan [2] showed that:

$$(0.1) \qquad \lim_{t \to \infty} t^{-d/(d+2)} \log \mathbb{P} \otimes P_0[T > t] = -c(d, \nu), \quad \text{with}$$

$$(0.2) \qquad c(d, \nu) = (\nu \omega_d)^{2/(d+2)} \left( \frac{d+2}{2} \right) \left( \frac{2\lambda_d}{d} \right)^{d/(d+2)}.$$

Here $\omega_d$, $\lambda_d$ stand respectively for the volume of the unit ball in $\mathbb{R}^d$, and the principal Dirichlet eigenvalue of $-\frac{1}{2}\Delta$ in the unit ball of $\mathbb{R}^d$. The constant $c(d, \nu)$ comes in fact from the minimization problem:

$$(0.3) \qquad c(d, \nu) = \inf_U \left\{ \nu|U| + \lambda(U) \right\},$$

where $U$ runs over the class of bounded open sets in $\mathbb{R}^d$, with negligible boundary. In (0.3), $|U|$ and $\lambda(U)$ denote respectively the volume of $U$ and the principal Dirichlet eigenvalue of $-\frac{1}{2}\Delta$ in $U$. The optimal choice of $U$ corresponds to a ball of radius

$$(0.4) \qquad R_0 = \left( \frac{2}{d} \frac{\lambda_d}{\nu \omega_d} \right)^{1/(d+2)}.$$

The lower bound part of (0.1) can in fact be obtained by assuming that $Z.$ does not leave until time $t$ the ball centered at the origin of radius $R_0 t^{1/(d+2)}$, and that no obstacle falls in this ball.

Alfred P. Sloan Fellow. Research partially supported by NSF grant DMS 8903858. Pres. addr.: Dept. Math., ETH-Zentrum, CH-8092 Zürich

In this paper, we are mainly concerned with the large $t$ behavior of the probability that Brownian motion surviving until time $t$, performs a "large excursion" at distance of order $t^{d/(d+2)}$ from the origin. We show that in dimension $d \geq 2$, for $x \geq 0$:

$$(0.5) \quad -(c(d,\nu) + k(d,\nu,a)x)$$
$$\leq \varliminf_{t\to\infty} t^{-d/(d+2)} \log \mathbf{P} \otimes P_0[\sup_{u\leq t}|Z_u| > xt^{d/(d+2)}, T > t]$$
$$\leq (c(d,\nu) + \nu\omega_{d-1}a^{d-1}x) ,$$

where the constant $k(d,\nu,a)$ is defined as:

$$(0.6) \qquad k(d,\nu,a) = \min_{r>0}(\nu\omega_{d-1}(a+r)^{d-1} + \sqrt{2\lambda_{d-1}}/r) .$$

In dimension $d = 1$, the situation is somewhat different, for these excursions are not so large, and we find:

$$(0.7) \qquad \lim_{t\to\infty} t^{-1/3} \log \mathbf{P} \otimes P_0\big[\sup_{u\leq t}|Z_u| \geq xt^{1/3} , T > t\big]$$
$$= -(\nu(x \vee 2R_0) + \frac{\pi^2}{2(x \vee 2R_0)^2}) .$$

These bounds suggest the presence of the large deviation property for the law of $t^{-d/(d+2)}\sup_{u\leq t}|Z_u|$ or $t^{-d/(d+2)}|Z_t|$ under the conditional law $\mathbf{P} \otimes P_0[\cdot/T > t]$. This can be checked in the case of dimension 1, which however is somewhat singular (see Remark 1.4).

The bounds (0.5), (0.7) have the following consequence. Eisele and Lang, motivated by the work of Grassberger and Procaccia [4], proved in [3] the existence of a change of regime in the long time behavior for the survival probability of Brownian motion with a constant drift $h$, evolving among Poissonian traps, as $|h|$ varies. Stated in terms of Wiener measure, they showed the existence of a constant $\alpha(d,\nu,a) \in [\nu\omega_{d-1}a^{d-1}, k(d,\nu,a)]$ such that

$$(0.8) \qquad \lim_{t\to\infty} \frac{1}{t} \log \mathbf{E} \otimes E_0[e^{h\cdot Z_t} , T > t] = 0 , \quad \text{when} \quad |h| \leq \alpha(d,\nu,a)$$
$$> 0 , \quad \text{when} \quad |h| > \alpha(d,\nu,a) .$$

The estimates (0.5) and (0.7) allow us to refine (0.8) for small $|h|$, that is $|h| < \nu\omega_{d-1}a^{d-1}$ when $d \geq 2$, and $|h| < \nu$, when $d = 1$. Indeed, in this case we show:

$$(0.9) \quad \lim_{t\to\infty} t^{-d/(d+2)} \log \mathbf{E} \otimes E_0[e^{h\cdot Z_t}, T > t]$$
$$= -c(1, \nu - |h|) , \quad \text{when} \quad d = 1 ,$$
$$= -c(d,\nu) , \quad \text{when} \quad d \geq 2 .$$

So unlike the case of dimension 1, when $d \geq 2$, the result does not depend on $h$, provided $0 \leq |h| < \nu \omega_{d-1} a^{d-1}$.

Let us give some indications on how the estimates (0.5), in the $d \geq 2$ situation, are derived.

The lower bound is the easier part. One considers a long cylindrical tube of radius $r$ and approximate length $x\,t^{d/(d+2)}$, with a sphere of radius $R_0 t^{1/(d+2)}$ attached to one end ($R_0$ is defined in (0.4)). We now require that no point of the cloud falls in a neighborhood of size $a$ of the union of the cylinder and the ball. We let Brownian motion start near the "free" end of the cylinder rush to the other end in a time of order $t^{d/(d+2)}$, and then rest until time $t$ in the ball of radius $R_0 t^{1/(d+2)}$. After some optimization this yields the lower bound part of (0.5).

It is worth mentioning the following point. A variation of the previous argument, with a cylinder of smaller size $t^\alpha$, $\alpha \in (0, d/(d+2))$ shows that the basic asymptotic result (0.1) remains unchanged if we replace the event $\{T > t\}$ by $\{T > t\} \cap \{|Z_t| > x t^\alpha\}$. This explains why it is delicate to obtain a confinement property of surviving Brownian motion in scale $t^{1/(d+2)}$. So far it is known to hold in the two dimensional case (see [1], [7]).

The upper bound part of (0.5) is more difficult. The crucial reduction step is derived in section II. It shows that to prove the upper bound, one can restrict the analysis to the following heuristic situation. The process does not exit before time $t$ a large box $T$ centered at the origin, of size $\sim$ const. $t^{(d+1)/(d+2)}$. This $T$ is chopped in subcubes of size $L t^{1/(d+2)}$. A finite number of these subcubes constitute "clearings" which can be used by the process as "resting places", whereas the other cubes, "the forest", constitute to a certain degree, a hostile environment for the process, which is killed there at a given rate. Adjusting parameters, one can in fact assume that the process does not perform too many excursions ($\leq \eta\, t^{d/(d+2)}$), in the forest, outside a neighborhood of size $t^{1/(d+2)}$ of the clearings, and does not spend too much time ($\leq \eta\, t$) in these excursions.

The notion of "clearing" we use is based on the technique of "enlargement of obstacles" and the notion of "good obstacles", i.e. well surrounded obstacles, developed in [5], [6], [7]. The reduction step allows one to separate the clearings which have a small size $\sim t^{1/(d+2)}$, where the process spends most of its time, from the forest, into which at some point, the process makes an excursion of size $t^{d/(d+2)}$. In this division, the clearings account for the $-c(d, \nu)$ term in the upper bound part of (0.5), whereas the excursions in the forest yield the $-\nu \omega_{d-1} a^{d-1} x$ term.

## I. Lower bounds

We consider random obstacles on $\mathbf{R}^d$, $d \geq 1$, which are translates of

$\overline{B}(0,a)$ the closed ball of radius $a > 0$, centered at the origin, at the points of a Poisson point measure $N(\omega, \cdot)$ with constant intensity $\nu > 0$, and law $P(d\omega)$. We are also given an independent canonical Brownian motion $Z_\cdot(w)$ on $C(\mathbf{R}_+, \mathbf{R}^d)$, and $P_0(dw)$ stands for Wiener measure. The entrance time of $Z_\cdot$ in the obstacles will be denoted by $T$. Our aim in this section is to derive asymptotic lower bounds as $t$ goes to infinity for the probability that Brownian motion has survived, i.e. has not entered the obstacles up to time $t$, and is located at a large distance from the origin. It will be convenient to introduce for $t > 0$:

$$(1.1) \quad \mu_t(dy) = \mathbf{P} \otimes P_0[Z_t^1 \in dy , T > t] = P_0[Z_t^1 \in dy , \exp\{-\nu|W_t^a|\}] ,$$

where $W_t^a = \bigcup_{0 \le s \le t} \overline{B}(Z_s, a)$ is the Wiener sausage of radius $a$ in time $t$ of $Z_\cdot$ and $Z^1$ is the first component of $Z_\cdot$. Let us recall that the constant $k(d, \nu, a)$ has been defined in (0.6) as $\inf_{r>0}(\nu\omega_{d-1}(a+r)^{d-1} + \sqrt{2\lambda_{d-1}}/r)$, when $d \ge 2$, and we set it equal to $\nu$ when $d = 1$. The infimum in the expression defining $k(d, \nu, a)$, when $d \ge 2$, is uniquely attained at the solution $r_0 > 0$ of the equation:

$$(1.2) \quad (d - 1)\nu\omega_{d-1}(a + r_0)^{d-2}r_0^2 = \sqrt{2\lambda_{d-1}} .$$

We now want to prove:

**Theorem 1.1.** *Suppose $d \ge 2$, and $x > 0$,*

(i)     if $\alpha < \dfrac{d}{d+2}$,    $\lim_{t\to\infty} t^{-d/(d+2)} \log(\mu_t(t^\alpha x, \infty)) = -c(d, \nu)$ ,

(ii)    if $\alpha = \dfrac{d}{d+2}$,    $\underline{\lim}_{t\to\infty} t^{-d/(d+2)} \log(\mu_t(t^{d/(d+2)}x, \infty))$

$$\ge -c(d, \nu) - k(d, \nu, a)x ,$$

(iii) if $\dfrac{d}{d+2} < \alpha < 1$ ,    $\underline{\lim}_{t\to\infty} t^{-\alpha} \log(\mu_t(t^\alpha x, \infty)) \ge -k(d, \nu, a)x$ ,

(iv) $\underline{\lim}_{t\to\infty} t^{-1} \log(\mu_t(tx, \infty))$

$$\ge -k(d, \nu, a)x , \quad \text{for } x \le \sqrt{2\lambda_{d-1}}/r_0$$

$$\ge -\min_{r>0}(\nu\omega_{d-1}(a + r)^{d-1}x + \frac{\lambda_{d-1}}{r^2}) - \frac{x^2}{2} \quad \text{otherwise.}$$

**Remark 1.2.**

1) The argument we use in the proof is valid in the one dimensional case as well, however it does not yield the correct behavior in case ii), as Theorem 1.3 will show.

2) As long as $\alpha < \dfrac{d}{d+2}$, both $E_0[Z_t^1 > t^\alpha x, \exp\{-\nu|W_t^a|\}]$ and $E_0[\exp\{-\nu|W_t^a|\}]$ behave as $\exp\{-c(d,\nu)t^{d/(d+2)}(1+o(1))\}$. This explains that the presence of a confinement property of surviving Brownian motion in scale $t^{1/(d+2)}$ (see [7] in the 2 dimensional case, or [1] for 2 dimensional random walks), involves more than principal logarithmic behavior for the probability of excursions of surviving Brownian motion.

3) Let us mention for the reader's convenience that when $\alpha > 1$, as will be clear from the proof, one obtains the same result as for Brownian motion in the absence of obstacles, namely:

$$\lim_{t\to\infty} t^{-(2\alpha-1)}\log(\mu_t(t^\alpha x,\infty)) = -\frac{x^2}{2}.$$

**Proof:** Pick $\beta \in (0,\alpha \wedge \frac{1}{d+2})$. For $t \geq 1$, we consider $U$ the open subset of $\mathbf{R}^d$, which is the union of the cylinder $C = (-t^\beta, t^\beta + xt^\alpha) \times B_{d-1}(r)$ ($B_{d-1}(r)$ stands for the open $d-1$ dimensional ball of radius $r$ centered at the origin), and of the $d$-dimensional ball $B$ centered at the point $(xt^\alpha, 0, \ldots, 0)$ with radius $R_0 t^{1/(d+2)}$. Let us recall here that $R_0$ defined in (0.4) is the radius of the ball such that

$$\nu|B(R_0)| + \lambda(B(R_0)) = c(d,\nu).$$

If $U^a$ denotes the $a$-neighborhood of $U$, we see that the volume of $U^a$ is smaller than

$$(1.3) \qquad v_t \stackrel{\text{def}}{=} \omega_d(R_0 t^{1/(d+2)} + a)^d + \omega_{d-1}(xt^\alpha + 2t^\beta + 2a)(a+r)^{d-1}.$$

Consequently, as $t$ goes to infinity,

$$(1.4) \qquad v_t \sim \omega_d R_0^d t^{d/(d+2)}, \qquad \alpha < \frac{d}{d+2}$$
$$\sim (\omega_d R_0^d + \omega_{d-1}(a+r)^{d-1}x)t^{d/(d+2)}, \qquad \alpha = \frac{d}{d+2},$$
$$\sim \omega_{d-1}(a+r)^{d-1}xt^\alpha, \qquad \alpha > \frac{d}{d+2}.$$

Let us introduce a function $s = s(t) \leq t$, which we will specify later on, as well as the cylinder $C_x \subset C$:

$$(1.5) \qquad C_x = (xt^\alpha, xt^\alpha + 1) \times B_{d-1}(r).$$

If $\theta_u$, $u \geq 0$ denotes the canonical shift on $C(\mathbf{R}_+, \mathbf{R}^d)$, and $T_C$ the exit time of $Z$ from the cylinder $C$, we have:

(1.6) $\mu_t(t^\alpha x, \infty) = P_0[Z_t^1 \geq xt^\alpha \ , \ \exp\{-\nu|W_t^a|\}]$
$$\geq \exp\{-\nu v_t\} P_0[T_C > s(t) \ , \ Z_{s(t)} \in C_x \ ,$$
$$Z_{t-s}^1 \circ \theta_s \geq Z_s^1 \ , \quad H_{(R_0 t^{1/(d+2)} - b)} \circ \theta_s > t - s] \ ,$$

where for any $c \geq 0$,

(1.7) $$H_c = \inf\{u \geq 0 \ , \ |Z_u - Z_0| \geq c\} \ ,$$

and $b = 1 + 2r$ is an upper bound on the diameter of $C_x$. It now follows that

(1.8) $\log \mu_t(t^\alpha x, \infty) \geq - \nu v_t + \log(P_0[T_C > s \ , \ Z_s \in C_x])$
$$+ \log(P_0[H_{R_0 t^{1/(d+2)} - b} > t - s \ , \ Z_{t-s}^1 \geq 0]) \ .$$

The second term in the right member of (1.8) equals
(1.9)
$$\log(P_0^{d-1}[H_r > s]) + \log(P_0^1[T_{(-t^\beta, t^\beta + xt^\alpha)} > s \ , \ Z_s \in (xt^\alpha \ , \ xt^\alpha + 1)]) \ .$$

Here, the superscript $(d-1)$ or 1 refers to the $d-1$ or 1 dimensional Wiener measure. In view of (1.8), (1.9), we now have a lower bound of $\log \mu_t(t^\alpha x, \infty)$ involving four terms. We will now specify our choice of the function $s(t)$, and then study the asymptotic behavior of each term for large $t$. We pick

(1.10) $s(t) = (\rho t^\alpha) \wedge t$ , $\quad$ when $\alpha < 1$ , with $\rho$ a positive constant,
$\qquad\quad = \rho t$ , $\qquad\qquad$ when $\alpha = 1$ , with the constant $0 < \rho \leq 1$ .

The asymptotic behavior of the first term $\nu v_t$ of the lower bound (1.8), (1.9) is provided in (1.4). For the second and fourth term using scaling and symmetry, we see that

(1.11) $$\lim_{t \to \infty} t^{-\alpha} \log(P_0^{d-1}[H_r > s]) = -\lambda_{d-1} \rho / r^2 \ , \quad \text{and}$$

$$\lim_{t \to \infty} t^{-d/(d+2)} \log(\tfrac{1}{2} P_0^d[H_1 > (t - s)/(R_0 t^{1/(d+2)} - b)^2])$$
(1.12)
$$= -\frac{\lambda_d}{R_0^2} \ , \qquad\qquad \text{if } \alpha < 1 \ ,$$

$$= -\frac{\lambda_d}{R_0^2}(1 - \rho) \ ; \qquad \text{if } \alpha = 1 \ .$$

Let us now study the third term. If $I$ stands for the interval $(xt^\alpha, 1 + xt^\alpha)$, for $t \geq 1$, the method of images allows us to see that

$$
\begin{aligned}
P_0^1[T_{(-t^\beta, t^\beta + xt^\alpha)} &> s \,, \, Z_s \in I] \\
&\geq P_0^1[Z_s \in I] - P_{-2t^\beta}^1[Z_s \in I] - P_{2xt^\alpha + 2t^\beta}^1[Z_s \in I] \\
&\geq (2\pi s)^{-1/2} \Big( \exp\{-\frac{(xt^\alpha + 1)^2}{2s}\} - \exp\{-\frac{(xt^\alpha + 2t^\beta)^2}{2s}\} \\
&\quad - \exp\{-\frac{(xt^\alpha + 2t^\beta - 1)^2}{2s}\} \Big) \,.
\end{aligned}
$$

It then follows that

$$
(1.13) \qquad \underline{\lim}_{t\to\infty} t^{-\alpha} \log P_0^1[T_{(-t^\beta, t^\beta + xt^\alpha)} > s \,, \, Z_s^1 \in I] \geq -\frac{x^2}{2\rho} \,.
$$

Collecting the asymptotic behavior of each term of the lower bound (1.8), (1.9), we see that when $\alpha < \dfrac{d}{d+2}$,

$$
(1.14) \qquad \underline{\lim}_{t\to\infty} t^{-d/(d+2)} \log \mu_t(t^\alpha x \,, \, \infty) \geq -\nu\omega_d R_0^d - \frac{\lambda_d}{R_0^2} = -c(d, \nu) \,.
$$

Our claim i) follows thanks to the natural upper bound (0.1). When $\alpha = d/(d+2)$, the left member of (1.14) is bigger than or equal to:

$$
-c(d, \nu) - \nu \, \omega_{d-1}(a + r)^{d-1} x - \lambda_{d-1}\frac{\rho}{r^2} - \frac{x^2}{2\rho} \,.
$$

Optimizing in $\rho$ and $r$, we find $\rho = xr/\sqrt{2\lambda_{d-1}}$, and $r = r_0$ (given by (1.2)). Our claim ii) follows. When $\alpha > \dfrac{d}{d+2}$, we now have

$$
(1.15) \ \underline{\lim}_{t\to\infty} t^{-\alpha} \log(\mu_t(t^\alpha x, \infty)) \geq -\nu\omega_{d-1}(a + r)^{d-1} x - \lambda_{d-1}\frac{\rho}{r^2} - \frac{x^2}{2\rho} \,.
$$

When $\dfrac{d}{d+2} < \alpha < 1$, optimizing over $\rho$, $r$, iii) follows. When $\alpha = 1$, we have the additional constraint $\rho \leq 1$, which leads to picking $\rho = \dfrac{xr_0}{\sqrt{2\lambda_{d-1}}}$, when $x \leq \sqrt{2\lambda_{d-1}}/r_0$, and $\rho = 1$ otherwise. Our claim iv) now follows. $\square$

In the one dimensional situation, the separation between scales $t^{1/(d+2)}$ and $t^{d/(d+2)}$ does not exist any more. If one sets $\ell_0 = 2R_0 = (\pi^2/\nu)^{1/3}$, one has in fact:

**Theorem 1.3.** *When $d = 1$,*

*i)* $\displaystyle\lim_{t\to\infty} t^{-1/3} \log(\mu_t(t^\alpha x, \infty)) = -c(1, \nu) = -\frac{3}{2}(\pi\nu)^{2/3}$ , $\qquad \alpha < \dfrac{1}{3}$ ,

*ii)* $\displaystyle\lim_{t\to\infty} t^{-1/3} \log(\mu_t(t^{1/3}x, \infty)) = -(\nu(x \vee \ell_0) + \frac{\pi^2}{2(x \vee \ell_0)^2})$

*iii)* $\displaystyle\lim_{t\to\infty} t^{-\alpha} \log(\mu_t(t^\alpha x, \infty)) = -\nu x$ , $\qquad\qquad \dfrac{1}{3} < \alpha < 1$ ,

*iv)* $\displaystyle\lim_{t\to\infty} t^{-1} \log(\mu_t(tx, \infty)) = -(\nu x + \frac{x^2}{2})$ .

**Proof:** Let us first treat the cases iii) and iv). One has the upper bound:

$$\mathbf{P} \otimes P_0^1 \left[ Z_t > xt^\alpha, \, T > t \right] \leq \mathbf{P} \left[ N([0, xt^\alpha]) = 0 \right] \times P_0^1 \left[ Z_t > xt^\alpha \right] ,$$

which immediately yields the upper estimates in iii) and iv). As for the lower bound part one uses (1.15), which in the one dimensional situation boils down to

$$\underline{\lim}_{t\to\infty} t^{-\alpha} \log(\mu_t(t^\alpha x, \infty)) \geq -\nu x - \frac{x^2}{2\rho} , \qquad \alpha > \frac{d}{d+2} ,$$

with the constraint $\rho \leq 1$ when $\alpha = 1$. This immediately provides the lower bound part of iii) and iv). The next observation is that i), (a similar remark could in fact be applied in the context of Theorem 1.1), is a consequence of ii). So there now remains to prove ii). Let us begin with the lower bound. Pick $\beta \in (0, 1/3)$, and consider the interval $J = (-t^\beta, t^\beta + (x \vee \ell_0)t^{1/3})$. One has:

$$(1.16) \quad \log(\mu_t(t^{1/3}x, \infty)) \geq -\nu(2a + 2t^\beta + (x \vee \ell_0)t^{1/3})$$
$$+ \log(P_0^1[T_J > t , \, Z_t > xt^{1/3}]) .$$

Using an eigenfunction expansion for the last term, it is standard to argue that $P_0^1[T_J > t , \, Z_t > xt^{1/3}]$ is equivalent to the principal term (with obvious notations):

$$\phi_1^J(0) \exp\{-\frac{\pi^2}{2|J|^2} t\} \int_{xt^{1/3}}^{(x\vee\ell_0)t^{1/3}+t^\beta} \phi_1^J(u) \, du$$

From the explicit behavior of $\phi_1^J$ it is now easy to deduce:

$$\underline{\lim}_{t\to\infty} t^{-1/3} \log(P_0^1[T_J > t , \, Z_t > xt^{1/3}]) \geq -\frac{\pi^2}{2(x \vee \ell_0)^2} .$$

This and (1.16) now proves the lower bound part of ii). For the upper bound part, it is enough to assume that $x > \ell_0$ in ii). Considering the points of the cloud falling immediately at the left and at the right of 0, we see that:

(1.17)   $\mathbf{P} \otimes P_0^1 [Z_t > x t^{1/3} \, , \, T > t]$

$$\leq \int_{u>0 \, , \, v \geq x} du \, dv \, \nu^2 t^{2/3} \, e^{-\nu(u+v)t^{1/3}} \, P_0^1 [T_{(-u,v)} > t^{1/3}]$$

Now

$$P_0^1 [T_{(-u,v)} > t^{1/3}] \leq \sup_{(0,1)} P_x^1 [T_{(0,1)} > \frac{t^{1/3}}{(u+v)^2}]$$

$$\leq (2\pi)^{-1} \int_0^1 dx \, P_x^1 [T_{(0,1)} > \frac{t^{1/3}}{(u+v)^2} - 1]$$

$$\leq (2\pi)^{-1} \exp \left\{ - \frac{\pi^2}{2} \left( \frac{t^{1/3}}{(u+v)^2} - 1 \right) \right\} ,$$

where the last inequality follows from the spectral theorem. If we use this estimate in (1.17), Laplace method now yields:

$$\overline{\lim}_{t \to \infty} t^{-1/3} \log \mu_t(t^{1/3}x, \infty) \leq - \inf_{\ell \geq x} (\nu\ell + \frac{\pi^2}{2\ell^2}) = \nu x + \frac{\pi^2}{2x^2} .$$

This finishes the proof of ii) and of Theorem 1.3.    □

**Remark 1.4.** It should be mentioned that the proof above works wi th trivial modifications if we replace the quantity $\mu_t(t^\alpha x, \infty)$ in the four cases of Theorem 1.3 by $\mathbf{P} \times P_0[\sup_{s \leq t} |Z_s| > x t^\alpha \, , \, T > t]$.

It is not difficult to see that $t^{-1/3} \sup_{s \leq t} |Z_s|$ satisfies a large deviation property under $\mathbf{P} \otimes P_0[\cdot/T > t]$, with rate function $I(x) = \nu y + \frac{\pi^2}{2y^2} - c(1, \nu)$, with $y = (2x) \wedge (\ell_0 \vee x)$, whereas $t^{-1/3}|Z_t|$ under the same conditional measure satisfies a large deviation property with rate function $I'(x) = \nu(x \vee \ell_0) + \frac{\pi^2}{2(x \vee \ell_0)^2} - c(1, \nu)$.    □

## II. Upper bound: a preliminary reduction step.

We suppose now that $d \geq 2$. We want to derive a preliminary estimate which will help us in proving upper bounds on the probability of existence prior to time $t$ of an excursion of surviving Brownian motion at a distance larger than const. $t^{d/(d+2)}$. This estimate will show that with no loss of generality we can assume that Brownian motion does not leave a suitable cube $\mathcal{T}$ of size const. $t^{(d+1)/(d+2)}$, that in this cube the obstacles form finitely many "clearings" of size $\sim t^{1/(d+2)}$. The estimate will also allow us to assume that the number of excursions of the process "in the forest", that is outside a neighborhood of size $t^{1/(d+2)}$ of the clearings, compared to $t^{d/(d+2)}$, as well as the fraction of time spent during these excursions are arbitrarily small. We will now make these notions precise.

First, it will be convenient to adopt respectively $t^{1/(d+2)}$ and $t^{2/(d+2)}$ as new space and time units. With this new choice of units, we now study a standard Brownian motion until time $s \overset{def}{=} t^{d/(d+2)}$, evolving among a Poisson cloud with intensity $\nu s$ of points $x_i$ of $\mathbf{R}^d$ surrounded by closed balls of radius $at^{-1/(d+2)} = as^{-1/d}$.

We now introduce a number $L > 0$, and an integer $N \geq 1$, and for $s \geq 1$ we define the cube $T$ of $\mathbf{R}^d$:

$$(2.1) \qquad T = \left( -N[s]L \,,\, N[s]L \right)^d , \; ([s] : \text{integer part of } s) .$$

For each multiindex $m \in [-N[s], N[s] - 1]^d$, $C_m$ will be the open subcube of $T$:

$$(2.2) \qquad C_m = \{z \in \mathbf{R}^d \,,\, m_i L < z_i < (m_i + 1)L \,,\, i = 1, \dots, d\} .$$

With no loss of generality we will assume that no point of the Poisson cloud has one of its coordinates which is an integer multiple of $L$.

We now introduce two numbers $b > a$, and $\delta > 0$. Following [5], we define a point $x_i \in C_m$ of the Poisson cloud to be a good point if for all closed balls $C = \overline{B}(x_i \,,\, 10^{\ell+1}\epsilon b)$, where $\epsilon \overset{def}{=} s^{-1/d}$, $0 \leq \ell$ and $10^{\ell+1}\epsilon b < L/2$,

$$(2.3) \qquad |C_m \cap C \cap ( \bigcup_{x_j \in C_m} \overline{B}(x_j, b\epsilon))| \geq \frac{\delta}{3^d}|C_m \cap C| .$$

We will denote by $G(m)$ the set of good points of $C_m$. Notice that for $r > 0$, and $z \in C_m$, using a homothety of ratio $1/3$ at $z$, one has:

$$|C_m \cap \overline{B}(z, r/3)| \geq 3^{-d}|C_m \cap \overline{B}(z, r)| .$$

It now follows from the covering Lemma 1.3 of [5], that for each $C_m$:

$$(2.4) \qquad |C_m \cap ( \bigcup_{\substack{x_i \notin G(m) \\ x_i \in C_m}} \overline{B}(x_i, b\epsilon))| \leq \delta|C_m| = \delta L^d .$$

In other words, the union of balls of radius $b\epsilon$ at bad points of $C_m$ cover a small fraction of the volume of $C_m$. We then chop identically each segment $[kL, (k+1)L]$ in at most $\left[ \frac{L}{b}\sqrt{d}\, s^{1/d} \right] + 1$ intervals of length $\frac{b}{\sqrt{d}}s^{-1/d}$, except maybe for the "last one". This yields closed boxes of diameter less than $b\epsilon = bs^{-1/d}$, in number less than $\left(\left[\frac{L}{b}\sqrt{d}\,s^{1/d}\right] + 1\right)^d$, whose union is $\overline{C}_m$.

We then introduce a number $r > 0$, and denote by $Cl_m$ the event that there is a "clearing of size $r$ in the cube $C_m$", that is:

$$(2.5) \qquad Cl_m = \left\{\omega \, , \, |\widetilde{U}_m(\omega)| \geq 2^{-d}|B(0,r)| = 2^{-d}\omega_d r^d\right\} \, ,$$

if $\widetilde{U}_m(\omega)$ is the random open set of $C_m$ obtained by taking the complement in $C_m$ of the closed boxes where a good point of the cloud falls. We also denote by $B(\omega)$ the union of all closed cubes $\overline{C}_m \subset \overline{T}$, where there is a clearing of size $r$:

$$(2.6) \qquad 1_{B(\omega)}(z) = \sum_m 1_{\overline{C}_m}(z) \cdot 1_{Cl_m}(\omega) \, .$$

We now pick a number $\ell > 0$. Let us mention that we will not need to let $\ell$ vary and we could very well pick $\ell = 1$, once and for all. We now define the successive excursions of the process $Z_.(w)$, at distance $\ell$ from $B(\omega)$:

$$D_1 = \inf\left\{v \geq 0 \, , \, Z_v \in (B^\ell)^c\right\} \leq \infty$$
$$R_1 = \inf\left\{v \geq D_1 \, , \, Z_v \in B\right\} = H_B \circ \theta_{D_1} + D_1 \leq \infty \, ,$$

and by induction for $n \geq 1$:

$$D_{n+1} = H_{(B^\ell)^c} \circ \theta_{R_n} + R_n \, ,$$
$$R_{n+1} = R_1 \circ \theta_{R_n} + R_n \, ,$$

here $B^\ell$ denotes the open neighborhood of point at distance less than $\ell$ of $B(\omega)$ (which is empty if $B$ is empty). We set $N_s$ to be the number of excursions completed by time $s$:

$$(2.7) \qquad \{N_s = k\} = \{R_k \leq s < R_{k+1}\} \, , \quad k \geq 0 \, ,$$

with the convention $R_0 = 0$. Finally, we define $L_s$ to be the fraction of time spent up to time $s$ in the excursions:

$$(2.8) \qquad L_s = \frac{1}{s}\sum_{i \geq 1}(R_i \wedge s - D_i \wedge s) \, .$$

We can now state the main object of this section:

**Theorem 2.1.** *For any $L > 0$, $\eta > 0$, $\ell > 0$:*
(2.9)

$$\overline{\lim}_{N \to \infty} \overline{\lim}_{r \to 0} \overline{\lim}_{n_0 \to \infty} \overline{\lim}_{b \to \infty, \delta \to 0} \overline{\lim}_{s \to \infty} \frac{1}{s} \log \left( \mathbf{P} \times P_0 [\{T > s\} \right.$$

$$\left. \cap \left[ \{T_T \le s\} \cup \{|B| > n_0 L^d\} \cup \{N_s \ge [\eta s]\} \cup \{L_s \ge \eta\} \right] \right]) = -\infty .$$

*Here $T_T$ denotes the entrance time in $T^c$.*

**Proof:** Our claim (2.9) will follow if we show that:

(2.10) $$\qquad \overline{\lim}_{N \to \infty} \overline{\lim}_{s \to \infty} \frac{1}{s} \log \left( P_0 [T_T \le s] \right) = -\infty ,$$

(2.11) $$\qquad \overline{\lim}_{n_0 \to \infty} \overline{\lim}_{b \to \infty, \delta \to 0} \overline{\lim}_{s \to \infty} \frac{1}{s} \log \mathbf{P} \left[ |B| > n_0 L^d \right] = -\infty$$

and with the notation $T \wedge T_T = \widetilde{T}$:

(2.12) $$\qquad \overline{\lim}_{r \to 0} \overline{\lim}_{b \to \infty, \delta \to 0} \overline{\lim}_{s \to \infty} \frac{1}{s} \log \sup_\omega P_0 \left[ \{\widetilde{T} > s\} \right.$$

$$\left. \cap \{N_s \ge [\eta s]\} \right] = -\infty ,$$

(2.13) $$\qquad \overline{\lim}_{r \to 0} \overline{\lim}_{b \to \infty, \delta \to 0} \overline{\lim}_{s \to \infty} \frac{1}{s} \log \sup_\omega P_0 \left[ \{\widetilde{T} > s\} \right.$$

$$\left. \cap \{N_s < [\eta s]\} \cap \{L_s \ge \eta\} \right] = -\infty ,$$

Let us start with the proof of (2.10). Using scaling, we immediately find:

$$P_0 [T_T \le s] \le d \, P_0^1 [ \sup_{0 \le u \le 1} |Z_u| \ge \frac{N}{\sqrt{s}} L[s] ] ,$$

where $P_0^1$ is the one dimensional Wiener measure. It follows that:

$$\overline{\lim}_{s \to \infty} \frac{1}{s} \log P_0 [T_T \le s] \le -\frac{N^2}{2} L^2 ,$$

which yields (2.10).

Let us now show (2.11). From the definition (2.3), we see that for any point $x_j \in C_m$, the property that $x_j$ is a good point of $C_m$ just depends on the restriction of the Poisson measure to $C_m$. From this observation, it follows that the events $Cl_m$ ("presence of a clearing of size $r$" in $C_m$), for $m \in [-N[s], N[s] - 1]^d$ are i.i.d. Let us now give an upper bound on $\mathbf{P}[Cl_m]$.

If $U_m(\omega)$ denotes the complement of the boxes in $C_m$ where some point falls (recall that for $\tilde{U}_m(\omega)$, it is a good point instead), we have since each box has diameter less than $b\epsilon$:

$$\tilde{U}_m \subseteq U_m \cup \Big( \bigcup_{\substack{x_i \notin G(m) \\ x_i \in C_m}} \overline{B}(x_i, bs^{-1/d}) \cap C_m \Big) .$$

Thanks to (2.4), this yields:

(2.14) $$|\tilde{U}_m| \leq |U_m| + \delta L^d .$$

This shows that

$$Cl_m = \{|\tilde{U}_m| > 2^{-d}|B(0,r)|\} \subseteq \{|U_m| \geq 2^{-d}|B(0,r)| - \delta L^d\} ,$$

and it follows that:

$$\begin{aligned} \mathbf{P}[Cl_m] &\leq \mathbf{P}[|U_m| \geq 2^{-d}|B(0,r)| - \delta L^d] \\ &\leq 2^{\left(\frac{L}{b}\sqrt{d}s^{1/d} + 1\right)^d} \exp\{-\nu s(2^{-d}|B(0,r)| - \delta L^d)\} . \end{aligned}$$

Now, there are at most $(2N\,s)^{dn_0}$ possible subsets of $n_0$ elements in the set of subcubes $C_m$ of $\mathcal{T}$. As a consequence,

$$\begin{aligned} &\mathbf{P}[|B| \geq n_0 L^d] \\ &\leq (2Ns)^{dn_0} 2^{\left((L/b)\sqrt{d}s^{1/d}+1\right)^d n_0} \exp\{-n_0\nu s(2^{-d}|B(0,r)| - \delta L^d)\} , \end{aligned}$$

so that

$$\overline{\lim}_{s\to\infty} \frac{1}{s} \log \mathbf{P}[|B| \geq n_0 L^d] \leq n_0\big(\log 2(\frac{L}{b}\sqrt{d})^d + \nu\delta L^d - \nu 2^{-d}|B(0,r)|\big) .$$

From this we obtain (2.11) immediately.

Let us now prove (2.12). We first introduce the constant $c_1$ defined as:

(2.15) $$c_1 = \frac{1}{2} \inf_{C \in \mathcal{C}} P_0[H_C < H_3] > 0 ,$$

where $H_3 = \inf\{v \geq 0, |Z_v - Z_0| \geq 3\}$, $H_C$ is the entrance time in $C$, and $\mathcal{C}$ is the class of compact subsets of $\overline{B}(0,2)$, such that $|C| \geq 2^{-d}(1-2^{-d})|B(0,2)|$. $c_1$ is easily seen to be positive, for instance by observing that the transition density $q_1(0,\cdot)$ at time 1 of Brownian motion kill ed outside $B(0,3)$ has a positive infimum on $\overline{B}(0,2)$. The notion of clearing we introduced will be of special help to us thanks to

**Lemma 2.2.** *For any $b > a$, $1 > \delta > 0$, $0 < r < L/4$, $N \geq 1$, there exists $s_0 > 0$ such that for $s \geq s_0$:*

$$(2.16) \qquad \inf_{\omega,\, z \in B(\omega)^c \cap T} P_z[\tilde{T} < H_{4r}] \geq c_1 .$$

**Proof:** First define the positive number:

$$(2.17) \quad \alpha(\delta, b, a) = \inf_{|z| \leq 1,\, C \in \mathcal{C}'} P_z[H_C < H_{10}] \times \inf_{|x| \leq b} P_z[H_{\overline{B}(0,a)} < H_{B(0,3b)^c}] ,$$

where $\mathcal{C}'$ is the class of compact subsets of $\overline{B}(0,1)$ with relative volume no less than $\delta/6^d$. Then set $m(\delta, b, a)$ to be the smallest integer such that:

$$(2.18) \qquad m \log(1 - \alpha) \leq -\log 2 .$$

Consider now $z \in \overline{C}_m$ a point at distance smaller or equal to $b\epsilon = bs^{-1/d}$ of a good point $x_i \in C_m$. Suppose $10^{m+1}b\epsilon + b\epsilon < L/2$, we have:

$$(2.19) \qquad P_z[H_{10^{m+1}b\epsilon + b\epsilon} < \tilde{T}] \leq P_z[H^{x_i}_{10^{m+1}b\epsilon} < T] ,$$

with $H^{x_i}_\rho = \inf\{u \geq 0 ,\ |Z_u - x_i| \geq \rho\}$. Since $x_i$ is a good point of $C_m$, we know that:

$$\left| C_m \cap C \cap \left( \bigcup_{x_j \in C_m} \overline{B}(x_j,\, bs^{-1/d}) \right) \right| \geq \frac{\delta}{3^d} |C_m \cap C| \geq \frac{\delta}{6^d} |C| ,$$

for all balls $C = \overline{B}(x_i, 10^{\ell+1}\epsilon b)$, with $0 \leq \ell$ and $10^{\ell+1}\epsilon b < L/2$. Now from the definition (2.17) of the constant $\alpha$, using scaling and Markov property, we see that

$$P_z[H^{x_i}_{10^{m+1}b\epsilon} < T] \leq (1 - \alpha)^m \leq \frac{1}{2} .$$

As a consequence, we see that for any $z$ in $T$ within distance $b\epsilon$ of some good point in some cube $C_m$, we have

$$(2.20) \qquad P_z[H_r > \tilde{T}] \geq P_z[H_{10^{m+1}b\epsilon + b\epsilon} \geq \tilde{T}] \geq 1 - (1 - \alpha)^m \geq \frac{1}{2} ,$$

provided $s$ is large enough so that

$$(2.21) \qquad 10^{m+1}b\epsilon + b\epsilon < r < L/4 .$$

Now, when $z \in \overline{C}_m \cap B(\omega)^c$, we know that $1_{Cl_m(\omega)} = 0$, and $|\tilde{U}_m| \leq 2^{-d}|B(0,r)|$. As a consequence, the intersection of $B(z, 2r)$ with the union of boxes in $\overline{C}_m$ containing some good point has a volume bigger than or equal to:

$$|B(z, 2r) \cap C_m| - 2^{-d}|B(0,r)| \geq (1 - 2^{-d})2^{-d}|B(0,2r)| \, , \quad \text{since } r < L/4 \, .$$

Consequently, using scaling and the definition (2.15) of $c_1$,

$$(2.22) \qquad P_z[H_{\overline{C}_m \setminus \tilde{U}_m} < H_{3r}] \geq 2c_1 \, , \quad \text{for } z \in \overline{C}_m \cap B(\omega)^c \, .$$

Combining (2.22) and (2.20), we see that when $s$ is large enough so that (2.21) holds, for $z \in T \cap B(\omega)^c$, $P_z[\tilde{T} < H_{4r}] \geq \frac{1}{2} \times 2c_1 = c_1$. This yields our claim. $\square$

Lemma 2.2 has the following consequence: for $b > a$, $1 > \delta > 0$, $0 < r < L/4$, $N \geq 1$, $s \geq s_0$ ($s_0$ appearing in Lemma 2.2), and $\rho > 0$:

$$(2.23) \qquad \sup_{\omega, \, z \in (B^\rho)^c \cap T} P_z[H_B < \tilde{T}] \leq (1 - c_1)^{[\rho/4r]} \, .$$

Indeed, if $z \in (B^\rho)^c \cap T$, the process has to exit successively at least $[\frac{\rho}{4r}]$ balls of radius $4r$ before entering $B$, so that (2.23) follows from (2.16) and a repeated use of strong the Markov property.

Let us now continue the proof of (2.12). We have:

$$\sup_\omega P_0[\{\tilde{T} > s\} \cap \{N_s \geq [\eta s]\}] \leq \sup_\omega P_0[R_{[\eta s]} < \tilde{T}] \, .$$

Observe now that for $k \geq 0$:

$$\begin{aligned} P_0[R_{k+1} < \tilde{T}] &\leq P_0[R_k < \tilde{T}, D_{k+1} < \tilde{T}, H_B \circ \theta_{D_{k+1}} < \tilde{T} \circ \theta_{D_{k+1}}] \\ &\leq P_0[R_k < \tilde{T}] \times \sup_{(B^\ell)^c \cap T} P_z[H_B < \tilde{T}] \\ &\leq \Big( \sup_{(B^\ell)^c \cap T} P_z[H_B < \tilde{T}] \Big)^{k+1} \, , \end{aligned}$$

using induction.

It now follows from (2.23), that when $s$ is large enough:

$$\sup_\omega P_0[\{\tilde{T} > s\} \cap \{N_s \geq [\eta s]\}] \leq (1 - c_1)^{[\ell/4r] \cdot [\eta s]}$$

from which we obtain (2.12) immediately.

Let us now prove (2.13). Pick $\lambda_0$ small enough so that

$$(2.24) \qquad P_0[\exp\{\lambda_0 H_1\}] \leq 1 + \frac{c_1}{2} .$$

Define $H^0 = 0$, and $H^{i+1} = H^i + H_{4r} \circ \theta_{H^i}$, the successive times when $Z$, travels at distance $4r$. Then choose $\lambda \leq \lambda_0/16r^2$ and $s \geq s_0$ given by Lemma 2.2, for $z \in (B^\ell)^c \cap T$, we have:

$$(2.25) \quad E_z[\exp\{\lambda R_1 \wedge \widetilde{T}\}]$$

$$= P_z[T = 0] + \sum_{k \geq 0} E_z[H^k < H_B \wedge \widetilde{T} \leq H^{k+1} , e^{\lambda(H_B \wedge \widetilde{T})}]$$

$$\leq 1 + \sum_{k \geq 0} E_z[H^k < H_B \wedge \widetilde{T} , e^{\lambda H^k}] \cdot E_0[e^{\lambda H_{4r}}] .$$

From scaling, $E_0[\exp\{\lambda H_{4r}\}] = E_0[\exp\{\lambda 16 r^2 H_1\}] \leq 1 + c_1/2$. Now for $k \geq 1$:

$$E_z[H^k < H_B \wedge \widetilde{T} , e^{\lambda H^k}] \leq E_z[H^{k-1} < H_B \wedge \widetilde{T} , e^{\lambda H^{k-1}}$$

$$(E_{Z_{H^{k-1}}}[H_{4r} < H_B \wedge \widetilde{T}] + E_{Z_{H^{k-1}}}[e^{\lambda H_{4r}}] - 1)]$$

$$\leq E_z[H^{k-1} < H_B \wedge \widetilde{T}, e^{\lambda H^{k-1}}](1 - \frac{c_1}{2}) ,$$

where we use the fact that $Z_{H^{k-1}} \in B(\omega)^c \cap T$, on the event $H^{k-1} < H_B \wedge \widetilde{T}$, together with (2.16). It now follows that for $\lambda \leq \lambda_0/16r^2$, $s \geq s_0$, $z \in (B^\ell)^c \cap T$, and any $\omega$:

$$(2.26) \qquad E_z[\exp\{\lambda R_1 \wedge \widetilde{T}\}] \leq 1 + (1 + \frac{c_1}{2}) \cdot \sum_{k \geq 0} (1 - \frac{c_1}{2})^k = 2 + \frac{2}{c_1} .$$

Picking $\lambda = \lambda_0/16r^2$, we find

$$(2.27) \quad \sup_\omega P_0[\{\widetilde{T} > s\} \cap \{N_s < [\eta s]\} \cap \{L_s \geq \eta\}]$$

$$\leq \sup_\omega \exp\{-\lambda \eta s\} \, P_0 \Big[ \exp \big\{ \lambda[(R_1 \wedge \widetilde{T}) \circ \theta_{D_1 \wedge \widetilde{T}}$$

$$+ (R_1 \wedge \widetilde{T}) \circ \theta_{D_2 \wedge \widetilde{T}} + \ldots + (R_1 \wedge \widetilde{T}) \circ \theta_{D_{[\eta s]} \wedge \widetilde{T}}\big\}\Big] .$$

Using now the strong Markov property, and the fact that (2.26) obviously holds when $z \notin T$, we see that the expression in (2.27) for $s \geq s_0$, is smaller than:

$$\exp\{-\lambda \eta s\} (2 + 2/c_1)^{\eta s + 1} ,$$

from which it follows that:

$$\overline{\lim}_{s \to \infty} \frac{1}{s} \log \sup_\omega P_0[\{\widetilde{T} > s\} \cap \{N_s < [\eta s]\} \cap \{L_s \geq \eta\}]$$

$$\leq \eta \Big( - \frac{\lambda_0}{16r^2} + \log (2 + 2/c_1) \Big) .$$

This immediately yields (2.13) and finishes the proof of Theorem 2.1. $\qquad \square$

## III. Upper bound

We suppose here, as in Section 2, that $d \geq 2$. Our main purpose is to derive an upper bound for the probability that surviving Brownian motion in time $t$ travels at distance of order $t^{d/(d+2)}$. This will provide a companion estimate to the lower bound of Theorem 1.1, ii).

**Theorem 3.1.**

$$(3.1)\quad \overline{\lim}_{t \to \infty} t^{-d/(d+2)} \log \mathbf{P} \otimes P_0[T > t\,,\, \sup_{u \leq t} |Z_u| > x\, t^{d/(d+2)}]$$

$$\leq -(c(d,\nu) + \nu \omega_{d-1} a^{d-1} x)\,.$$

**Proof:** Using the same scaling argument and same notations as in Section 2, we want to show that

$$(3.2)\qquad \overline{\lim}_{s \to \infty} s^{-1} \log \mathbf{P} \otimes P_0[T > s\,,\, \sup_{u \leq s} |Z_u| > x s^{(d-1)/d}]$$

$$\leq -(c(d,\nu) + \nu \omega_{d-1} a^{d-1} x)\,,$$

where we recall that $s = t^{d/(d+2)}$ and $\mathbf{P}$ now denotes a Poisson point process of intensity $\nu s$, the points of the random cloud being surrounded by closed balls of radius $a s^{-1/d}$, constituting the obstacles. Thanks to Theorem 2.1, we see that for a given $0 < \eta < 1$, and $L > 0$, we can pick $N(\eta, L)$, $r(\eta, L)$, $\eta_0(\eta, L)$, $b(\eta, L)$, $\delta(\eta, L)$ so that the probability of the event which appears in (2.9), decays at a faster exponential rate in $s$ than $-(c(d,\nu) + \nu \omega_{d-1} a^{d-1} x)$. So, (3.2) will follow if we show that

$$(3.3)\qquad \overline{\lim}_{L \to \infty}\, \overline{\lim}_{\eta \to 0}\, \overline{\lim}_{s \to \infty} \frac{1}{s} \log \mathbf{P} \otimes P_0[\tilde{T} > s\,,$$

$$|B| \leq n_0 L^d\,,\, N_s < [\eta s]\,,\, L_s \leq \eta \sup_{v \leq s} |Z_v| > x s^{(d-1)/d}]$$

$$\leq -(c(d,\nu) + \nu \omega_{d-1} a^{d-1} x)\,.$$

Observe that the condition $L_s \leq \eta < 1$, forces $B(\omega)$ not to be empty. The number of nonempty subsets $\mathcal{M}$ of $[-N[s], N[s] - 1]^d$, with no more than $n_0$ elements is smaller than $(2Ns)^{dn_0} \times n_0$. It is consequently enough to show that:

$$(3.4)\qquad \overline{\lim}_{L \to \infty}\, \overline{\lim}_{\eta \to 0}\, \overline{\lim}_{s \to \infty} \sup_{1 \leq |\mathcal{M}| \leq n_0(\eta, L)} \frac{1}{s} \log \mathbf{P} \otimes P_0[\tilde{T} > s\,,$$

$$N_s < [\eta s]\,,\, L_s \leq \eta\,,\, \sup_{u \leq s} |Z_u| > x s^{(d-1)/d}] \leq -(c(d,\nu) + \nu \omega_{d-1} a^{d-1} x)\,,$$

where now $N_s$ denotes the number of excursions between $B = \cup_\mathcal{M} \overline{C}_m$ and $(B^\ell)^c$, as in (2.7) and $L_s$ is the fraction of time spent up to time $s$, coming back from these excursions as in (2.8). Observe that the restrictions of the Poisson cloud to $B^{2\ell}$ and $(B^{2\ell})^c$, are independent. We denote by $\mathbf{P}_{B^{2\ell}}$ and $\mathbf{P}_{(B^{2\ell})^c}$ their respective laws. Conditioning on the restriction to $B^{2\ell}$ of the cloud, we see that the probability under study in (3.4) equals:

$$(3.5) \quad \mathbf{P}_{B^{2\ell}} \otimes P_0[\widetilde{T} > s \ , \ N_s < [\eta s] \ , \ L_s \leq \eta \ , \ \sup_{u \leq s} |Z_u| > x s^{(d-1)/d} \ ,$$

$$\exp\{-\nu s |W_s^{a s^{-1/d}} \cap (B^{2\ell})^c|\}] \ ,$$

where $\widetilde{T} = T_T \wedge T$, and $T$ denotes now the entrance time in the obstacles coming from the cloud with law $\mathbf{P}_{B^{2\ell}}$. It will now be convenient to derive:

**Lemma 3.2.** *Let* $\phi \in C([0,T], \mathbf{R}^d)$, $\phi(0) = 0$. *For* $\rho > 0$,

$$(3.6) \quad |W^\rho(\phi) \cap (B^{2\ell})^c| \geq \omega_{d-1} \rho^{d-1} \left( \sup_{[0,T]} |\phi| - n_0(L + 4\ell + 2\rho)\sqrt{d} \right) - n_0 \omega_d \rho^d \ .$$

**Proof:** With no loss of generality, possibly reducing the interval $[0,T]$, we assume that $|\phi(T)| = \max_{[0,T]} |\phi|$. Define for $m \in \mathcal{M}$, $\widetilde{C}_m \supset C_m^{2\ell}$ to be the closed cube with same center and parallel to $C_m$ of side length $L + 4\ell + 2\rho$, and set

$$\widetilde{B} = \bigcup_\mathcal{M} \widetilde{C}_m \supset B^{2\ell + \rho} \ .$$

We then define $s_1 = \inf\{u \in [0,T] \ , \ \phi(u) \in \widetilde{B}\}$ ($s_1 = \infty$, if the previous set is empty). Then there is one or maybe several $m \in \mathcal{M}$ such that $\phi(s_1) \in \widetilde{C}_m$. Define $t_1$ to be the supremum of the $v \in [s_1, T]$ where $\phi(u)$ belongs to the union o f these $\widetilde{C}_m$. Continue then by defining $s_2 = \inf\{u \geq t_1, \ \phi(u) \in \widetilde{B}\}$, and construct a finite sequence, $0 \leq s_1 \leq t_1 \leq s_2 \leq t_2 \leq \cdots \leq s_k \leq t_k \leq T$, with $0 \leq k \leq n_0$, such that for $u \notin \bigcup_{1 \leq i \leq k}[s_i, t_i]$, $\phi(u) \notin \widetilde{B}$, and $\phi(s_i)$, $\phi(t_i)$ belong to the same $\widetilde{C}_m$, $m \in \mathcal{M}$. We can now define the new trajectory $\psi$, which agrees with $\phi$ outside $\bigcup_{1 \leq i \leq k}[s_i, t_i]$, and is the linear interpolation of $\phi(s_i)$ and $\phi(t_i)$ on each interval $[s_i, t_i]$. Since $W^\rho(\phi_{|[0,T] \setminus \cup_i[s_i,t_i]}) \leq (B^{2\ell})^c$, we have

$$|W^\rho(\phi) \cap (B^{2\ell})^c| \geq |W^\rho(\phi_{|[0,T] \setminus \cup_i[s_i,t_i]})|$$
$$\geq |W^\rho(\psi)| - n_0((L + 4\ell + 2\rho)\sqrt{d}\,\omega_{d-1}\rho^{d-1} + \omega_d \rho^d)$$
$$\geq \omega_{d-1}\rho^{d-1}(|\psi(T)| - n_0(L + 4\ell + 2\rho)\sqrt{d}) - n_0 \omega_d \rho^d \ ,$$

which proves our claim (3.6) since $|\psi(T)| = |\phi(T)| = \max_{[0,T]} |\phi|$. $\quad\square$

If we now apply Lemma 3.2 to (3.5), we find that the expression in (3.5) is smaller than:

$$\mathbf{P}_{B^{2\ell}} \otimes P_0[\widetilde{T} > s \,,\, N_s < [\eta s] \,,\, L_s \leq \eta] \exp\{-\nu s \omega_{d-1} a^{d-1}$$
$$(x - 2n_0 \sqrt{d}\,(L + 4\ell + 2as^{-1/d})\,s^{-(d-1)/d}) - 2n_0 \omega_d a^d\} \,.$$

So our claim (3.4) will follow if we show that:

$$(3.7) \quad \overline{\lim}_{L \to \infty} \; \overline{\lim}_{\eta \to 0} \; \overline{\lim}_{s \to \infty} \sup_{1 \leq |\mathcal{M}| \leq n_0(\eta, L)} \frac{1}{s} \log \mathbf{P}_{B^{2\ell}} \otimes P_0[\widetilde{T} > s,$$
$$N_s < [\eta s], \; L_s \leq \eta] \leq -c(d, \nu) \,.$$

Let us introduce the torus of size $L$, $T_L = (\mathbf{R}/L\mathbf{Z})^d$, and denote by $\mathrm{proj}_{T_L}$ the canonical projection on $T_L$. We have:

$$|W_s^{as^{-1/d}} \cap B^{2\ell}| \geq |\mathrm{proj}_{T_L}(W_s^{as^{-1/d}} \cap B^{2\ell})| \,,$$

where we use $|\cdot|$ to denote the usual volume on $T_L$ as well. As soon as $as^{-1/d} < \ell$, the last quantity is bigger than

$$|\mathrm{proj}_{T_L}(W_{[0,D_1)}^{as^{-1/d}} \cup W_{[R_1, D_2)}^{as^{-1/d}} \ldots \cup W_{[R_{N_s}, s \wedge D_{N_s})}^{as^{-1/d}})| \,.$$

For $as^{-1/d} < \ell$, we find

$$\mathbf{P}_{B^{2\ell}} \otimes P_0[\widetilde{T} > s \,,\, N_s < [\eta s] \,,\, L_s \leq \eta]$$
$$= P_0[\exp\{-\nu|W_s^{as^{-1/d}} \cap B^{2\ell}|\} \,,\, N_s < [\eta s] \,,\, L_s \leq \eta \,,\, T_T > s]$$
$$\leq P_0[\exp\{-\nu|\mathrm{proj}_{T_L}(W_{[0,D_1)}^{as^{-1/d}} \cup W_{[R_1, D_2)}^{as^{-1/d}} \cup \ldots \cup W_{[R_{N_s}, s \wedge D_{N_s})}^{as^{-1/d}})|\} \,,$$
$$(3.8) \qquad N_s < [\eta s] \,,\, L_s \leq \eta \,,\, T_T > s] \,.$$

Let us now denote by $\mathbf{P}^L$ the law of the random point process obtained by picking a Poisson configuration of points with intensity $\nu s$ in $(0, L)^d$ and extending it to the whole of $\mathbf{R}^d$ by periodicity. As before let us define the obstacles as closed balls of radius $as^{-1/d}$ centered at each point of the periodic configuration. If $T$ denotes as before the entrance time in the obstacles, we see that (3.8) equals

$$(3.9) \quad \mathbf{P}^L \otimes P_0[T_T > s, \; T \geq D_1, \; T \circ \theta_{R_1} \geq D_1 \circ \theta_{R_1}, \; \ldots,$$
$$T \circ \theta_{R_{N_s}} \geq (s - R_{N_s}) \wedge D_1 \circ \theta_{R_{N_s}}, \; N_s < [\eta s], \; L_s \leq \eta] \,.$$

Denote by $\lambda_L(\omega)$ the principal Dirichlet eigenvalue for the generator of the self adjoint trace class semigroup on $L^2(T_L)$, corresponding to Brownian motion on $T_L$ killed on the obstacles on $T_L$, projection of the periodic obstacles on $\mathbf{R}^d$. Observe that on the event whic h appears in (3.9),

$$T \wedge D_1 + T \wedge D_1 \circ \theta_{R_1} + \cdots + T \wedge D_1 \circ \theta_{R_{N_s}} \geq (1 - \eta)s$$

It follows that for $M > 0$, and $\rho > 0$, (3.9) is smaller than:

$$\mathbf{P}^L \otimes P_0[T_T > s, \; \exp\{(\lambda_L \wedge M - \rho)_+ (T \wedge D_1 + T \wedge D_1 \circ \theta_{R_1} + \cdots$$
$$+ T \wedge D_1 \circ \theta_{R_{N_s}} - (1 - \eta)s)\} \, , \; N_s < [\eta s]]$$
$$\leq \mathbf{P}^L[\exp\{-(\lambda_L \wedge M - \rho)_+ (1 - \eta)s\}$$
$$P_0[\exp\{(\lambda_L \wedge M - \rho)_+ \sum_0^{[\eta s]} T \wedge D_1 \circ \theta_{R_i} \, 1(R_i < \infty)\}]]$$
$$\leq \mathbf{P}^L[\exp\{-(\lambda_L \wedge M - \rho)_+ (1 - \eta)s\}$$
$$(\sup_{z \in \mathbf{R}^d} P_z[\exp\{(\lambda_L \wedge M - \rho)_+ T\}])^{\eta s + 1}] \, .$$

If now $P_z^L$ denotes the law of Brownian motion on $T_L$ starting from $z$, using obvious notations, we have

$$\sup_{\mathbf{R}^d} P_x[\exp\{(\lambda_L \wedge M - \rho)_+ T\}] = \sup_{T_L} P_z^L[\exp\{(\lambda_L \wedge M - \rho)_+ T^L\}] \, .$$

This last quantity, by inequality (1.22) of [5] is smaller than $e^M (1 + c_L + c_L \frac{M}{\rho})$, where $c_L$ stands for the supremum of the transition density at time one, with respect to the normalized volume measure of Brownian motion on $T_L$. It is may be helpful to mention that the notations of [5] are somewhat different and $T^L$ here corresponds to $T_b$ in [5], and $\lambda_L$ to $\lambda_b$. We now see that (3.9) is smaller than

$$f(L, \eta, M, \rho, s)$$
$$\overset{def}{=} \exp\{(2M\eta + \rho)s + M\}(1 + c_L + c_L \frac{M}{\rho})^{\eta s + 1} \mathbf{P}^L[\exp\{-(\lambda_L \wedge M)s\}] \, .$$

It now follows that the left member of (3.7) is smaller than:

$$\varlimsup_{L \to \infty} \varlimsup_{M \to \infty, \rho \to 0} \varlimsup_{\eta \to 0} \varlimsup_{s \to \infty} \frac{1}{s} \log(f(L, \eta, M, \rho, s))$$
$$= \varlimsup_{L \to \infty} \varlimsup_{M \to \infty} \varlimsup_{s \to \infty} \frac{1}{s} \log \mathbf{P}^L[\exp\{-(\lambda_L \wedge M)s\}] = -c(d, \nu) \, ,$$

The last equality comes from the upperbound we derive for $\mathbf{P}^L[exp\{-(\lambda_L \wedge M)s\}]$ in [5], see what follows there (1.39), Theorem 2.2 and Lemma 3.3.

We should also mention that in the notations of [5], $\lambda_L$ plays the role of $\lambda_K$. T his proves our claim (3.7) and finishes the proof of Theorem 3.1.   □
As a counterpart to the lower bounds of Theorem 1.1, we have

**Corollary 3.3.** *Suppose $d \geq 2$, and $x > 0$,*

$$(3.10) \quad \overline{\lim}_{t\to\infty} t^{-d/(d+2)} \log(\mu_t(t^{d/(d+2)}x, \infty))$$
$$\leq -\left(c(d,\nu) + \nu\omega_{d-1}a^{d-1}x\right) ,$$

$$(3.11) \quad \overline{\lim}_{t\to\infty} t^{-\alpha} \log(\mu_t(t^\alpha x, \infty)) \leq -\nu\omega_{d-1}a^{d-1}x , \quad \frac{d}{d+2} < \alpha < 1 ,$$

$$(3.12) \quad \overline{\lim}_{t\to\infty} t^{-1} \log(\mu_t(tx, \infty)) \leq -\left(\frac{x^2}{2} + \nu\omega_{d-1}a^{d-1}x\right) .$$

**Proof:** (3.10) is an immediate consequence of Theorem 3.1. The proof of (3.11) and (3.12) is straightforward, it relies on the already observed fact that:

$$(3.13) \qquad |W_t^a(Z)| \geq \nu\omega_{d-1}a^{d-1}|Z_t - Z_0| ,$$

from which we deduce $\mu_t(t^\alpha x, \infty) \leq \exp\{-\nu\omega_{d-1}a^{d-1}xt^\alpha\} P_0[Z_t^1 > xt^\alpha]$. (3.11) and (3.12) now follow easily.   □

**Remark 3.4.**
1) It should be observed that (3.10) does not hold in the one dimensional case (see Theorem 1.3).
2) From Theorem 1.1 ii) and Theorem 3.1, we see that, when $d \geq 2$:

$$-k(d,\nu,a)x \leq \overline{\lim}_{t\to\infty} \; t^{-d/(d+2)} \log \mathbf{P} \otimes P_0[\sup_{s\leq t} |Z_s| \geq xt^{d/(d+2)}/T > t]$$
$$\leq -\nu\omega_{d-1}a^{d-1}x .$$

This strongly suggests the possibility of a large deviation property for the law of $t^{-d/(d+2)} \sup_{s\leq t} |Z_s|$, conditional on survival up to time $t$.   □

## IV. An application

We are now going to apply our results to the study of the large $t$ behavior of

$$E_0[\exp\{h \cdot Z_t - \nu|W_t^a|\}] , \quad \text{when} \quad |h| < \nu\omega_{d-1}a^{d-1} .$$

This quantity was investigated by Eisele and Lang, in their study of survival probability for Brownian motion with a constant drift evolving among Poissonian obstacles. They showed that
(4.1)

$$\lim_{t\to\infty} \frac{1}{t} \log E_0[\exp\{h \cdot Z_t - \nu|W_t^a|\}] = 0 , \quad \text{when} \quad |h| < \alpha(d,\nu,a)$$
$$> 0 , \quad \text{when} \quad |h| > \alpha(d,\nu,a) ,$$

for a critical value $\alpha(d, \nu, a)$ in the interval $[\nu\omega_{d-1}a^{d-1}, k(d, \nu, a)]$, where $k(d, \nu, a)$ is defined in (0.6). We are now going to apply the results of section III, in the $d \geq 2$ case, and section I in the $d = 1$ case, to refine (4.1) when $|h| < \nu\omega_{d-1}a^{d-1}$.

**Theorem 4.1.** *When* $d = 1$, *and* $|h| < \nu$:

(4.2)
$$\lim_{t\to\infty} t^{-1/3} \log E_0[\exp\{hZ_t - \nu|W_t^a|\}] = -c(1, \nu - |h|) = -\frac{3}{2}(\pi(\nu - |h|))^{2/3} .$$

*When* $d \geq 2$, *and* $|h| < \nu\omega_{d-1}a^{d-1}$:

(4.3)
$$\lim_{t\to\infty} t^{-d/(d+2)} \log E_0[\exp\{h \cdot Z_t - \nu|W_t^a|\}] = -c(d, \nu) .$$

**Proof:** It is enough to study $E_0[\exp\{hZ_t^1 - \nu|W_t^a|\}]$ for $h > 0$. This quantity is bigger than

(4.4)
$$\int_0^\infty \exp\{hx\} d\mu_t(x) = E_0[\exp\{-\nu|W_t^a|\}] + h\int_0^\infty e^{hx}\mu_t(x, \infty)\,dx ,$$

and anyway smaller than:

$$2E_0[\exp\{-\nu|W_t^a|\}] + h\int_0^\infty e^{hx}\mu_t(x, \infty)\,dx .$$

It follows that we simply have to study the second term to the right of (4.4). It is the sum of

$$a_1 \overset{\text{def}}{=} ht^{d/(d+2)} \int_0^A \exp\{t^{d/(d+2)}hx\}\,\mu_t(t^{d/(d+2)}x, \infty)\,dx \quad \text{and}$$

$$a_2 \overset{\text{def}}{=} h\int_{At^{d/(d+2)}}^\infty \exp\{hx\}\,\mu_t(x, \infty)\,dx , \quad \text{with} \quad A \text{ a positive constant.}$$

Using (3.13), we see that $a_2$ is smaller than

$$\int_{At^{d/(d+2)}}^\infty dx\, h\exp\{x(h - \nu\omega_{d-1}a^{d-1})\},$$

so that

(4.5)
$$\overline{\lim}_{t\to\infty} t^{-d/(d+2)} \log a_2 \leq -A(\nu\omega_{d-1}a^{d-1} - h) .$$

We see our claims (4.2), (4.3) will be proved if we show that for large enough $A$:

(4.6)
$$\overline{\lim}_{t\to\infty} t^{-d/(d+2)} \log a_1 \leq -c(d, \nu) , \quad \text{when } d \geq 2 ,$$

(4.7)        $\lim_{t \to \infty} t^{1/3} \log a_1 = -c(1, \nu - |h|)$ ,    when $d = 1$ .

Observe that $\mu_t(t^{d/(d+2)}x, \infty)$ is a decreasing function of $x$. Using Riemann sums, and Laplace method, we see that when $d = 2$, thanks to (3.10),

$$\overline{\lim}_{t \to \infty} t^{-d/(d+2)} \log a_1 \leq \sup_{x \in [0,A]} \{hx - c(d, \nu) - \nu \omega_{d-1} a^{d-1} x\} = -c(d, \nu) ,$$

since $h < \nu \omega_{d-1} a^{d-1}$. This proves (4.6). On the other hand when $d = 1$, using Theorem 13, we find

$$\lim_{t \to \infty} t^{-1/3} \log a_1 = \sup_{x \in [0,A]} \left\{ hx - \nu(x \vee \ell_0) - \frac{\pi^2}{2(\ell_0 \vee x)^2} \right\} = -c(1, \nu - h) ,$$

provided $A$ is bigger than $[\pi^2/(\nu - h)]^{1/3}$. This yields (4.7) and finishes the proof of our claim.    □

## Bibliography

[1] BOLTHAUSEN, E.: "Localization of a two dimensional random walk with an attractive path interaction", preprint.

[2] DONSKER, M. D., VARADHAN, S.R.S.: "Asymptotics for the Wiener sausage", Comm. Pure Appl. Math., 28, 525-565, (1975).

[3] EISELE, T., LANG, R.: "Asymptotics for the Wiener sausage with drift", Prob. Th. Rel. Fields, 74, 1, 125-140, (1987).

[4] GRASSBERGER, P., PROCACCIA, I.: "Diffusion and drift in a medium with randomly distributed traps", Phys. Rev., A 26, 3686-3688, (1982).

[5] SZNITMAN, A.S.: "Lifschitz tail and Wiener sausage I", in J. Funct. Anal., 94, 223-246 (1990).

[6] —————: "Long time asymptotics for the shrinking Wiener sausage", Comm. Pure Appl. Math., 43, 809-820, (1990).

[7] —————: "On the confinement property of two dimensional Brownian motion among Poissonian obstacles", to appear in Comm. Pure Appl. Math.

Printed in the United States
By Bookmasters

Printed in the United States
By Bookmasters